Theoretical Spectroscopy of Transition Metal and Rare Earth Ions

Theoretical Spectroscopy of Transition Metal and Rare Earth Ions

From Free State to Crystal Field

Mikhail G. Brik | Chong-Geng Ma

JENNY STANFORD
PUBLISHING

Published by

Jenny Stanford Publishing Pte. Ltd.
Level 34, Centennial Tower
3 Temasek Avenue
Singapore 039190

Email: editorial@jennystanford.com
Web: www.jennystanford.com

British Library Cataloguing-in-Publication Data
A catalogue record for this book is available from the British Library.

Theoretical Spectroscopy of Transition Metal and Rare Earth Ions: From Free State to Crystal Field

ISBN 978-981-4800-56-3 (Hardcover)
ISBN 978-0-429-27875-4 (eBook)

Contents

Preface

In the life of every scientist, there was a time of a start of the career ("once upon a time..."). This is usually during the university years, MSc and PhD studies, when the research interests are formed and shaped. Very often, this happens under the influence of professors, supervisors, older friends, and books. That time is of paramount importance for a young person, since the foundations of the future career steps, skills, and knowledge are implicitly created. As a rule, that period is the time of really hard work in the laboratories and libraries, search for various good books and articles to learn new things, accept and understand new concepts, "digest" them, and use them in the future research.

Also, in the life of every scientist, there is a time when he looks back upon his younger (or better—in a way) years. Then the understanding comes that some things were not done in the best way. For example, some books could have been missed for some reasons, or the time was not organized efficiently. And then the thought is coming to help young researchers make their scientific path smoother. When both authors were going through their PhD studies—chronologically and geographically it was in different times and countries—they felt a huge gap between the university courses on atomic physics and quantum mechanics and specialized books on atomic spectroscopy, theory of crystal field, and group theory. Many different books had to be consulted at that time to make a (more or less) clear picture of why different crystals and different impurity ions behave in remarkably different ways, to understand their main spectroscopic properties, to describe and quantify them and, finally, to start real research on the basis of that acquired knowledge. Going through many books on the same subject has a clear and enriching advantage of looking at the same topic from the perspectives of different authors. The price to be paid for that—especially if we speak about quantum mechanics and spectroscopy—is fighting with different conventions, notations, systems of units, numerical factors, normalizations, etc.

In an attempt to fill in this gap between the university courses on physics and research on the optical properties of the transition metal and lanthanide ions, we wrote this book on theoretical spectroscopy of these ions in a free state and in crystals. We hope that having such a single book with description of the energy levels' formation (in a free state), their splitting (in crystal fields), group theory in its application to spectroscopy, and examples of the application of all that to real optical materials would help young researchers to understand these topics clearly. We tried to write the book in as simple way as possible (although, of course, a certain level of quantum mechanics and calculus knowledge would be required to feel comfortable when reading the book). The book contains many examples of the application of those theoretical concepts to the real physical systems and those examples are based on the original authors' publications. It is hoped that the book will be a bridge to more advanced quantum-mechanical books, many of those are cited here or mentioned as a literature for further reading.

We also believe that the book will be helpful to the experimental scientists to refresh (or to get familiar with) basic theoretical concepts that may be useful for a proper interpretation of the experimental spectroscopic results.

Writing this book was an awesome adventure of going back in time to our PhD years—trying to recollect what the most difficult topics were for us and at the same time contacting our current students to understand what the most difficult topics are for them. The content of the book is based on the lectures on spectroscopy of the transition metal and rare earth ions, which the authors have given (and continue to give) at the University of Tartu in Estonia and Chongqing University of Posts and Telecommunications in China. Besides, as a guest professor, one of the authors (MGB) taught a number of relevant topics at Kyoto University in Japan, Lanzhou University and South China University of Technology in China, and Jan Długosz University in Częstochowa, Poland.

We also would like to express our sincere gratitude to all our numerous colleagues and friends worldwide. We simply cannot mention their names here because of the lack of space on the book pages—but we truly remember all of them! Fruitful and stimulating discussions with them on a variety of topics—science, teaching, culture, art, sport, history, literature, food, etc.—were extremely interesting and enjoyable! Science is an amazing thing: In addition to real pleasure of being involved into solving interesting scientific problems, it gives a person a unique chance to travel a lot, to be exposed to different cultures, to see exciting places all over the globe, and to appreciate breathtaking rare beauty of our world—this must be cherished!

Being unable to name here all our friends and colleagues, we wish, however, to mention our first teachers—they introduced us to the field of spectroscopy of the transition metal and rare earth ions and they made us in the way we are now.

MGB gratefully acknowledges very important influence of Dr. V. V. Zhorin and late Professors A. G. Avanesov, V. A. Lebedev, and V. F. Pisarenko from his *alma mater* (Kuban State University, Krasnodar, Russia). He is also much indebted to his mother and his late father, who was a school physics teacher. Many thanks are due to Professor N. M. Avram (West University of Timisoara, Romania) for long friendship and first international visit to his place, back in September of 2000.

CGM gratefully acknowledges the help on his earlier scientific career offered by Professors S.-D. Xia, M. Yin, C.-K. Duan (University of Science and Technology of China, Hefei, China), P. A. Tanner (City University of Hong Kong, Hong Kong, China), and M. F. Reid (University of Canterbury, Christchurch, New Zealand).

Many thanks are due to Sarabjeet Garcha, Arvind Kanswal, and all Jenny Stanford staff involved for their guidance during all stages of the preparation of this book. We thank Prof. Martin Nikl (Institute of Physics, Czech Academy of Sciences, Prague) for introducing us to Jenny Stanford Publishing.

We also acknowledge the financial support, which made this book possible. MGB thanks the Chinese National Recruitment Program of High-end Foreign Experts (Grant Nos. GDW20145200225 and GDT20185200479), the Chongqing Recruitment Program for 100 Overseas Innovative Talents (Grant No. 2015013), the Programme for the Foreign Experts offered by Chongqing University of Posts and Telecommunications (Grant No. W2017011), Ministry of Education and Research of Estonia, Project PUT430, Estonian Research Council Grant PUT PRG111, European Regional Development Fund (Project TK141), Guest Professorship at Kyoto University (Prof. S. Tanabe Laboratory, October 2017—January 2018), and MOBILITY FZU project from Operational Programme: Research, Development and Education, MEYS, no. CZ.02.2.69/0.0/0.0/16_027/0008215 (stay at the Institute of Physics, Czech Academy of Sciences, group of Prof. M. Nikl). CGM thanks the Innovation and Entrepreneurship Programs for Returned Overseas Chinese Scholars offered by Chinese Ministry and Chongqing Bureau of Human Resources and Social Security (Grant Nos. [2014] 167 and CX2018125), respectively, Scientific and Technological Research Program of Chongqing Municipality Education Commission (Grant No. KJ1600415) and Wenfeng High-End Talents Project (Grant No. W2016-01) offered by Chongqing University of Posts and Telecommunications.

At last, but by no means at least, we would like to thank sincerely our families—for their everlasting patience, support, and understanding during our whole scientific careers! We owe them a lot, if not everything!

And—with all that said—*lectori benevolo salutem!*

Mikhail G. Brik

Chong-Geng Ma

November 2019

Chongqing–Tartu–Częstochowa–Kyoto–Prague–Chongqing

Chapter 1

*Fiat lux,** or What Are the Main Optical Applications of the *d* and *f* Ions

Numerous fascinating optical devices have firmly entered our everyday life. It is now unthinkable to spend even a short time without those already ordinary and/or yet fancy things, which just a few decades ago could only have been a result of rich imagination of a novelist writing the scientific fiction novels or stories. Simple or sophisticated lamps and light emitting diodes (LED) create comfortable conditions for work and relaxation, help in making proper atmosphere in theaters and museums, or are used in breathtaking light shows. TV sets, computers, and mobile phones with large and small colorful screens are now inseparable parts of entertainment and serious work. Lasers producing emission in various spectral ranges are widely applied in the telecommunications industry, medicine, research, etc. Glow-in-the-dark safety signs indicate escape routes in the public areas, and the toys with long persistent phosphors are funny things for kids to play with. Nanoparticles emitting infrared light are used for non-invasive bioimaging and thermometry. The list of these applications can be continued and extended, but this is not the main aim of this book. Rather than discussing and describing the applications of the optical materials, the reader's attention will be brought to the so-

*Fiat lux—Let there be light (Latin).

Theoretical Spectroscopy of Transition Metal and Rare Earth Ions: From Free State to Crystal Field
Mikhail G. Brik and Chong-Geng Ma

ISBN 978-981-4800-56-3 (Hardcover), 978-0-429-27875-4 (eBook)
www.jennystanford.com

called activators or active ions, responsible for the absorption and emission of light. Many people are quite advanced users of those above-mentioned devices, but very few think about the physical principles of their operation and those substances producing bright colored emissions that make watching modern TVs or playing with modern computers/mobile phones so amusing.

It took people really a long time to understand what the nature of light is, how it interacts with matter, how it can be absorbed and re-emitted, and what those processes in physical bodies are that correspond to the light absorption and/or emission.

Since the times when Niels Bohr developed his theory of the hydrogen atom in the second decade of the 20th century, it has been recognized that the atoms are characterized by discrete energy level schemes with energy levels numbered by the so-called principal quantum number *n* (Fig. 1.1) and the light is absorbed and emitted when the electrons in the atoms or ions "jump" from one allowed orbit into another (Fig. 1.2).

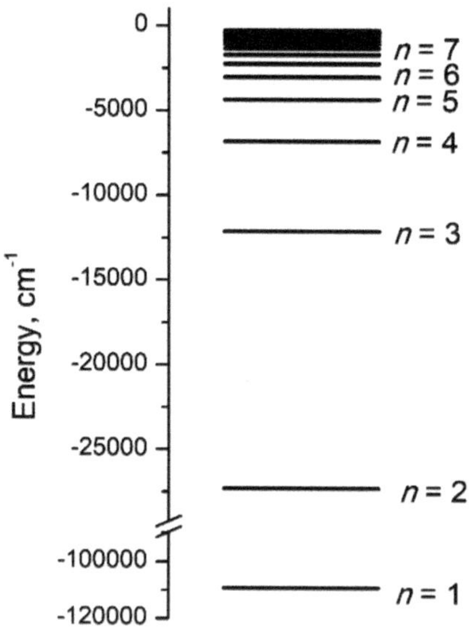

Figure 1.1 Discrete energy level scheme of the hydrogen atom (see Chapter 3 for more details).

Further development of quantum mechanics and spectroscopy allowed explaining the order of the chemical elements in the

Periodic table; it was also understood that the filling of the electron orbitals in atoms occurs in a certain order and there are well-defined strict limitations on how the electrons can be distributed through the available states in atoms. It was shown that the partially filled electron shells have a large number of the allowed states that can be involved into the absorption/emission transitions. As a result, the optical spectra of the ions with those partially filled electron shells exhibit remarkable diversity from the point of view of possible transitions, their intensities and spectral ranges covered by those transitions.

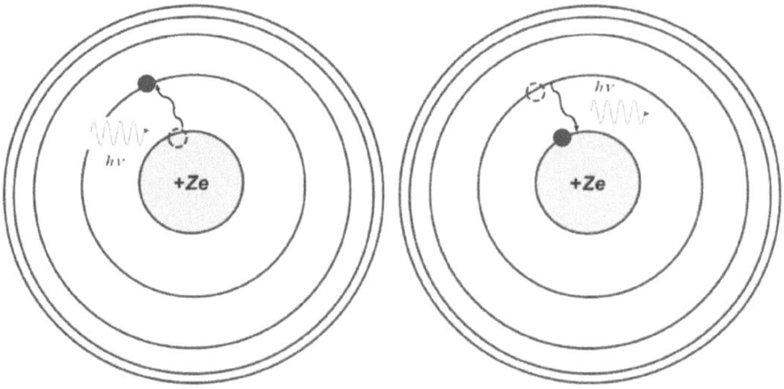

Figure 1.2 Schematic representation of absorption (left) and emission (right) of a photon with energy *hv* by the hydrogen-like atom with the nucleus charge +*Ze*. The electron's orbits are shown in a simplified way.

Among various chemical elements in the Periodic table, there are several specific groups with the unfilled *d* and *f* electron shells. Namely these elements are mostly suitable for the optical applications, since they have many—sometimes up to several hundreds and even thousands—energetic states, which span a very wide energy range from infrared to vacuum ultraviolet, thus creating practically countless possibilities for fascinating potential applications.

The first such group is the transition metal ions with the 3*d* unfilled shell, from Ti to Zn. The most known representatives from this group are the Cr^{3+}, Mn^{4+}, Mn^{2+}, and Ni^{2+} ions. In the first ever built laser, ruby laser $Al_2O_3{:}Cr^{3+}$,[1] the red emission coming from the Cr^{3+} ions is utilized. The Mn^{4+} ions are widely used for creating red phosphors for white LED,[2–5] and the Ni^{2+} ions are the active ions for getting tunable laser generation in the infrared spectral range.[6] The spectral properties of the 3*d* ions strongly depend on the crystalline host they are placed in. Figure 1.3, which shows the emission spectra of the Ni^{2+} ions in several crystals, serves as a good illustration to this statement—the emission maximum is shifted from about 1100 nm in $YAlO_3$ to about 1700 nm in MgF_2—so, the same 3*d* ion emits in very different ways in different materials.

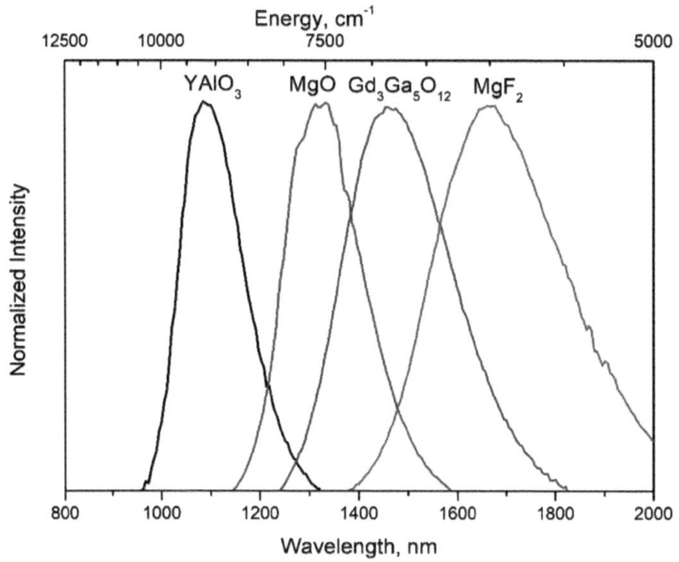

Figure 1.3 Normalized emission spectra of the Ni^{2+} ions in $YAlO_3$, MgO, $Gd_3Ga_5O_{12}$ and MgF_2 (after 6).

The groups of ions with the unfilled 4*d* (from Y to Cd) and 5*d* (from Lu to Hg) electron shells are somewhat similar to the 3*d* ions, although due to the increased atomic numbers the spin-orbit interaction effects are more important for them and the optical absorption spectra are shifted towards higher energies.

Another fascinating group of elements is formed by the lanthanide ions, from La to Lu. They all have the 4*f* unfilled shell, and chemically they are very similar. According to Pimentel and Sprately,[7] "Lanthanum has only one important oxidation state in aqueous solution, the +3 state. With few exceptions, this tells the whole boring story about the other 14 Lanthanides."

Luckily, the whole story of the lanthanides is not so boring at all, especially from the spectroscopic and optical points of view. Their emission spectra cover a wide range from infrared to ultraviolet (Fig. 1.4) and, remarkably, the positions of the absorption/emission maxima are practically unchangeable from one crystalline host to another, what is in striking contrast to the case of the 3*d* ions. Another quote of a British chemist Sir William Crookes from his address to the British Association in 1887 (at that time nobody would even think of lasers, LED, phosphor materials, etc.!) is a kind of a look through the magic crystal ball into the future with foreseen outstanding importance of the lanthanides for science and technology: "These elements (rare earths) perplex us in our researches, baffle us in our speculations and haunt us in our dreams. They stretch like an unknown sea before us—mocking, mystifying and murmuring strange revelations and possibilities."

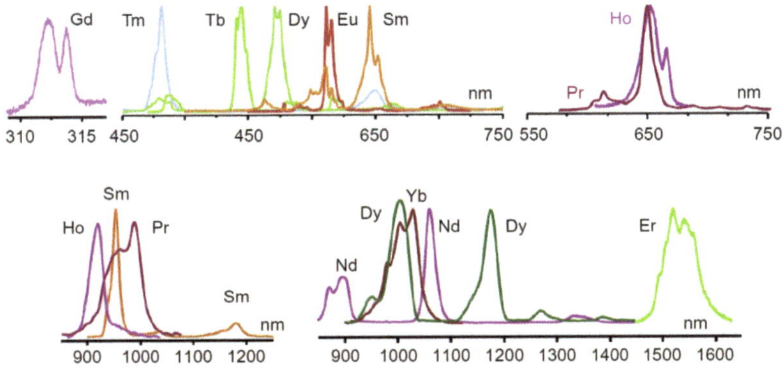

Figure 1.4 Normalized emission spectra of the trivalent lanthanide ions. Reproduced with permission from Ref. 10. All spectra are normalized to their maxima for convenience.

The areas of applications of the lanthanide ions are extremely diverse[8–10] and include lasers, telecommunications industry, lighting, security markers (to prevent banknotes counterfeiting, for example), luminescent biomarkers, non-invasive thermometry, persistent phosphors for the safety signs to the escape roots, scintillators for detecting X-ray and γ-radiation, solar energy converters, etc. Each of the above-mentioned application fields represents a wide attractive and quickly developing area of research, thousands of scientific papers and hundreds of scientific books focused on those topics are published every year in the world.

However, all these examples—no matter how diverse they are—have one common thing: At their cores there are always activators, those transition metal and lanthanide ions. The light absorption and emission occur between the states of those impurity ions, which do or sometimes do not interact strongly with the nearest crystalline environment. How these states are formed, what are the main physical mechanisms underlying their behavior, how they can be described, understood, quantified, calculated—all these questions constitute the main content of this book.

A few remarks regarding the structure of the whole book can be appropriate now. The book starts with very basic description of interaction of light with matter (Chapter 2), continues with the description of theory of atom of hydrogen (Chapter 3), which is necessary to make the stage for a quantified discussion of the ions with many-electron configurations (Chapter 4). After that, the spectroscopic properties of these ions in a free state are described in detail in the book (Chapters 5 and 6). A separate chapter is devoted to the group theory (Chapter 7), since this theory gives a very convenient and extremely beautiful method to analyze the spectra of the impurity ions in crystals employing the concept of symmetry. These seven chapters are a good preparation to describe the spectroscopic properties of the impurity ions in crystals (Chapter 8, theory of crystal field and Chapter 9, electron-vibrational interaction). Finally, Chapter 10 finishes the book with selected examples of a joint application of the first-principles and semi-empirical methods to the description of the optical properties of doped crystals, containing impurity ions. Many of the examples considered in the book are from the authors' original publications. The CGS system units are used in the book, except for some equations in Chapter 10 (Judd–Ofelt theory), when the SI units are employed.

Finally, let us repeat once more "*Fiat lux!*" and wish the reader a pleasant and interesting journey through the pages of this book.

Bibliography

1. Maiman TH (1960). Stimulated optical radiation in ruby. *Nature* **160** 493–494.

2. Ye, S, Xiao, F, Pan, YX, Ma, YY, Zhang, QY (2010). Phosphors in phosphor-converted white light-emitting diodes Recent advances in materials, techniques and properties. *Mater Sci Eng R Rep* **71**(1) 1–34.
3. George NC, Denault KA, Seshadri R (2013). Phosphors for solid-state white lighting. *Ann Rev Mater Res* **43** 481–501.
4. Xia ZG, Liu QL (2016). Progress in discovery and structural design of color conversion phosphors for LEDs. *Prog Mater Sci* **84** 59–117.
5. Lin CC, Meijerink A, Liu RS (2016). Critical red components for next-generation white LEDs. *J Phys Chem Lett* **7**(3) 495–503.
6. Kück S (2001). Laser-related spectroscopy of ion-doped crystals for tunable solid-state lasers, *Appl Phys B* **72**(5), 515–562.
7. Pimentel CG, Sprately RD (1971). *Understanding Chemistry*. Holden-Day Inc., San Francisco.
8. Bünzli JCG, Piguet C (2005). Taking advantage of luminescent lanthanide ions. *Chem Soc Rev* **34** 1048–1077.
9. Eliseeva SV, Bünzli JCG (2010). Lanthanide luminescence for functional materials and bio-sciences. *Chem Soc Rev* **39** 189–227.
10. Bünzli JCG (2016). Lanthanide Luminescence: From a Mystery to Rationalization, Understanding, and Applications. In: Jean-Claude G, Bünzli, Vitalij K Pecharsky, eds, *Handbook on the Physics and Chemistry of Rare Earths, vol. 50*, Amsterdam: Elsevier, pp. 141–176.

Chapter 2

Basic Processes of Interaction of Radiation with Matter

2.1 Introduction

The elementary processes of the radiation absorption and emission in a generalized quantum-mechanical system with a discrete energy level scheme will form the main subject of this chapter. The whole consideration will be based on a mixed description: The absorbed/emitted radiation will be treated classically, as a monochromatic plane wave, whereas the atomic system that can absorb/emit such radiation will be described quantum-mechanically. Although such approach may seem to be somewhat inconsistent, it allows one, nevertheless, to establish basic properties of the emission and absorption transitions, the shapes and widths of the emission spectral lines, time-dependent behavior of populations of individual energy levels. It should be emphasized that the presented description is general, and there is no need to specify exactly the origin of the considered quantum-mechanical system. The obtained results are valid for any atom or ensemble of atoms.

Theoretical Spectroscopy of Transition Metal and Rare Earth Ions: From Free State to Crystal Field
Mikhail G. Brik and Chong-Geng Ma

ISBN 978-981-4800-56-3 (Hardcover), 978-0-429-27875-4 (eBook)
www.jennystanford.com

2.2 Absorption/Emission Transitions and Einstein Coefficients

Let us start our discussion with the simplest case possible—a two-level system with energy levels E_2 and E_1 ($E_2 > E_1$). The E_1 level is called the ground state, whereas the E_2 level is referred to as an excited state. If such a system is illuminated by a beam of electromagnetic radiation, whose photons have the energy $h\nu = E_2 - E_1$, then the following three processes can take place: (i) induced absorption, (ii) induced emission and (iii) spontaneous emission (see Fig. 2.1).

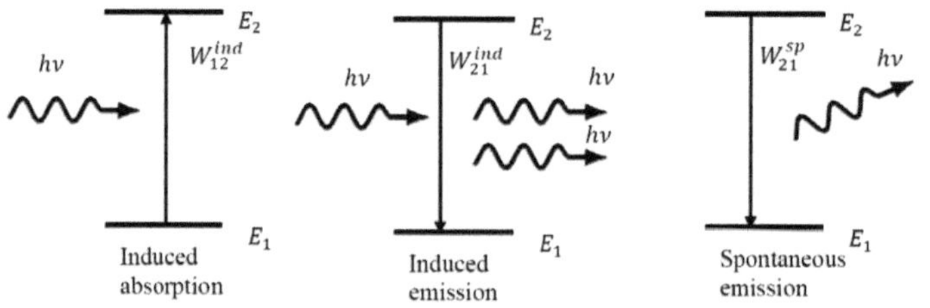

Figure 2.1 Possible transitions in a two-level system with energy levels E_2 and E_1 ($E_2 > E_1$). The absorption and emission transitions are shown by the up- and down- ward arrows, respectively.

As a result of induced absorption, the atom "jumps" from the ground state to the excited state, and the number of photons in the incident beam is decreased by one.

The second process, induced emission, requires an atom to be already in the excited state. One of the photons in the incident beam "kicks away" the atom from the excited state to the ground state with a subsequent emission of the photon $h\nu = E_2 - E_1$. The number of the photons in the incident beam is increased then by one. It is essential that in the case of induced emission the emitted photon and the photon, which has caused the induced emission transition, are identical, i.e., they have the same energy, polarization, direction of propagation and phase.

Finally, the third process, spontaneous emission, also implies an atom to be in the excited state. However, no incident radiation beam is needed, and the atom "falls down" after a certain time interval (called the lifetime of this excited state) to the ground state without any external reason. In this case the photon $h\nu = E_2 - E_1$ is emitted, whose phase and direction of propagation are absolutely arbitrary. The spontaneous emission is a purely quantum mechanical phenomenon and it cannot be interpreted from the classical physics point of view. The induced transitions can be understood by invoking the classical mechanics concepts such as an interaction between a harmonic oscillator and an external force.

Existence of these types of transitions was suggested by A. Einstein in 1916. Since all three of them imply involvement of the photons—either absorbed or emitted—these transitions are also called radiative. In molecules or in crystals, however, another type of transitions can occur—the so-called non-radiative transitions, when the electrons in atoms change their electronic states without absorption or emission of a photon, but on account of exchange of energy with vibrations of the nearest neighboring atoms. We shall not consider such processes in this chapter, but will return to this point later, when discussing the spectra of impurities in crystals in Chapter 9.

The induced transitions have the following important properties:

1. the probability of the induced transitions is not zero only if the incident photons have the energy $h\nu = E_2 - E_1$ equal to the difference of energies of two states involved into a transition (the resonance condition);
2. the photons emitted in the induced emission transitions are coherent;
3. the probability W_{12}^{ind} and W_{21}^{ind} of the induced transitions (per unit of time) is directly proportional to the spectral volume density of the electromagnetic field energy ρ_ν:

$$W_{12}^{\text{ind}} = B_{12}\rho_\nu, \tag{2.1}$$

$$W_{21}^{\text{ind}} = B_{21}\rho_\nu, \tag{2.2}$$

where B_{12} and B_{21} are the Einstein coefficients for the induced absorption and induced emission, respectively. (The order of indices in both subscripts is important and determines the direction of the transition. In the general case, $B_{12} \neq B_{21}$.)

In the spontaneous emission processes the emitting atom can be thought of as a classical oscillating dipole, whose amplitude decreases due to the energy loss through the radiation. The probability dw^{sp} of a spontaneous transition from the excited state to a lower state during the time interval dt can be written as

$$dw^{sp} = A_{21}dt, \tag{2.3}$$

with A_{21} being the Einstein coefficient for the spontaneous emission.

A deeper insight into the properties of the induced and spontaneous transitions can be gained if we consider an ensemble of particles, and each particle can occupy any of the above-mentioned two energy levels. If this ensemble is in the state of thermodynamic equilibrium at absolute temperature T, the numbers of particles n_1 and n_2 at each level follow the Boltzmann's statistics:

$$\frac{n_2}{g_2} = \frac{n_1}{g_1}\exp\left(-\frac{E_2 - E_1}{kT}\right), \tag{2.4}$$

where g_1 and g_2 are the statistical weights or degeneracies of the considered states, and k is the Boltzmann's constant. If the thermodynamic equilibrium is maintained, the energy of the whole ensemble is not changed, which means that the number of the absorption transitions should be equal to the number of the emission transitions for the same time interval. The number N_{12} of the "1→2" transitions can be determined as a product of the number of particles n_1 in the state "1" and the probability $W_{12}{}^{\text{ind}}$ of such a transition. Since the absorption can be realized only on account of the induced absorption transitions, we can write

$$N_{12} = n_1 W_{12}{}^{\text{ind}} = n_1 B_{12}\rho_\nu, \tag{2.5}$$

where Eq. (2.1) is used. The emission processes, on the contrary, can be a sum of the induced and spontaneous emission transitions. Then for the total number of the "2→1" transitions, we will have (Eqs. (2.2) and (2.3) were used)

$$N_{21} = n_2 W_{21} = n_2\left(W_{21}{}^{\text{ind}} + A_{21}\right) = n_2\left(B_{21}\rho_\nu + A_{21}\right). \tag{2.6}$$

Since the equilibrium condition implies $N_{21} = N_{12}$, the following expression for the spectral volume density of the electromagnetic field energy ρ_ν can be obtained from Eqs. (2.4)–(2.6):

$$\rho_\nu = \frac{A_{21}}{B_{21}} \frac{1}{\frac{B_{12}}{B_{21}} \frac{g_1}{g_2} \exp\left(\frac{E_2 - E_1}{kT}\right) - 1}. \tag{2.7}$$

A further very important step was made by A. Einstein, who postulated that the radiation emitted and absorbed by a quantum-mechanical system in thermodynamic equilibrium, should be described by the Planck's distribution of the black body radiation energy:

$$\rho_\nu = \frac{8\pi\nu^2}{c^3} \frac{h\nu}{\exp\left(\frac{h\nu}{kT}\right) - 1}, \tag{2.8}$$

where h and c are Planck's constant and speed of light, respectively, and ν stands for the radiation frequency. The comparison of Eqs. (2.7) and (2.8) gives the following results:

$$h\nu = E_2 - E_1, \tag{2.9}$$

$$B_{12} g_1 = B_{21} g_2, \tag{2.10}$$

$$A_{21} = \frac{8\pi\nu^2}{c^3} h\nu B_{21}. \tag{2.11}$$

The first Eq. (2.9) is just Bohr's postulate: The energy emitted (or absorbed) by an atom is equal to the energy difference between two stationary states involved into a transition. Equation (2.10) shows equal probabilities of the induced absorption and induced emission, and the Einstein coefficients for these two processes are equal to each other $B_{12} = B_{21}$ only if $g_1 = g_2$. Finally, Eq. (2.11) links together the Einstein coefficients for the induced emission and spontaneous emission.

Taking the frequency of the visible light as $\sim 10^{14}$ Hz, it is possible to estimate that $A_{21} \sim 6.2 \times 10^{-16} B_{21}$, which means that the Einstein coefficient for the spontaneous emission is much smaller than the Einstein coefficient for the induced emission. If the spontaneous emission probability is zero for a particular transition (this may happen if the transition between a certain pair of energy levels is forbidden by the selection rules—this will be discussed later), the induced emission transition between these states is also impossible. This conclusion has a great importance for laser applications.

2.3 Natural Width of the Emission Lines and Their Different Profiles

As follows from Eq. (2.9), the emission or absorption transitions between two states are possible if the energy of the photons involved into such a transition is equal to the energy difference between these states. If this resonance condition is fulfilled, the absorption/emission lines in the experimental spectra would appear as infinitely narrow peaks, which is not the real case. All spectral lines are characterized by a definite non-zero width. Why does it happen? One of the possibilities to explain this phenomenon is to refer to one of the main principles of quantum mechanics—the Heisenberg's uncertainty relations. Those physical quantities, whose quantum-mechanical operators do not commute, cannot be measured at the same time. The pairs of such quantities that cannot be determined simultaneously are, for example, the momentum p and coordinate r, the energy E and time t. The uncertainties in determining both E and t are linked by the following inequality:

$$\Delta E \cdot \Delta t \geq \hbar \ , \tag{2.12}$$

where $\hbar = h/2\pi$. The energy levels are not infinitely sharp. Considering the energy uncertainty for both states involved into a transition and the fact that the transition can take place for any values of energy within those uncertainties, we already arrive at somewhat non-zero width of the emission line (Fig. 2.2, where different lengths of the downward arrows symbolize different energies of the emission transition within the energy uncertainties of both ground and excited states). Equation (2.12) in terms of the frequency ν and time t looks as follows:

$$\Delta\nu \geq \frac{1}{2\pi \cdot \Delta t}. \tag{2.13}$$

If Δt is the uncertainty of the excited state lifetime, then, as follows from Eq. (2.13), the smaller is Δt, the greater is the uncertainty in frequency, or the wider the corresponding emission band is. Therefore, since the lifetime of any excited state is always finite (due to the spontaneous emission an atom in any excited state sooner or later relaxes to the lower states until it reaches the ground state), the spectral lines have a definite non-zero width. The intensities of the absorption or emission transitions are not zero when $h\nu \neq E_2 - E_1$, which means that the transitions can take place not at exactly resonant frequency, but in its vicinity as well.

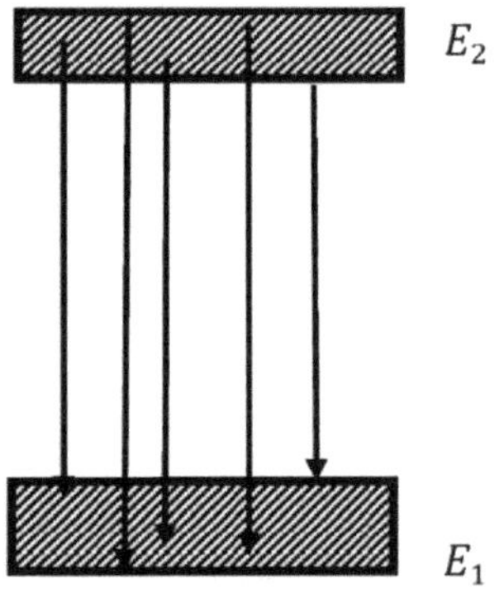

Figure 2.2 Illustration of a finite width of the emission line by uncertainties in the determination of energy of both levels involved into a transition.

Mathematical derivation of distribution of the emission intensity in the atomic spectra can be accomplished using a model of the damped harmonic oscillator, whose equation is as follows:

$$\ddot{x} + \gamma \dot{x} + \omega_{01}{}^2 x = 0, \tag{2.14}$$

where $\omega_{01} = \dfrac{E_2 - E_1}{\hbar}$ is the resonance frequency, a dot above x means the time derivative, and γ denotes the damping force coefficient. Following Bohr's theory, the emission of the photon is followed by a transition of an electron from an orbit with a greater radius to an orbit with a smaller radius, which exactly can be understood as a decreased amplitude of the harmonic oscillator. The solution of this differential equation has the following form

$$x(t) = A\exp\left(-\frac{\gamma t}{2}\right) \cdot \exp(i\omega_1 t),\ \omega_1 = \sqrt{\omega_{01}{}^2 - \left(\frac{\gamma}{2}\right)^2}, \tag{2.15}$$

with A staying for the amplitude of the oscillations and $i = \sqrt{-1}$. After a Fourier transform of this time-dependent function $x(t)$, we get the frequency-dependent function $g(\omega)$:

$$g(\omega)=\frac{1}{\sqrt{2\pi}}\int_{-\infty}^{\infty}x(t)\cdot\exp(-i\omega t)dt=\frac{A}{\sqrt{2\pi}}\cdot\frac{\frac{\gamma}{2}+i(\omega_1-\omega)}{(\omega_1-\omega)^2+\left(\frac{\gamma}{2}\right)^2}. \tag{2.16}$$

This is a complex function, which is also called a form-factor, and the intensity distribution $I(\omega)$ in the emission spectrum is given by

$$I(\omega)=g(\omega)g^*(\omega)=\frac{A^2}{2\pi}\cdot\frac{1}{(\omega_1-\omega)^2+\left(\frac{\gamma}{2}\right)^2}, \tag{2.17}$$

where the complex conjugation is denoted by a star. Equation (2.17) is the Lorentzian function, and its general shape is depicted in Fig. 2.3.

The emission band, whose shape is described by such a function, is said to have a Lorentzian profile. The spontaneous emission lines have such profile. It is easy to see that function (2.17) has a maximum at $\omega = \omega_1$. To distinguish quantitatively between the lines with different widths, a concept of the full width at half maximum (FWHM) is introduced. It is defined as the frequency interval, in which the emission intensity decreases to one half from its maximum value I_{max} (Fig. 2.3). It can be shown in a straightforward way that in the case of the Lorentzian function (2.17) the FWHM is equal to γ.

The natural spectral width of the emission lines can be roughly estimated as an inverse value of the excited state lifetime (see Eq. (2.13)). For example, if the excited state lifetime is 10 ns, then the corresponding emission line width will be approximately 16 MHz, or about 10^{-8} from the visible light frequency.

So, the finite lifetime of any excited level leads to the broadening of the emission lines of the transitions from that level. Such broadening is referred to as the homogeneous broadening. If an ensemble of identical atoms is considered and if, for simplicity, each atom emits radiation with the same spectral shape and FWHM, it is impossible to ascribe a definite spectral component $g(\omega)$ with intensity $I(\omega)$ to a definite atom. In this case, such a broadening is identical for all atoms in the ensemble.

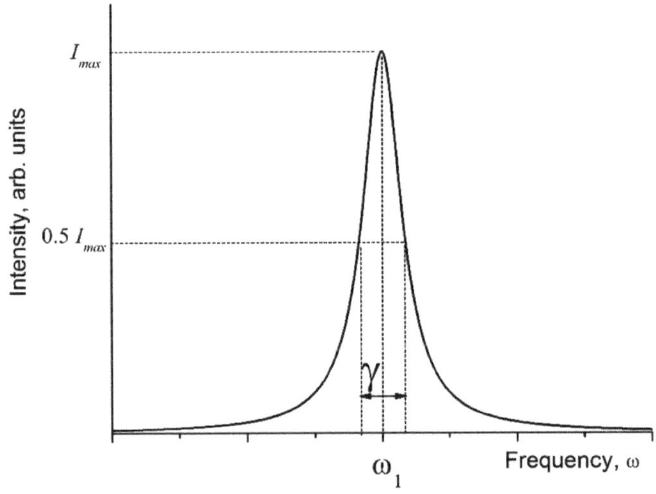

Figure 2.3 Lorentzian profile of the emission line.

However, it is not easy to create such an ensemble. In real atomic ensembles, there can be several sorts of atoms of different chemical elements, for example, or the individual conditions for every atom can be different from each other. Then the spectral lines will be inhomogeneously broadened. The simplest example of the inhomogeneous broadening is the Doppler broadening of the emission lines of gas molecules. This effect can also be quantified.

All molecules in gases are never at rest but participate in thermal motion. Therefore, the emission/absorption processes will be taking place in the moving molecules or atoms, and the Doppler effect should be considered. If a molecule moves at a velocity u with respect to an observer (or to a spectral detector) and emits radiation at the same time, the observer will detect a shift in the frequency of the received radiation if compared to the frequency of the emitted one. The registered frequency v will be

$$v = v_0 \left(1 \pm \frac{u}{c}\right), \tag{2.18}$$

where v_0 is the frequency of radiation when the emitter is at rest. The positive/negative signs correspond to the motion of the emitting molecule to/from the observer, and c is the speed of light. Even if all molecules in the considered emitting ensemble are identical, their velocities that are described by a Maxwell distribution for an ideal gas are different. Therefore, the registered frequency will be varied for all the particles depending on the direction and magnitude of their velocity.

If the velocity distribution function is denoted as $p(u)$, then the form-factor $q(\nu)$ of the spectral line will be proportional to $p(u)$:

$$q(\nu)d\nu = p(u)du, \tag{2.19}$$

with $d\nu$ and du being the frequency and the velocity intervals, respectively. The $p(u)$ function is the Maxwell velocity distribution for an ideal gas:

$$p(u) = \left(\frac{m}{2\pi kT}\right)^{1/2} \exp\left(-\frac{mu^2}{2kT}\right), \tag{2.20}$$

where m is the mass of an individual molecule, and T is the absolute temperature. Since $\nu = \nu_0\left(1+\frac{u}{c}\right)$ (here for certainty we assume that the emitting molecule moves to the observer), then $u = c\frac{\nu-\nu_0}{\nu_0}$ and $du = \frac{c}{\nu_0}d\nu$. Combining these expressions with Eqs. (2.19) and (2.20), we get for the form-factor

$$q(\nu) = \frac{c}{u_0\nu_0}\frac{1}{\sqrt{\pi}}\exp\left(-\frac{c^2}{u_0{}^2}\left(\frac{\nu-\nu_0}{\nu_0}\right)^2\right). \tag{2.21}$$

This is the Gaussian function. As in the previous case of the Lorentzian profile, it is also possible to determine the FWHM, which turns out to be $2\nu_0\frac{u_0}{c}\sqrt{\ln 2}$, where $u_0 = \left(\frac{2kT}{m}\right)^{1/2}$. A rough estimation of the FWHM value in this case gives about 10^{-6} from the visible light frequency, which is about two orders of magnitude greater than the natural line width caused by the spontaneous emission.

Figure 2.4 shows the comparison between these two line shapes—the Lorentzian (solid lines) and the Gaussian (dashed lines). In Fig. 2.4a, both functions are normalized to have the same area under the curves, whereas in Fig. 2.4b both plotted functions have the same maximum value. The Gaussian function drops off faster than the Lorentzian when moving to the left or to the right from the central frequency, at which both functions reach their maxima.

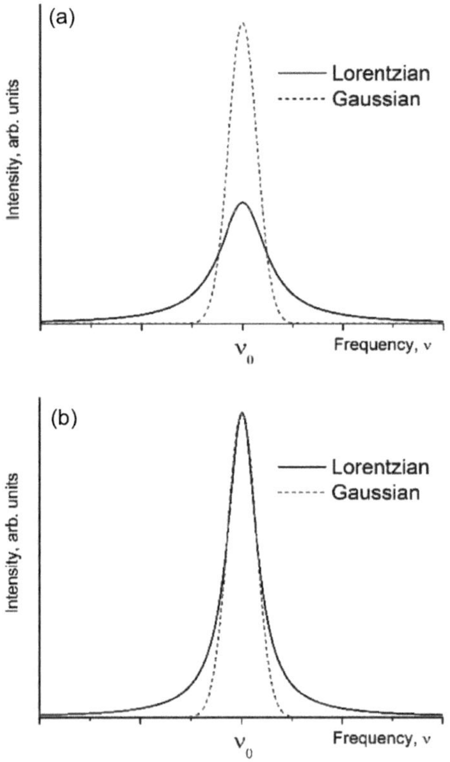

Figure 2.4 Comparison between the Gaussian (dashed lines) and Lorentzian (solid line) line shapes. The functions have the same area (a) or the same maxima (b).

2.4 Two-Level System, Rabi Oscillations

Let us consider again a two-level quantum mechanical system shown in Fig. 2.1. For the sake of simplicity, let us also assume that there is only one atom in this system. In this paragraph we shall analyze the absorption and emission transitions between these two states, under conditions of a continuous illumination by the monochromatic light with frequency ω. The wave functions of the considered states $|\Psi_1\rangle$ and $|\Psi_2\rangle$ should be the solutions of the time-dependent Schrödinger equation

$$i\hbar\frac{\partial\Psi}{\partial t}=\widehat{H}\Psi. \tag{2.22}$$

The coordinate and time variables can be separated (the ϕ_1, ϕ_2 functions depend only on the spatial coordinates):

$$\Psi_1=\phi_1\exp\left(i\frac{E_1t}{\hbar}\right),\Psi_2=\phi_2\exp\left(i\frac{E_2t}{\hbar}\right). \tag{2.23}$$

These functions should be orthonormalized:

$$\int\left|\Psi_1\right|^2dV=\int\left|\Psi_2\right|^2dV=1,\quad\int\Psi_1{}^*\Psi_2dV=\int\Psi_2{}^*\Psi_1dV=0. \tag{2.24}$$

For our discussion there is no need to indicate the details of the Hamiltonian $\widehat{H}$; the main requirement is that the $|\Psi_1\rangle$, $|\Psi_2\rangle$ functions have the properties determined by Eqs. (2.23) and (2.24).

If the system is subjected to an additional interaction $\widehat{H}'$ (called a perturbation) with its environment, the new Hamiltonian of a system becomes $\widehat{H}+\widehat{H}'$. By this perturbation in this particular case we shall understand interaction of an atom with an incident electromagnetic wave. As a result, a new mixed state $|\Psi\rangle$ can be formed, and this state is a superposition of both $|\Psi_1\rangle$ and $|\Psi_2\rangle$:

$$|\Psi\rangle=a|\Psi_1\rangle+b|\Psi_2\rangle, \tag{2.25}$$

where a and b are the complex coefficients satisfying the following condition: $|a|^2+|b|^2 = 1$. The probabilities to find the system in the first or in the second state are given by the $|a|^2$ or $|b|^2$ numbers, respectively. It is natural to assume that at the initial moment $t = 0$ these coefficients have the following values: $a = 1$, $b = 0$, which means that initially the atom is in the ground state. After the perturbation is "switched on," both coefficients a, b become the time-dependent functions. Successful description of the time-dependent state of the system implies knowledge of the variation of both a and b coefficients with time.

At first, Eq. (2.25) should be substituted into the perturbed time-dependent Schrödinger equation:

$$i\hbar\frac{\partial\Psi}{\partial t}=\left(\widehat{H}+\widehat{H}'\right)\Psi. \tag{2.26}$$

Using the fact that $i\hbar\frac{\partial\Psi_1}{\partial t}=\widehat{H}\Psi_1$, $i\hbar\frac{\partial\Psi_2}{\partial t}=\widehat{H}\Psi_2$, we easily get

$$i\hbar\Psi_1\frac{\partial a}{\partial t}+i\hbar\Psi_2\frac{\partial b}{\partial t}=a\widehat{H}'\Psi_1+b\widehat{H}'\Psi_2, \tag{2.27}$$

As the next step, we multiply both parts of Eq. (2.27) by Ψ_2^* from the left side and integrate over all coordinates to arrive at the following expression:

$$i\hbar\frac{\partial b}{\partial t}=a(t)\exp\left(\frac{i(E_1-E_2)t}{\hbar}\right)\int\phi_2^*\widehat{H}'\phi_1 dV+b(t)\int\phi_2^*\widehat{H}'\phi_2 dV. \tag{2.28}$$

If Eq. (2.27) will be multiplied by Ψ_1^* and also integrated over all coordinates, a similar equation for $\frac{\partial a}{\partial t}$ can be obtained. Now it is time to specify the perturbation operator $\widehat{H}'$, which can be chosen as a dipole operator of interaction of an atom with an incident monochromatic electromagnetic wave with frequency ω:

$$\widehat{H}'=-e\vec{\mathbf{r}}\cdot\vec{\mathbf{E}}(\exp(-i\omega t)+\exp(i\omega t))=-\mu E(\exp(-i\omega t)+\exp(i\omega t)). \tag{2.29}$$

Here e denotes the electron charge, $\vec{\mathbf{r}}$ and $\vec{\mathbf{E}}$ stand for the electron's radial coordinate and electric field vector of the incident wave, respectively, and the atom's dipole moment is μ.

At first, we consider a special case of a short-lived perturbation, which means that the atom is exposed to a very short pulse of radiation. Then the values of $a(t)$ and $b(t)$ can be replaced by their values at the initial moment of time $a(0) = 1$, $b(0) = 0$, which transforms Eq. (2.28) to

$$\frac{\partial b}{\partial t}=-\frac{1}{i\hbar}\langle\mu\rangle E\left(e^{i(\omega_0+\omega)t}+e^{-i(\omega_0-\omega)t}\right), \tag{2.30}$$

where $\langle\mu\rangle=\int\phi_2^{*}\mu\phi_1 dV$ is the matrix element of the dipole operator, and $\omega_0=\frac{E_2-E_1}{\hbar}$ is the resonant frequency of the transition between these two states.

The next simplification is made by involving the so-called "rotating wave approximation." Both frequencies ω_0 and ω are of the same order of magnitude. Therefore, the first exponent in the right-hand side of Eq. (2.30) oscillates at much higher frequency than the second one. When integrating over time, the first exponent averages to zero, which allows to simply drop it off. Finally, we get a very simple differential equation

$$\frac{\partial b}{\partial t}=-\frac{1}{i\hbar}\langle\mu\rangle Ee^{-i(\omega_0-\omega)t}, \tag{2.31}$$

which can be easily integrated over time from 0 to t (which corresponds to the moment when the perturbation will be switched off) to yield

$$b(t)=\frac{\langle\mu\rangle E}{\hbar}\frac{e^{-i(\omega_0-\omega)t}-1}{\omega-\omega_0}. \tag{2.32}$$

The $b(t)$ coefficient appears to be a complex function, and the square of its absolute value returns the probability to find the system after time t in the $|\Psi_2\rangle$ state. Physically it is equal to the probability of the "$1\rightarrow 2$" transition:

$$|b(t)|^2=\left(\frac{\langle\mu\rangle E}{\hbar}\right)^2\cdot\frac{\sin^2\left(\frac{\omega_0-\omega}{2}t\right)}{\left(\frac{\omega_0-\omega}{2}\right)^2}. \tag{2.33}$$

As seen from Eq. (2.33), if the frequency ω of the incident radiation is far from the resonance frequency ω_0, then the probability of the "$1\rightarrow 2$" transition rapidly goes to zero. Figure 2.5 shows a general shape of the function (2.33).

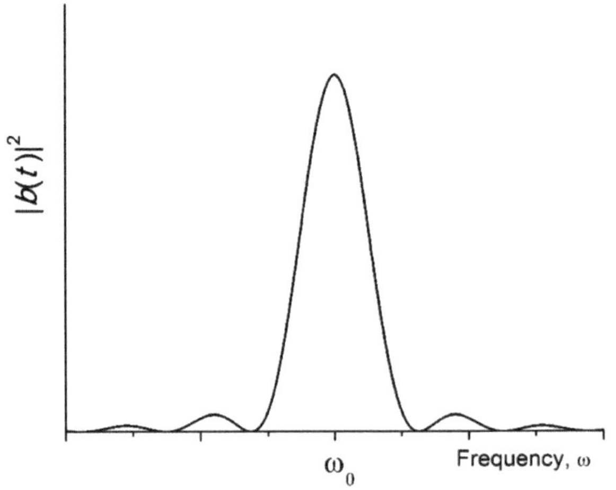

Figure 2.5 A plot of Eq. (2.33).

Although the $b(t)$ function has a pronounced resonant character and reaches its maximum at $\omega = \omega_0$, it is, nevertheless, not absolutely zero if the resonance condition is not fulfilled (Fig. 2.5). There exists a non-zero probability of absorption of a photon with the off-resonance energy. Therefore, to find a total probability Π of the "1 $\rightarrow$ 2" transition, Eq. (2.33) has to be integrated over whole range of frequencies. First, the square of the electric field vector E should be replaced by the following expression $E^2 = \frac{8\pi\rho_\nu}{3}$ (which relates the spectral volume density of electric field to the electric field vector). Then we have

$$\Pi = \int_{-\infty}^{\infty} |b(t)|^2 d\nu = \left(\frac{\langle\mu\rangle}{\hbar}\right)^2 \frac{8\pi}{3} \int_{-\infty}^{\infty} \rho_\nu \frac{\sin^2(\nu_0 - \nu)\pi t}{\pi^2(\nu_0 - \nu)^2} d\nu, \qquad (2.34)$$

where $\omega = 2\pi\nu$ is used instead of ν. Since the $\frac{\sin^2 x}{x^2}$ function has a sharp maximum, the frequency-dependent function ρ_ν (see Eq. (2.8)) can be moved out of integral and replaced by its value at $\nu = \nu_0$.

The remaining integral $\int_{-\infty}^{\infty} \frac{\sin^2 x}{x^2} dx$ is equal to π. Finally, we arrive at the following result:

$$\Pi = \frac{8\pi}{3}\left(\frac{\langle\mu\rangle}{\hbar}\right)^2 \rho_\nu(\nu_0)t. \tag{2.35}$$

If we divide Eq. (2.35) by t, we get the probability of the "$1 \rightarrow 2$" transition per a unit of time (per second), which can be also expressed in terms of the Einstein coefficient for the induced absorption B_{12}(see Eq. (2.1)). This allows to obtain a remarkable result, namely, that the Einstein coefficient of the induced absorption is proportional to the square of the absolute value of the electric dipole operator matrix element calculated between the wave functions of the states involved into a transition:

$$B_{12} = \frac{8\pi}{3}\left(\frac{\langle\mu\rangle}{\hbar}\right)^2. \tag{2.36}$$

If the perturbation is not short-lived and if the frequency ω of the incident radiation is close (but may not be equal) to the resonance frequency, the problem of analysis of time-evolution of the system becomes more complicated. Let us introduce the parameter $\delta = \omega_0 - \omega$, which shows how far from resonance the frequency of the incident monochromatic wave is. Near the resonance frequency $\delta << \omega_0$, ω. In a complete analogy with the previous case, we can obtain the equations for both $\frac{\partial a}{\partial t}$ and $\frac{\partial b}{\partial t}$ (the rotating wave approximation should be used again):

$$i\hbar\frac{\partial a}{\partial t} = -b(t)e^{i\delta t}\int \Psi_1{}^* \mu E \Psi_2 dV, \tag{2.37}$$

$$i\hbar\frac{\partial b}{\partial t} = -a(t)e^{-i\delta t}\int \Psi_2{}^* \mu E \Psi_1 dV. \tag{2.38}$$

For a convenience, we introduce here a short-hand notation $F = -\int \phi_2 * \mu E \phi_1 dV$ and write down two equations for the unknown functions $a(t)$ and $b(t)$ (for simplicity, the argument t is dropped in these equations):

$$i\hbar\frac{\partial a}{\partial t} = bF^* e^{i\delta t}, \tag{2.39}$$

$$i\hbar\frac{\partial b}{\partial t} = aFe^{-i\delta t}. \tag{2.40}$$

Expressing a from Eq. (2.40) and substituting it into Eq. (2.39), we arrive at the second-order homogeneous differential equation for the time-dependent b coefficient:

$$\frac{\partial^2 b}{\partial t^2} + i\delta \frac{\partial b}{\partial t} + b\frac{F^2}{\hbar^2} = 0. \tag{2.41}$$

The linear combination of the function b itself, its first and second derivatives equals to zero. This can be realized if and only if the unknown function $b(t)$ is an imaginary exponent. Therefore, the general form of the solution of this equation will be as follows:

$$b = Re^{-i\bar{\omega} t}, \tag{2.42}$$

where B is a constant. Introducing notation

$$\Omega = \sqrt{\frac{\delta^2}{4} + \frac{F^2}{\hbar^2}}, \tag{2.43}$$

we find that

$$b(t) = A\exp\left\{-i\left(\frac{\delta}{2} - \Omega\right)t\right\} + B\exp\left\{-i\left(\frac{\delta}{2} + \Omega\right)t\right\}. \tag{2.44}$$

Since $a = \frac{i\hbar}{F} e^{i\delta t} \frac{\partial b}{\partial t}$ (Eq. (2.40)), we get

$$a(t) = \frac{\hbar}{F} A\left(\frac{\delta}{2} - \Omega\right)\exp\left\{i\left(\frac{\delta}{2} + \Omega\right)t\right\} + \frac{\hbar}{F} B\left(\frac{\delta}{2} + \Omega\right)\exp\left\{i\left(\frac{\delta}{2} - \Omega\right)t\right\}. \tag{2.45}$$

Finally, the general expression for the wave function of the mixed state $|\Psi\rangle = a|\Psi_1\rangle + b|\Psi_2\rangle$ will be written as

$$|\Psi\rangle = |\Psi_1\rangle \frac{\hbar}{F} \left\{A\left(\frac{\delta}{2} - \Omega\right)\exp\left\{i\left(\frac{\delta}{2} + \Omega\right)t\right\} + B\left(\frac{\delta}{2} + \Omega\right)\exp\left\{i\left(\frac{\delta}{2} - \Omega\right)t\right\}\right\} + |\Psi_2\rangle\left\{A\exp\left\{-i\left(\frac{\delta}{2} - \Omega\right)t\right\} + B\exp\left\{-i\left(\frac{\delta}{2} + \Omega\right)t\right\}\right\}. \tag{2.46}$$

The unknown so far values of the constants of integration A and B can be found from the initial conditions. As we assumed earlier, at

$t = 0$ $|\Psi\rangle = |\Psi_1\rangle$. This condition is equivalent to the following system of two linear equations:

$$A + B = 0, \tag{2.47}$$

$$A\frac{\hbar}{F}\left(\frac{\delta}{2} - \Omega\right) + B\frac{\hbar}{F}\left(\frac{\delta}{2} + \Omega\right) = 1. \tag{2.48}$$

The solution is

$$A = -B = \frac{F}{2\hbar\Omega}. \tag{2.49}$$

The square of the absolute value of the coefficient at $|\Psi_2\rangle$ gives us the probability to find the atom in the second state. After simple algebraic transformations, we get the final expression for the probability of the "$1 \rightarrow 2$" transition:

$$W = \frac{F^2}{2\hbar^2\Omega^2}(1 - \cos 2\Omega t). \tag{2.50}$$

It oscillates with frequency 2Ω between 0 and its maximal value $\frac{F^2}{\hbar^2\Omega^2}$ (Fig. 2.6).

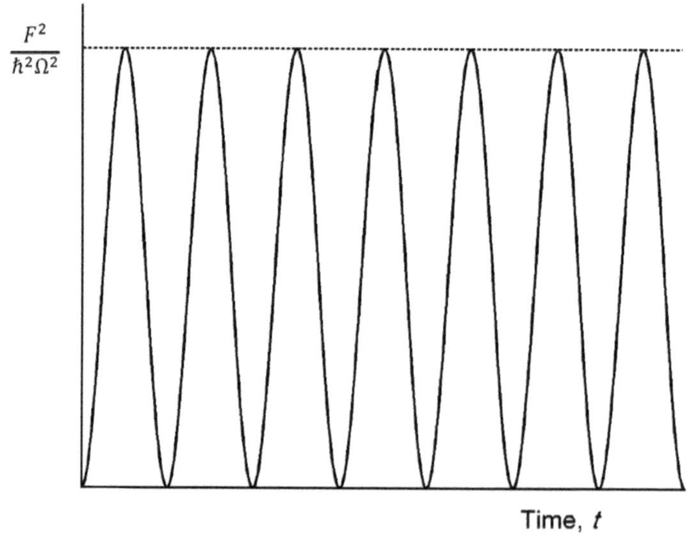

Figure 2.6 A plot of Eq. (2.50).

Let us introduce a new parameter

$$\Omega_R = \frac{\langle \mu E \rangle}{\hbar}, \tag{2.51}$$

which has dimension of frequency and is called the "Rabi frequency" (after the Nobel Prize winner I. I. Rabi (1898–1988)). Then $\Omega = \sqrt{\frac{\delta^2}{4} + \frac{\langle \mu E \rangle^2}{\hbar^2}} = \sqrt{\frac{\delta^2}{4} + \Omega_R{}^2}$ and

$$W = \frac{2\Omega_R{}^2}{\delta^2 + 4\Omega_R{}^2}\left(1 - \cos\left(\delta^2 + 4\Omega_R{}^2\right)^{1/2} t\right). \tag{2.52}$$

Now two interesting limiting cases can be considered. The first one is when $\delta << \Omega_R$, which corresponds either to almost exact resonance $\delta \sim 0$ or very intensive incident field with very large values of module of the electric field vector E. The second situation is the opposite inequality $\delta >> \Omega_R$, which means that the incident radiation frequency is very far from resonance or intensity of the incident radiation is very low. In the first case we can drop δ from Eq. (2.52) to have

$$W = \frac{1}{2}(1 - \cos 2\Omega_R t),\ 0 \le W \le 1. \tag{2.53}$$

This equation can be understood as follows: If the resonance condition is almost fulfilled or the incident radiation is very intensive (even if the incident radiation frequency is greatly detuned from the resonance), the probability of the "1 → 2" transition (or, what is the same, the population of the upper level) oscillates between 0 and 1. A remarkable result is that very strong field (even far from the resonance!) can always populate the upper level with a probability of 100%.

In the second case Eq. (2.52) is reduced to:

$$W = \frac{2\Omega_R{}^2}{\delta^2}(1 - \cos \delta t),\ 0 \le W << 1. \tag{2.54}$$

In other words, in the far-from-resonance case or in very weak fields the population of the upper level can never reach unity. However, certain non-zero probability of the absorption of light in this case does exist, and this is another illustration—now physically based and mathematically derived—of non-zero widths of the spectral lines.

It is instructive to consider numerically several cases with different combinations of δ and Ω_R. Figure 2.7 shows two plots of Eq. (2.52) for exact resonance and low field (Fig. 2.7a) and for a small detuning and intensive field (Fig. 2.7b).

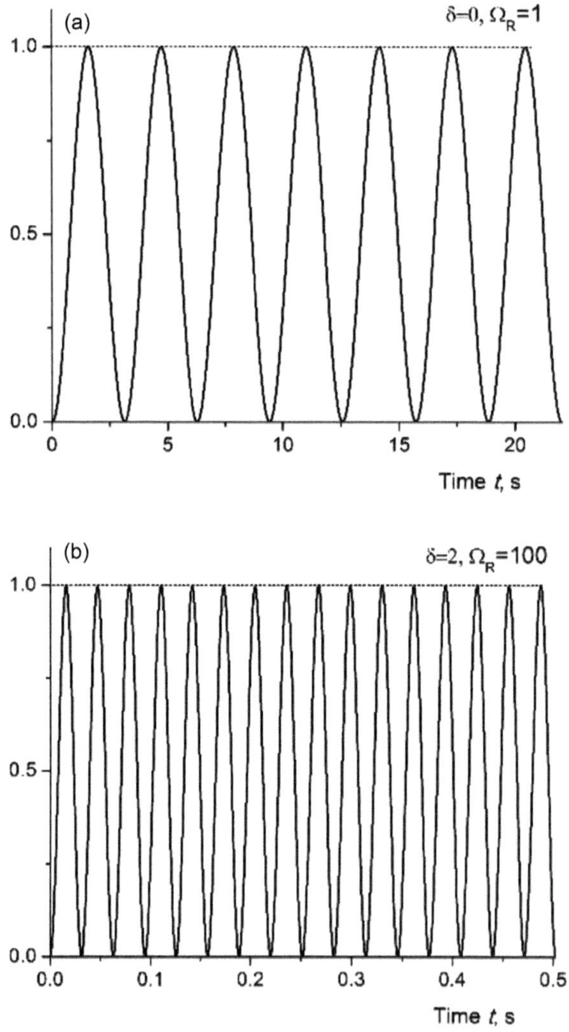

Figure 2.7 Plots of Eq. (2.52) for different δ and Ω_R (small detuning).

In both cases the population of the upper state oscillates between 0 and 1, but in the second case the frequency of these oscillations is considerably higher. So, a very intensive field makes population oscillations much faster.

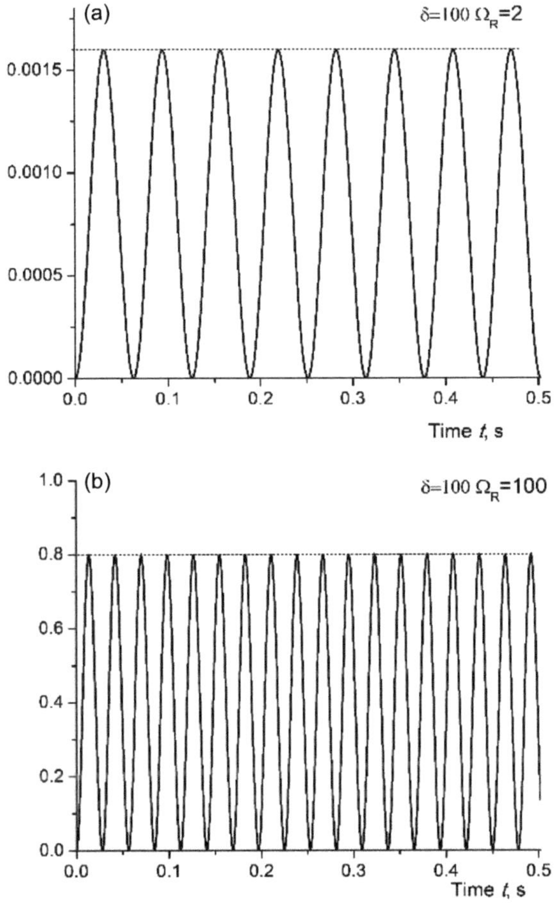

Figure 2.8 Plots of Eq. (2.54) for different δ and Ω_R (large detuning). Compare the vertical axes scales in both graphs.

Figure 2.8 shows an opposite situation: a large detuning in the weak (Fig. 2.8a) and intensive (Fig. 2.8b) fields. In both cases, the population of the upper state does not reach unity. It should be noticed, however, that increasing intensity of the incident field increases the upper state population drastically. With Ω_R increasing, the probability of the "1 → 2" transition will tend to 1. If we average then the population of both states, we get 1/2, which means that the population is divided equally between these two states. A similar result will be obtained later for a large number of atoms in a two-level system.

2.5 Positive and Negative Absorption, Population Inversion, Saturation, Absorption Cross Section

When an electromagnetic wave propagates through a medium, some part of the wave's energy is absorbed. If I_0 is the intensity of the incident radiation before it enters a considered medium, then the intensity I of the beam of radiation passed through this medium is given by

$$I = I_0 e^{-\alpha l}, \tag{2.55}$$

where α is the absorption coefficient and l is the distance traveled by radiation in the medium. It can be easily derived, if we assume that decrease of the intensity of light $-dI$ is proportional to the distance dx traveled by light in the medium and intensity of light I: $-dI = I\alpha dx$. Equation (2.55) is known as the Bouguer–Lambert–Beer law; it is valid in the case of low intensities of the incident radiation.

The absorption coefficient α is a characteristic property of the medium; it depends on the wavelength—that is the reason why different materials have different colors. If, for example, a glass absorbs red light, then it will appear to have blue color in the passing light. On the microscopic scale, absorption of radiation is reduced to the transitions between pairs of energy levels of a discrete atomic spectrum. The quantitative treatment of the absorption processes can be based on the harmonic oscillator model.

An atom in the field of the electromagnetic wave with frequency ω can be effectively described as a harmonic oscillator, whose oscillations are affected by an external force $qE_0 \exp(i\omega t)$ (interaction of the charge of atom q with the field of the incident plane monochromatic wave), damping force $b\dot{x}$ (energy losses) and restoring force *kx*. The equation of motion for such an oscillator is well known:

$$m\ddot{x} + b\dot{x} + kx = qE_0 \exp(i\omega t). \qquad (2.56)$$

Introducing new notations $\gamma = \frac{b}{m}$, $\omega_0^2 = \frac{k}{m}$, we can write down the solution of Eq. (2.56) as

$$x = \frac{qE_0}{m\left(\omega_0^2 - \omega^2 + i\omega\gamma\right)} \exp(i\omega t). \qquad (2.57)$$

The oscillations of a charge *q* produce the oscillating dipole moment

$$p = qx = \frac{q^2 E_0 \exp(i\omega t)}{m(\omega_0^2 - \omega^2 + i\omega\gamma)}. \qquad (2.58)$$

As a result, in the sample with *N* atoms a microscopic polarization $P = Nqx$ will be formed. On the other hand, there exists a relation between the microscopic polarization and electric field vector *E*: $P = \varepsilon_0(\varepsilon - 1)E$, where $\varepsilon_0 = 8.85 \times 10^{-12}$ F/m is the vacuum permittivity and ε is the relative dielectric permittivity of a medium. From the Maxwell theory, it is known that the refractive index of a medium $n = \sqrt{\varepsilon\mu}$, with μ being a permeability of a medium. Since for the dielectric materials $\mu \approx 1$, then $n = \sqrt{\varepsilon}$. Combining all these equations together, we get

$$\varepsilon = 1 + \frac{Nq^2}{m\varepsilon_0\left(\omega_0^2 - \omega^2 + i\omega\gamma\right)}. \qquad (2.59)$$

Estimation of the second term (with *q* being not more than a few elementary charges, *N* of the order of the Avogadro number, and light frequency to be about 10^{14} Hz—as for the visible light) yields its value much less then unity. Then using a series expansion $\sqrt{1+x} \approx 1 + \frac{x}{2}$, which is valid for small *x*, we get

$$n = \sqrt{\varepsilon} = 1 + \frac{Nq^2}{2m\varepsilon_0\left(\omega_0^2 - \omega^2 + i\omega\gamma\right)}. \qquad (2.60)$$

The refractive index n turns out to be a complex function $n = n' - i\kappa$, with

$$n' = \mathrm{Re}(n) = 1 + \frac{Nq^2}{2m\varepsilon_0}\frac{\omega_0^2 - \omega^2}{\left(\omega_0^2 - \omega^2\right)^2 + \omega^2\gamma^2}, \tag{2.61}$$

and

$$\kappa = -\mathrm{Im}(n) = -\frac{Nq^2}{2m\varepsilon_0}\frac{\gamma\omega}{\left(\omega_0^2 - \omega^2\right)^2 + \omega^2\gamma^2}. \tag{2.62}$$

The physical meaning of the imaginary part of the refractive index can be understood as follows: If the electromagnetic wave passes through a medium, its frequency does not change, but the wavelength become shorter $\lambda = \lambda_0/n$ (where n is a refractive index); it corresponds to the increased wave vector $k = k_0 n = k_0(n' - i\kappa)$. Inserting this expression into the general form of a plane monochromatic wave propagating along the z direction $E = E_0 \exp(i(\omega t - kz))$ we get

$$E = E_0 \exp(-\kappa k_0 z)\exp\left(i\left(\omega t - k_0 n' z\right)\right). \tag{2.63}$$

Since the intensity of the radiation is proportional to the electric vector square, then

$$I = I_0 \exp(-2\kappa k_0 z), \tag{2.64}$$

which is equivalent to Eq. (2.55), if we assume $\alpha = 2\kappa\, k_0$. So, the absorption coefficient α is proportional to the imaginary part of the refractive index of the medium.

Figure 2.9 shows the frequency dependence of the real and imaginary parts of the refractive index. With increasing frequency, the value of Re(n) increases (Fig. 2.9a)—this is the so called normal dispersion. In the vicinity of the ω_0 frequency, the value of Re(n) decreases; this is the range of the anomalous dispersion. It is related to the light absorption. Finally, when moving further from ω_0, the normal dispersion behavior is restored. The imaginary part of the refractive index (Fig. 2.9b) has a maximum when $\omega = \omega_0$ (maximum of absorption), and rapidly falls when moving away from ω_0.

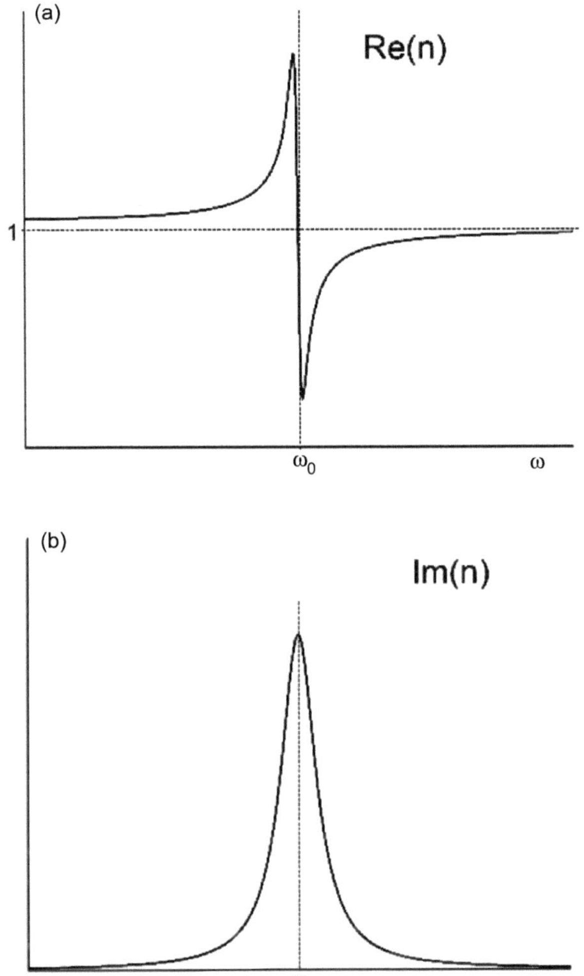

Figure 2.9 Real Re(n) (a) and imaginary Im(n) (b) parts of the refractive index (see Eqs. (2.61) and (2.62)). At $\omega = \omega_0$ the real part of the refractive index is unity (there is no refraction at all), and the imaginary part (which is proportional to the absorption coefficient) has a maximum value.

Intensity of radiation passing through a two-level system can be decreased due to the "$1 \to 2$" absorption transition and can be increased because of the "$2 \to 1$" induced emission transitions. At this point, we do not take into account the spontaneous emission transitions, since, as we have already seen, their probabilities are negligibly small if compared with those of the induced transitions. These two processes of induced absorption and emission compete with each other, constantly changing the number of atoms populating the first and the second energy levels, denoted by n_1 and n_2, respectively. Then the change dI of intensity I of radiation travelling the distance dx through such a system will be expressed by the following equation:

$$-dI = (n_1 B_{12} - n_2 B_{21}) \frac{I}{c} h\nu S(\nu) dx . \tag{2.65}$$

Here $\rho = \frac{I}{c}$ is the spectral density of the electromagnetic field energy, $S(\nu)$ is a spectral distribution of frequencies of the incident radiation (we remind here that absorption of radiation is selective and depends on its frequency), $h\nu$ is the energy of an individual quantum, and the first and the second terms in the parenthesis are related to the "$1 \rightarrow 2$" absorption and "$2 \rightarrow 1$" emission transitions, respectively. Using the properties of the Einstein coefficients ($B_{12} = \frac{g_1}{g_2} B_{21}$, $B_{21} = \frac{c^3}{8\pi h\nu^3} A_{21}$), the last expression can be modified as

$$-dI = \frac{c^2 A_{21}}{8\pi\nu^2} n_1 \frac{g_2}{g_1} \left(1 - \frac{g_1}{g_2} \frac{n_2}{n_1} \right) S(\nu) I dx . \tag{2.66}$$

On the other hand, for low intensities of the incident radiation decrease of the radiation intensity is proportional to the distance travelled by this radiation in a given medium $-dI = \alpha I dx$, with α being the frequency-dependent absorption coefficient of such medium; it has a dimension of inverse length. Then we get

$$\alpha = \frac{c^2 A_{21}}{8\pi\nu^2} n_1 \frac{g_2}{g_1} \left(1 - \frac{g_1}{g_2} \frac{n_2}{n_1} \right) S(\nu) . \tag{2.67}$$

One of special features of this equation is that it links the macroscopic characteristics of a medium, such as its absorption coefficient, with the microscopic ones, such as the probabilities of transitions and populations of the considered energy levels. It should be emphasized that the absorption coefficient depends not only on the incident radiation frequency, but on the number of atoms in the lower and upper states and the statistical weights of these levels. Under normal conditions the number of atoms in the lower state always exceeds the number of atoms in the upper state: $n_1 >> n_2$. We assume for simplicity $g_1 = g_2$ (in real atomic systems even if $g_1 < g_2$, the second term in the parenthesis is still much less than unity because of a great difference between n_1 and n_2). Then the value of α determined by Eq. (2.67) is positive (positive absorption), which means decrease of intensity of radiation after passing the sample. Such a conclusion is in line with the Bouguer–Lambert–Beer law $I = I_0e^{-\alpha x}$.

If the number of atoms n_2 is increased because of induced absorption, then it may happen that $g_1n_2 = g_2n_1$ and $\alpha = 0$. This result means that the radiation beam does not lose its intensity after passing the sample, no matter what the radiation frequency is. This case is called the saturation of absorption. Such situation can be realized if considerably large number of atoms determined by the $g_1n_2 = g_2n_1$ condition is already excited to populate the upper level.

Although the radiation intensity before and after the sample in this case is not changed at all, it should not be, however, understood that there are absolutely no transitions between these two considered energy levels. The induced absorption and emission transitions do occur, but their numbers are exactly equal, keeping the radiation intensity at the same level.

Finally, under some special circumstances, the absorption coefficient in Eq. (2.67) can become negative, which implies that $g_1n_2 >> g_2n_1$. This will be realized when a considerable part of atoms has been already excited before the sample is irradiated. Then $\alpha < 0$ (negative absorption), and a remarkable situation is realized: The intensity of radiation is increased after passing through the sample. This happens because the number of the induced emission transition significantly exceeds the number of induced absorption transition. The state of a system, when the number of atoms at the upper energy level is greater than at the lower level, is called the population inversion. Creating population inversion is necessary for laser performance.

The increased emission intensity in this case should not be interpreted as a gain of energy. The enhanced emission is due to the previously arranged excitations of atoms and occurs on account of energy already absorbed and stored in the system.

Sometimes the states with population inversion are spoken of as the states with the negative temperature. This may seem to be in a contradiction with the definition of the thermodynamic temperature, which cannot be less than zero. The term "negative temperature" in its application to the system with inversion of population is just a formal description, which becomes clear after looking at the Boltzmann's distribution equation $n_2 = n_1 \exp\left(-\frac{E_2 - E_1}{kT}\right)$. If $E_2 > E_1$, then $n_2 > n_1$ if $T < 0$. Of course, it is impossible to reach negative absolute temperatures, and we emphasize that no special meaning (other than $n_2 > n_1$) is associated with the "negative temperature" term.

It is convenient to introduce a new physical quantity σ called the "absorption cross section" and defined by the following expression:

$$\alpha = \sigma\left(n_1 - \frac{g_1}{g_2} n_2\right). \tag{2.68}$$

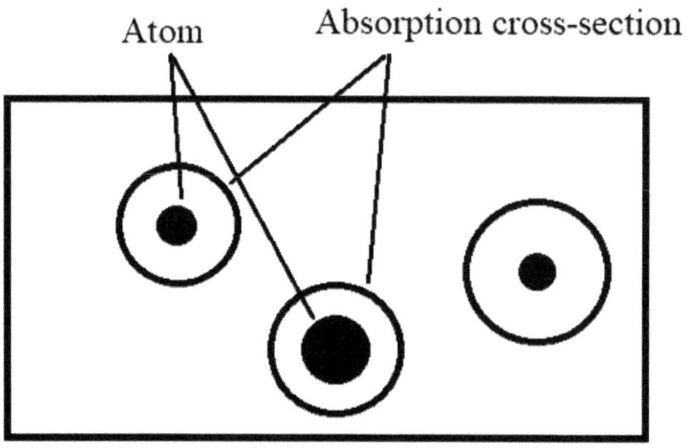

Figure 2.10 Physical meaning of the absorption cross section (see the text for the details).

It is widely used in spectroscopy to describe different absorption abilities of atoms. The absorption cross section has the dimension of area and can be thought of as an effective disk with an absorbing atom at the center (Fig. 2.10). This disk is in some sense a target, which absorbs the incident photons. If the photons passing near an absorbing atom miss these "disks," there is no absorption. The absorption cross section can be also thought of as an effective area around an absorbing atom that extracts power from the incident light beam. Transitions between different energy levels can have different values of the absorption cross sections.

2.6 Absorption of Radiation in an Ensemble of Particles and Two Schemes of Laser Generation

Let us consider an ensemble of particles (atoms or molecules), each of which can have only two energy levels 1 and 2 (Fig. 2.11). For the sake of simplicity, the energy levels are assumed to be non-degenerate, $g_1 = g_2 = 1$. Let the total number of particles be denoted as N. If the number of particles in the first (ground) and the second (excited) levels are N_1 and N_2, respectively, then, obviously, $N_1 + N_2 = N$. Since there are many particles, they can interact, at least, by colliding with each other and exchanging energy in the result of such collisions. The transitions between the considered states in both directions can be induced not only by the incident radiation (the solid arrows in Fig. 2.11), but also by elastic and inelastic collisions between these particles shown by the dashed arrows in the same figure.

At the stationary conditions the number of the up-ward transitions, both radiation- and collision-induced, should be equal to the number of the down-ward transitions.

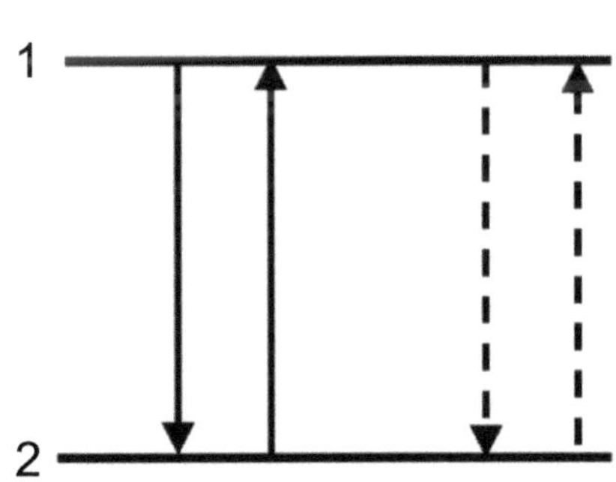

Figure 2.11 Possible types of transitions in a two-level system of N particles. See the text for further details.

Let us consider how the population of both levels can vary in time. The population of the ground level can decrease because of the absorption processes, when the particles get excited to the second level. On the other hand, the number of particles in the ground level can increase because of the emission processes, or de-activation of the excited particles. The probability of absorption is proportional to the number of particles in the ground state, whereas the probability of emission is proportional to the number of particles in the excited state. Considering all possible radiative and non-radiative processes, the expression for $\frac{dN_1}{dt}$ (the change of population of the first level) can be written as follows:

$$\frac{dN_1}{dt} = -B_{12}\rho N_1 + B_{12}\rho N_2 + (A_{21} + C_{21})N_2 - C_{12}N_1, \qquad (2.69)$$

where the first term is responsible for a decrease of population in the first level due to the induced absorption, the second term describes population increase on account of induced emission, the third term represents population increase because of spontaneous emission (Einstein coefficient A_{21}) and collision induced emission with the proportionality coefficient C_{21}. The last term accounts for a decrease of population due to the collision induced absorption with the proportionality coefficient C_{12}. At the stationary conditions $\frac{dN_1}{dt} = -\frac{dN_2}{dt} = 0$. Introducing a new short-hand notation $R_2 = A_{21} + C_{21}$, $R_1 = C_{12}$ and keeping in mind that $N_1 + N_2 = N$, it is easy to find the difference of the energy levels populations $\Delta N = N_1 - N_2$ as

$$\Delta N = N\frac{R_2 - R_1}{2B_{12}\rho + R_2 + R_1}. \qquad (2.70)$$

If ρ is increased (i.e., the electromagnetic radiation of higher intensity is used), then $\Delta N \to 0$. This means that the total number of particles will be distributed equally between these two states, and each of them will accommodate exactly one half of particles, $N/2$. Since $\alpha = \sigma(N_1 - N_2)$ (if $g_1 = g_2 = 1$, see Eq. (2.68)), the absorption coefficient under such conditions becomes zero. Therefore, a two-level system in a very strong electromagnetic field becomes completely transparent for incident radiation; the radiation propagates through such a system without losing energy.

It also follows from this result that in a two-level system it is simply impossible to create a situation, when $N_2 > N_1$. In other words, an increase of intensity of radiation, also known as the radiation amplification, after passing through a two-level system cannot be achieved under any circumstances.

However, the whole situation changes considerably if there are additional energy levels in the system that can be connected with the energy levels 1 and 2 by means of the collision-induced transitions (Fig. 2.12). The energy intervals between these additional energy levels and the levels denoted by 1, 2 should be small enough, and for sure they should be smaller than the energy interval between the levels 1 and 2.

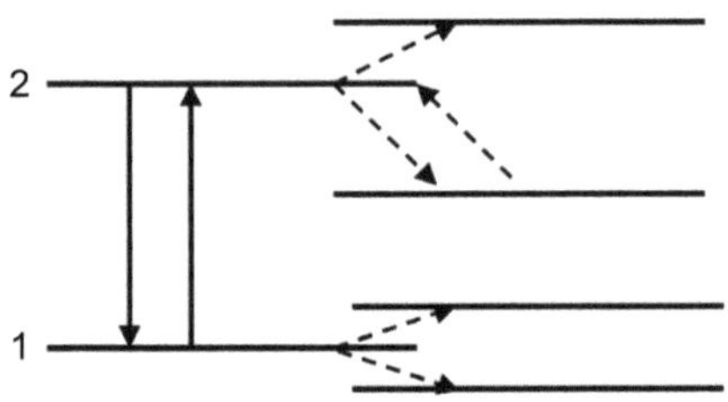

Figure 2.12 Possible types of transitions in a system of N particles distributed through several energy levels. The collision-induced transitions are shown by the dashed arrows. The emission and absorption transitions are shown by the vertical arrows. See the text for further details.

The kinetic equations for the populations of the energy levels 1 and 2 in such a system would be different. For the change of population of level "1" we have

$$\frac{dN_1}{dt} = -B_{12}\rho N_1 + B_{12}\rho N_2 - R_1 N_1 + \sum_k R_{k1} N_k, \tag{2.71}$$

the first and the second terms describe the induced absorption and emission, respectively, and the third and the fourth terms account for all de- and re-populating processes, respectively. All other levels in the system are denoted by a subscript k, the sum includes all transitions to the first level from all k states located above the ground state. The R_1 entry is a proportionality coefficient for the collision-induced depopulation processes from the level 1, whereas the R_{k1} entries stand for the proportionality coefficients of the collision-induced processes that can increase the first level population on account of the transitions from the higher located levels with number k. A similar equation for level "2" will be

$$\frac{dN_2}{dt} = -B_{12}\rho N_2 + B_{12}\rho N_1 - R_2 N_2 + \sum_k R_{k2} N_k. \qquad (2.72)$$

Note the change of index "1" into "2" in the R_2, R_{k2} entries. For the sake of simplicity, we introduce a new notation for all repopulating processes as $\sum_k R_{k1} N_k = C_1$, $\sum_k R_{k2} N_k = C_2$. Again, for the stationary conditions $\frac{dN_1}{dt} = -\frac{dN_2}{dt} = 0$. A very important difference from the previous case is that $N_1 + N_2 \neq N$, since some particles can occupy other energy levels as well. Solving Eqs. (2.71) and (2.72) under stationary conditions yields the following expressions for the N_1, N_2 populations:

$$N_1 = \frac{B_{21}\rho(C_1 + C_2) + R_2 C_1}{B_{21}\rho(R_1 + R_2) + R_2 R_1}, \qquad (2.73)$$

$$N_2 = \frac{B_{12}\rho(C_1 + C_2) + R_1 C_2}{B_{12}\rho(R_1 + R_2) + R_1 R_2}. \qquad (2.74)$$

For certain combinations of the energy levels one can get the situation when $N_2 > N_1$ (this depends on the numerical values of the coefficients involved into the last equations). Therefore, to achieve the population inversion, there should be at least three energy levels in a system. This is the basic principle of the laser operation.

Two principal schemes (shown in Fig. 2.13) for getting population inversion in the laser media are used.

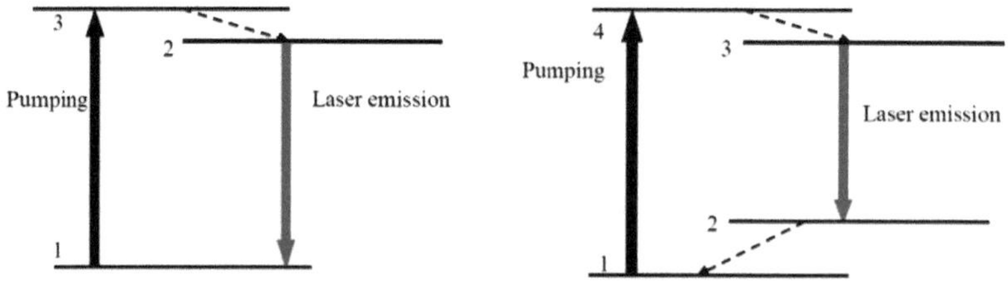

Figure 2.13 The three-level (left) and four-level (right) schemes of laser operation. Fast non-radiative relaxation transitions are shown by the dashed arrows.

In the three-level system the atoms are excited from the ground level 1 to some excited level 3, which is connected to the lower level 2 by fast non-radiative relaxation, and the energy difference between the levels 3 and 2 will transform into the heat or atomic vibrations. Level 2 can accommodate atoms, if it has a considerably long lifetime, and then at some moment of time the population of this level will exceed the ground state population, thus creating conditions for getting the amplified induced (stimulated) radiation.

In the four-level system the atoms are excited from the ground level 1 to the excited level 4, from which they relax non-radiatively to the level 3. This level is connected by a radiative transition with the level 2, which, in its turn, is connected by very fast non-radiative relaxation with the ground level, so we may assume the level 2 to be always empty. Then exciting even one atom to the level 3 creates inversion of population between the levels 3 and 2. So, the four-level system provides an easier way to get the population inversion, than the three-level system.

Finally, let us recall here that the word "laser" itself is an abbreviation of the following word combination: Light Amplification by Stimulated Emission of Radiation, and this emphasizes importance of the induced transitions for the laser performance.

Bibliography

1. Demtröder W (1981, 1996, 2003). *Laser Spectroscopy: Basic Concepts and Instrumentation* (various editions). Springer-Verlag Berlin Heidelberg.
2. Einstein A (1916). Strahlungs-Emission und -Absorption nach der Quantentheorie. *Verhandlungen der Deutschen Physikalischen Gesellschaft* **18** 318–323.
3. Karlov NV (1993). *Lectures on Quantum Electronics*. CRC Press.
4. Scully MO, Zubairy MS (1997). *Quantum Optics*. Cambridge University Press.
5. Svelto O (2010). *Principles of Lasers*, 5th Ed. Springer, New York.

Chapter 3

The Theory of Atom of Hydrogen

3.1 Introduction

In the previous chapter, the basic processes of light absorption and emission in simple quantum mechanical systems were considered. Very important concepts of the induced and spontaneous transitions were introduced; general results were obtained for the probabilities of those transitions. At the same time, we did not specify yet what the nature of the energy levels involved in these transitions was. It was not mentioned at all what the main characteristics of those levels are, how their energetic positions can be obtained or how the energy levels can be described in a unified way. Moreover, a mixed approach, in which the atom was considered quantum-mechanically, and the incident radiation was treated as a monochromatic wave, was employed. A deeper quantum mechanical analysis of the energetic spectra of free atoms/ions will constitute the main content of this and next chapters. Certain quantum mechanical ideas and mathematical tools will be used for a quantitative description of the atomic states.

Theoretical Spectroscopy of Transition Metal and Rare Earth Ions: From Free State to Crystal Field
Mikhail G. Brik and Chong-Geng Ma

ISBN 978-981-4800-56-3 (Hardcover), 978-0-429-27875-4 (eBook)
www.jennystanford.com

3.2 Experimental Spectroscopic Results Known by the End of the 19th Century–Beginning of the 20th Century

The second half of the 19th century was marked by fast development of the experimental spectroscopy as a separate part of physics. It was preceded by pioneering works by J. von Fraunhofer (discovery of the dark lines in the solar spectrum, which later were explained as the absorption lines of various atoms), G. Kirchhoff and R. Bunsen (the laws of thermal radiation, discovery of several chemical elements, etc.). By that time, considerable amount of information about absorption and emission spectra of various elements and compounds had been collected. In particular, it was well established that the spectrum of atomic hydrogen in the visible region consists of four very sharp lines with the wavelengths of 410, 434, 486, 656 nm. However, no clear understanding of physical reasons for those experimental results was developed. In an attempt of finding any explanation (or, at least, description of this spectrum), a Swiss mathematician and physicist J. J. Balmer noticed in 1885 that these wavelengths fit the following expression:

$$\lambda = B\left(\frac{n^2}{n^2-4}\right), \tag{3.1}$$

where B is a constant equal to 364.56 nm, and n is an integer number equal to 3, 4, 5, 6 for the 656, 486, 434, and 410 nm lines, correspondingly. Although this equation describes the series of the visible hydrogen spectrum very well, it does not explain at all the physical origin of these spectral lines.

Three years later, in 1888, a Swedish physicist J. R. Rydberg obtained a general expression, which describes the wavenumbers (which is an inverse of the wavelength) of lines in the spectra of hydrogen and alkali metals. For the reasons that will be clear later, we point out here that these elements have only one outer electron above the completely filled electron shells (this resembles the electron configuration of hydrogen). For hydrogen, this expression is as follows:

$$\frac{1}{\lambda} = R_{\mathrm{H}}\left(\frac{1}{n_1^{\,2}} - \frac{1}{n_2^{\,2}}\right), \tag{3.2}$$

where R_H = 10967758.341 ± 0.001 m^{-1} is the Rydberg constant for hydrogen, and n_1 and n_2 are integer numbers. The wavelengths of the above-mentioned four lines in the visible spectrum of hydrogen, which form the so-called Balmer series, can be obtained from Eq. (3.2) by taking $n_1 = 2$ and $n_2 = 3, 4, 5, 6$.

Equation (3.2) is another form of the following so-called combination principle proposed in 1888 by a Swiss physicist W. Ritz: The frequencies of the spectral lines of any element are either the sums or the differences of frequencies of two other lines.

Later, other regular spectral series, or groups of sharp lines, were discovered in the hydrogen spectrum. They lie in the UV and infrared regions and are named after their discoverers. The Lyman series discovered in 1906 fits Eq. (3.2) if $n_1 = 1$ and $n_2 = 2, 3, \ldots$; the Paschen series discovered in 1908 corresponds to $n_1 = 3$ and $n_2 = 4, 5, \ldots$; the Brackett series observed for the first time in 1922 can be obtained by putting $n_1 = 4$ and $n_2 = 5, 6, \ldots$; the Pfund series (1924) can be built up if $n_1 = 5$ and $n_2 = 6, 7, \ldots$; and the Hamphreys series (the last one designated by the personal name in 1953) corresponds to n_1 = 6 and $n_2 = 7, 8, \ldots$. Obviously, the spectral series for other values of n_1 can be obtained; in fact, they were detected, but the tradition to name the spectral series in the hydrogen spectra after a scientist, who observed it for the first time, is no longer in use.

It should be pointed out that the Rydberg equation (3.2) does not explain the origin of the spectral series, although it describes the observed results with high accuracy. The world should have waited until 1913, when a Danish physicist N. Bohr proposed an entirely new theory for a description and partial explanation of the spectral properties of the hydrogen atom. His theory was based on two postulates, the so-called Bohr postulates, and both of them were impossible to be understood from the point of view of classical physics. These postulates are as follows:

1. The electrons travel (in accordance with the classical laws of motion) in orbits that have discrete quantized speeds, and, therefore, quantized energies. Not every classical orbit is possible, but only specific ones, at certain specific distances from the nucleus. The allowed orbits are defined by the expression $L = n\hbar$ ($n = 1, 2, 3\ldots$), where L stands for the orbital momentum of an electron, and $\hbar = 1.05 \times 10^{-34}$ J·s is Planck's constant

2. The electrons do not continuously lose energy as they travel (as would be expected according to the classical electrodynamics of an accelerated charge). They can only absorb and emit energy by jumping from one allowed orbit to another; the emitted (absorbed) energy is equal to the difference of energies for the orbits involved: $h\nu = \hbar\omega = E_1 - E_2$ (here another Planck's constant $h = 2\pi\hbar$ is used).

Being based on these postulates and employing classical physics, N. Bohr was able to derive explicit expressions for the energy levels of the hydrogen atom and the wavelengths of possible transitions (this derivation will not be repeated here, since it can be found even in the school textbooks on physics). The obtained results agreed perfectly with the experimental data, and in 1922 N. Bohr was awarded the Nobel Prize in physics for his pioneering works.

Bohr's theory—though it was a real breakthrough in developing our knowledge about the fundamental properties of matter—was still not very consistent. It was, in fact, "a happy alliance" between the classical physics and new quantum-mechanical postulates. Obvious intrinsic inconsistency of Bohr's theory, on one side, and its remarkable success in explanation of the observed spectra, on the other side, turned out to be a strong driving force for creating a new theory, with new physical concepts and a new mathematical apparatus. Such a theory was developed in the late 1920s and was called quantum (or wave) mechanics. Its application to the description of the hydrogen atom finally provided physicists with strictly consistent explanation of the spectral series and N. Bohr's postulates.

3.3 Solution of the Schrödinger Equation for the Hydrogen-Like Atom

The hydrogen-like atom (or an ion) has only one electron, and all others have been removed by various ways of ionization. An essential feature of these systems is the presence of one electron only. The Coulomb interaction between this electron and the nucleus is spherically symmetrical. Later, we shall see that the situation becomes considerably more complicated in the case of multielectron atoms.

In the present paragraph, we shall be concerned with finding a solution of the following physical problem: determination of a spectrum (possible energetic states and corresponding wave functions) of an electron with the mass m and the charge $-e$ revolving around a nucleus with the mass M and the charge $+Ze$ (Fig. 3.1).

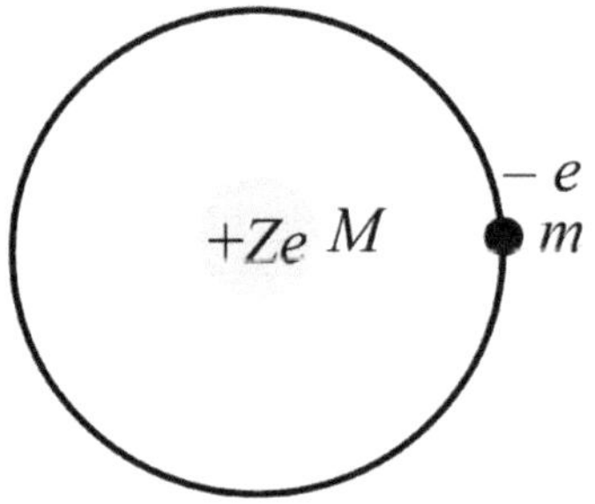

Figure 3.1 A schematic representation of a hydrogen-like ion.

Basically, such a problem as known from the classical mechanics can be reduced to a problem of a particle with an effective mass $\mu = \dfrac{mM}{m+M}$ revolving around a center of mass of this system in a Coulomb potential $U = -\dfrac{Ze^2}{r}$ produced by a nucleus (r is the distance between an electron and a nucleus). Since even for the hydrogen atom $M >> m$ ($M/m \approx 1840$) and the M/m ratio becomes even much greater for heavier ions, we shall assume $\mu = m$.

From the quantum-mechanical point of view, the problem will be solved if the eigenvalues and eigenfunctions of the following Hamiltonian $\widehat{H} = -\dfrac{\hbar^2}{2m}\Delta - \dfrac{Ze^2}{r}$ would be found. Here the first term is the kinetic energy of an electron with $\Delta = \dfrac{\partial}{\partial x^2} + \dfrac{\partial}{\partial y^2} + \dfrac{\partial}{\partial z^2}$ being the Laplace operator, and the second term is the potential energy of Coulomb interaction between the electron and nucleus. The so-called stationary or time-independent Schrödinger equation should be solved:

$$\left(-\frac{\hbar^2}{2m}\Delta - \frac{Ze^2}{r}\right)\Psi = E\Psi, \tag{3.3}$$

where E and Ψ are an eigenvalue (energy of the electron) and an eigenfunction, respectively.

Since the considered system possesses the spherical symmetry, it is natural to use the spherical coordinates r,θ,φ with the origin of the system of reference placed at the nucleus position. It should be, however, kept in mind that such an assumption is true for a free atom or ion only, in the absence of the external electric and/or magnetic fields. In the spherical coordinates Laplace's operator Δ has the following form:

$$\Delta=\frac{1}{r^2}\frac{\partial}{\partial r}\left(r^2\frac{\partial}{\partial r}\right)+\frac{1}{r^2\sin\theta}\frac{\partial}{\partial\theta}\left(\sin\theta\frac{\partial}{\partial\theta}\right)+\frac{1}{r^2\sin^2\theta}\frac{\partial^2}{\partial\phi^2}. \quad (3.4)$$

A very important consequence of the spherical symmetry of the considered system is that the wave function Ψ can be represented as a product of two functions, each of which depends on the radial coordinate r and angular coordinates θ,φ only:

$$\Psi(r,\theta,\varphi) = R(r)Y(\theta,\varphi). \quad (3.5)$$

Such a "trick," widely known as the separation of variables, has a clear physical reasoning: There is no selected direction in an isolated atom, and the basic properties of such an atom (the energy, for example) can depend only on the distance r between the electron and nucleus and are independent of the spherical angles.

The next step is to substitute Eq. (3.5) into Eq. (3.3). Dividing both parts by the product $R(r)Y(\theta,\varphi)$ and really separating the variables by grouping all the r-dependent terms in the left-hand side, and all the θ,φ -dependent terms in the right-hand side, we arrive at the following expression (the arguments of the R and Y functions are omitted for the sake of brevity, but it is implied that they depend on r and θ,φ, respectively):

$$\frac{1}{R(r)}\frac{\partial}{\partial r}\left(r^2\frac{\partial R(r)}{\partial r}\right)+r^2\left(\lambda+\frac{2\alpha}{r}\right)$$
$$=-\left\{\frac{1}{Y(\theta,\phi)\sin\theta}\frac{\partial}{\partial\theta}\left(\sin\theta\frac{\partial Y(\theta,\phi)}{\partial\theta}\right)+\frac{1}{Y(\theta,\phi)\sin^2\theta}\frac{\partial^2 Y(\theta,\phi)}{\partial\phi^2}\right\}$$

(3.6)

For the sake of brevity, the following notations were introduced in Eq. (3.6): $\frac{2m}{\hbar^2}E = \lambda$; $\frac{Ze^2m}{\hbar^2} = \alpha$. Since both parts of Eq. (3.6) depend on entirely different variables but are identically equal at the same time, they both should be equal to some numerical constant denoted as σ. Let us consider the "angular" equation (the right-hand side of Eq. (3.6)) first:

$$\frac{1}{\sin\theta}\frac{\partial}{\partial\theta}\left(\sin\theta\frac{\partial Y}{\partial\theta}\right)+\frac{1}{\sin^2\theta}\frac{\partial^2 Y}{\partial\phi^2}+\sigma Y=0. \tag{3.7}$$

Equation (3.7) is an equation for the spherical functions. More details can be found in any textbook on the mathematical physics or special functions. We just mention here that the spherical functions are the homogenous polynomials satisfying the Laplace's equation $\Delta u = 0$. The explicit form of this equation in the spherical system of coordinates is

$$\frac{\partial}{\partial r}\left(r^2\frac{\partial u}{\partial r}\right)+\frac{1}{\sin\theta}\frac{\partial}{\partial\theta}\left(\sin\theta\frac{\partial u}{\partial\theta}\right)+\frac{1}{\sin^2\theta}\frac{\partial^2 u}{\partial\phi^2}=0. \tag{3.8}$$

Here we postulate that the Laplace's equation in the Cartesian coordinates $\frac{\partial^2 u}{\partial x^2}+\frac{\partial^2 u}{\partial y^2}+\frac{\partial^2 u}{\partial z^2}=0$ has a solution in the form of a polynomial $u(x, y, z)$ of a degree l. If, then, the Cartesian coordinates are transformed into the spherical ones, the solution u will be transformed as $u = r^l Y_{lm}(\theta,\varphi)$, where $Y_{lm}(\theta,\varphi)$ is a spherical function with l being an integer number like 0, 1, 2, ..., and m taking all integer values from $-l$ to $+l$ with step of one. Below the explicit expressions for a few first spherical functions are given:

$$\begin{aligned}
&Y_{00}=\sqrt{\frac{1}{4\pi}} && Y_{22}=\sqrt{\frac{15}{32\pi}}\sin^2\theta e^{2i\varphi}\\
&Y_{11}=-\sqrt{\frac{3}{8\pi}}\sin\theta e^{i\varphi} && Y_{21}=-\sqrt{\frac{15}{8\pi}}\sin\theta\cos\theta e^{i\varphi}\\
&Y_{10}=\sqrt{\frac{3}{4\pi}}\cos\theta \quad , && Y_{20}=\sqrt{\frac{5}{16\pi}}\left(3\cos^2\theta-1\right)\quad ,\\
&Y_{1-1}=\sqrt{\frac{3}{8\pi}}\sin\theta e^{-i\varphi} && Y_{2-1}=\sqrt{\frac{15}{8\pi}}\sin\theta\cos\theta e^{-i\varphi}\\
& && Y_{2-2}=\sqrt{\frac{15}{32\pi}}\sin^2\theta e^{-2i\varphi}
\end{aligned} \tag{3.9}$$

Substitution of $u = r^l Y_{lm}(\theta,\varphi)$ into Eq. (3.8) yields:

$$l(l+1)Y + \frac{1}{\sin\theta}\frac{\partial}{\partial\theta}\left(\sin\theta\frac{\partial Y}{\partial\theta}\right) + \frac{1}{\sin^2\theta}\frac{\partial^2 Y}{\partial\phi^2} = 0\,. \tag{3.10}$$

Comparison of Eq. (3.10) with (3.7) returns immediately the constant σ value as

$$\sigma = l(l+1). \tag{3.11}$$

With this value of σ the "radial" equation or the left-hand side of Eq. (3.6) will look like

$$\frac{d^2R}{dr^2} + \frac{2}{r}\frac{dR}{dr} + \left(\lambda + \frac{2\alpha}{r} - \frac{l(l+1)}{r^2}\right)R = 0\cdot \tag{3.12}$$

Again, the argument r of the R function is omitted. It is instructive to replace the α and λ constants in the last equation by their explicit expressions defined above:

$$\frac{d^2R}{dr^2} + \frac{2}{r}\frac{dR}{dr} + \frac{2m}{\hbar^2}\left(E + \frac{Ze^2}{r} - \frac{\hbar^2}{2m}\frac{l(l+1)}{r^2}\right)R = 0\,. \tag{3.13}$$

The last equation describes a motion of an electron in the central-symmetric field with an effective potential

$$U_{\text{eff}}(r) = -\frac{Ze^2}{r} + \frac{\hbar}{2m}\frac{l(l+1)}{r^2}\,.$$

The last term, which is an addition to the Coulomb potential energy, describes a kinetic energy of an electron revolving around a nucleus. In what follows, the new designations will be used: $\beta^2 = -\frac{2m}{\hbar^2}E;\ q = \frac{2mZe^2}{\hbar^2}$. Then Eq. (3.13) will be re-written as

$$\frac{d^2R}{dr^2} + \frac{2}{r}\frac{dR}{dr} + \left(\frac{q}{r} - \beta^2 - \frac{l(l+1)}{r^2}\right)R = 0. \tag{3.14}$$

Finding a solution of this equation is not an easy task at all, and this can be done in several steps. First of all, from Eq. (3.14), some properties of the asymptotic behavior of the R function can be derived. In the limit of very large values $r \to \infty$, Eq. (3.14) is reduced to a very simple form

$$\frac{d^2R}{dr^2} - \beta^2 R = 0, \tag{3.15}$$

which has two solutions: $R = e^{\pm\beta r}$. Since the wave function should be finite at infinity, to ensure a proper normalization condition, which in the spherical coordinates looks like $\int_0^\infty R(r)^2 r^2 dr = 1$, the solution with a positive argument of the exponent should be neglected. Therefore, the R function should contain a factor $e^{-\beta r}$:

$$R(r) = e^{-\beta r} f(r), \tag{3.16}$$

with $f(r)$ being a new unknown function, which from now on we shall try to determine. Substituting Eq. (3.16) into Eq. (3.14) gives us the differential equation for the $f(r)$ function:

$$\frac{d^2 f}{dr^2} + 2\frac{df}{dr}\left(\frac{1}{r} - \beta\right) + f\left(\frac{q-2\beta}{r} - \frac{l(l+1)}{r^2}\right) = 0. \tag{3.17}$$

The proposed solution of this equation will be represented as a series in powers of r:

$$f(r) = r^\gamma \sum_{k=0}^{\infty} a_k r^k. \tag{3.18}$$

with unknown coefficients a_k. The main idea of such a representation is that it reduces the differential equation Eq. (3.17) of the second order with varying coefficients to the linear equations for the coefficients a_k. The pre-factor r^γ before the summation sign is needed to make the function be zero at $r = 0$ and any values of the coefficient a_0 (corresponding to the first term of the sum). Physically the zero-th value of the electron's wave function at $r = 0$ means impossibility for an electron to fall onto the nucleus. After substituting Eq. (3.18) into Eq. (3.17), we get

$$\begin{aligned}&\sum_k a_k r^{k+\gamma-2}\{(k+\gamma)(k+\gamma-1)+2(k+\gamma)-l(l+1)\}\\&+\sum_k a_k r^{k+\gamma-1}\{-2(k+\gamma)\beta+(q-2\beta)\}=0\end{aligned}. \tag{3.19}$$

Equation (3.19) represents a polynomial, which is equal to zero for all values of r. This can happen if and only if all coefficients at all consecutive powers of r ($r^0, r^1, r^2, \ldots$) are simultaneously equal to zero. Let us find out what the coefficient at the lowest power of r is. The lowest power of r (equal to $\gamma - 2$) occurs when $k = 0$, and the corresponding coefficient is $\gamma(\gamma + 1) - l(l + 1)$. Equating it to zero, we get two possible values of γ: $\gamma = l, \gamma = -l - 1$. The second root should be neglected, since it leads to a singularity of the $f(r)$ function at $r = 0$. Hence, the value of γ is determined to be l, and with this value of γ Eq. (3.19) will look like

$$\sum_k a_k r^{k+l-2}\{(k+l)(k+l+1)-l(l+1)\} + \sum_k a_k r^{k+l-1}\{-2(k+l)\beta+(q-2\beta)\}=0. \tag{3.20}$$

Collecting the coefficients at the same powers of r, the following recurrent relation for a_k can be obtained:

$$a_{k+1} = a_k \frac{2\beta(k+l+1)-q}{(l+k+1)(l+k+2)-l(l+1)}, \tag{3.21}$$

which allows to calculate all coefficients, if the first coefficient a_0 is known. At some value of q the numerator in Eq. (3.21) will be zero and the series expansion (3.18) will be terminated, i.e. all coefficients after that one with a definite number will be zero. Let us denote the last non-zero coefficient as a a_{n_r}. Then $a_{n_r+1} = 0$ and

$$q = 2\beta(n_r + l + 1). \tag{3.22}$$

An essential feature of Eq. (3.22) is that all numbers in the parenthesis are integer numbers, and their sum is, obviously, an integer number too. Then, coming back to the earlier introduced notations for β and q, we immediately obtain a remarkable result for the discrete energy levels of an electron in the hydrogen-like atom (see also Fig. 1.1):

$$E = -\frac{mZ^2e^4}{2\hbar^2(n_r+l+1)^2}. \tag{3.23}$$

This result is identical with that one of Bohr's theory if

$$n = n_r + l + 1. \tag{3.24}$$

Now those n^2 entries in the Balmer and Ritz equations become clear: The energy of the hydrogen atom depends on $1/n^2$, and the emitted/absorbed photon energy is equal to the difference of energies of two states.

The quantum number $n = 1, 2, 3, \ldots$ chosen in such a way is called the principal quantum number; it defines the energy of the electron's bound states in the hydrogen-like atom. Equation (3.24) also determines the allowed interval for the values of l. The minimum value of n_r is zero (this means that we have only one term $r^l a_0$ in

Eq. (3.18)) and it corresponds to the maximum value of $l = n - 1$. Increasing n_r decreases the values of l, unless it becomes zero. So, the allowed values of l for a given value of n are all integer numbers from 0 to $n - 1$. The quantum number l is called an orbital, or azimuthal, quantum number. For each value of l there exist $2l + 1$ values of m, which is called a magnetic quantum number, varying from $-l$ to $+l$ with a step of unity. As seen from Eq. (3.23), the energy of the stationary states of a hydrogen-like atom depends on the principal quantum number n only; the states with the same n but different l all have the same energy. In other words, all states turn out to be degenerate; for a given value of n there are $\sum_{l=0}^{n-1}(2l+1) = n^2$ states with the same energy (at this point we do not include into account an additional two-fold degeneracy, which is due to existence of the spin momentum). Basically, Eqs. (3.16), (3.18), and (3.21) determine the general form of the radial wave function $R(r)$, which has to be normalized to unity as $\int_0^\infty R(r)^2 r^2 dr = 1$. It is possible to generalize the expression for $R(r)$ using the special functions of mathematical physics. The general form of the radial wave function for the hydrogen-like atom is

$$R_{nl}(r) = -\sqrt{\frac{(n-l-1)!}{(n+l)!^3\,2n}}\left(\frac{2Z}{na_0}\right)^{3/2}\exp\left(-\frac{Zr}{na_0}\right)\left(\frac{2Zr}{na_0}\right)^l L_{n+l}^{2l+1}\left(\frac{2Zr}{na_0}\right), \tag{3.25}$$

with

$$L_n^m(x) = (-1)^m \frac{n!}{(n-m)!} e^x x^{-m} \frac{d^{n-m}}{dx^{n-m}}\left(e^{-x}x^n\right) \tag{3.26}$$

being the generalized (or associated) Laguerre polynomials. $a_0 = \hbar^2/me^2 = 0.529 \times 10^{-8}$ cm or 0.529 Å (with 1 Å = 10^{-8} cm = 10^{-10} m) is the atomic unit of length. It is the radius of the first orbit of an electron in the hydrogen atom. The $R_{nl}(r)$ functions are orthogonal and normalized:

$$\int_0^\infty R_{nl}(r)R_{n'l}(r)r^2 dr = \delta_{nn'}, \tag{3.27}$$

and $\delta_{nn'}$ is the Kronecker symbol ($\delta_{nn'} = 1$ if $n = n'$ and 0 otherwise). We also give here a general expression to generate the explicit form of a spherical function with any indices l and m:

$$Y_{lm}(\theta,\phi) = (-1)^m \sqrt{\frac{2l+1}{4\pi}\frac{(l-m)!}{(l+m)!}} P_l^m(\cos\theta)e^{im\phi}, \text{ for } m \geq 0 \quad (3.28)$$

$$Y_{lm}(\theta,\phi) = (-1)^{|m|} Y_{l|m|}^{\ *}(\theta,\phi), \text{ for } m < 0, \quad (3.29)$$

and

$$P_l^m(x) = \frac{1}{2^l l!}(1-x^2)^{m/2}\frac{d^{l+m}}{dx^{l+m}}(x^2-1)^l \text{ for } m \geq 0 \quad (3.30)$$

are the associated Legendre polynomials. The complete wave function Υ_{nlm} will be defined as a product of the radial part (3.25) and the corresponding spherical function Y_{lm}, defined in Eqs. (3.28) and (3.29). It is also seen that the quantum numbers n, l, m determine the expression for the total wave function. Although the general form of the wave functions seems to be rather complicated, the explicit form of the first wave functions corresponding to the lowest energy levels for the atom of hydrogen is relatively simple. As an example, several wave functions are given below (r is measured in units of a_0 and $Z = 1$):

$$\Psi_{100} = \frac{1}{\sqrt{\pi}}\exp(-r)$$

$$\Psi_{200} = \frac{1}{4\sqrt{2\pi}}(2-r)\exp\left(-\frac{r}{2}\right)$$

$$\Psi_{210} = \frac{1}{4\sqrt{2\pi}} r\exp\left(-\frac{r}{2}\right)\cos\theta$$

$$\Psi_{211} = \frac{1}{8\sqrt{\pi}} r\exp\left(-\frac{r}{2}\right)\sin\theta e^{i\phi}$$

$$\Psi_{21-1} = \frac{1}{8\sqrt{\pi}} r\exp\left(-\frac{r}{2}\right)\sin\theta e^{-i\phi}.$$

The physical meaning of the wave function is as follows: The square of the absolute value of the wave function Ψ_{nlm} at a given point is the probability density to find an electron at that point. Figures 3.2–3.4 show the radial part of the probability $|\Psi_{nlm}|^2 r^2$ for several hydrogen wave functions. These plots (obtained with the help of codes from the F. Y. Wang's book "Physics with Maple" cited at the end of this chapter) in the polar coordinates are the cross sections of the surfaces corresponding to a particular value of probability. The $|\Psi_{nlm}|^2 r^2$ function has no φ term; in all figures the z-axis in the space corresponds to the horizontal axis in the figure, and the angle θ is measured with respect to the horizontal axis. These plots illustrate that the electron density distribution is quite different in different states.

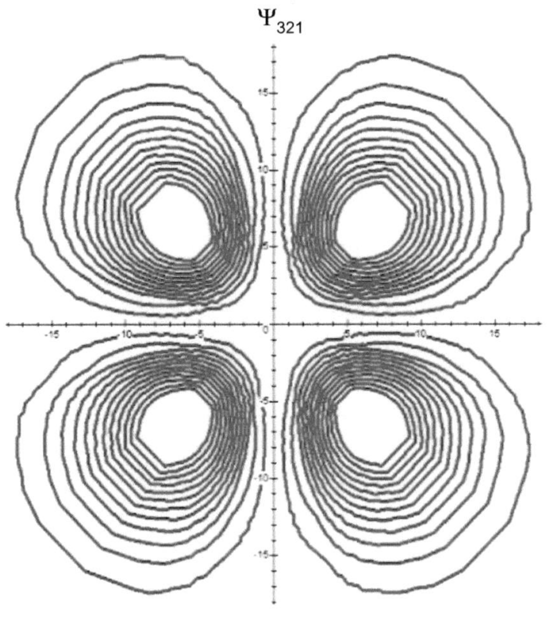

Figure 3.2 Cross section of the electron density distribution in the Ψ_{321} state.

After the wave functions of the hydrogen-like atom are obtained, it is possible to find the averaged value of any physical quantity A in a particular state by performing the following integration:

$$\langle A \rangle = \int \Psi_{nlm}{}^* \hat{A} \Psi_{nlm} d\tau , \tag{3.31}$$

where $\hat{A}$ is the quantum-mechanical operator corresponding to this quantity, and $d\tau$ is the volume element in the spherical coordinates. Special attention should be paid to the coordinate operator r and its powers, since it determines the energies of various states and different kinds of interactions inside the atom. Without derivation, we give here the following expressions (r^2 appears in all integrands, since the integration is performed in the spherical coordinates), which shall be used later:

$$\langle r\rangle = \int \Psi_{nlm}{}^{*} r \Psi_{nlm} r^2 dr = \frac{1}{2}\left(3n^2 - l(l+1)\right)\frac{a_0}{Z}, \quad (3.32)$$

$$\langle r^2\rangle = \int \Psi_{nlm}{}^{*} r^2 \Psi_{nlm} r^2 dr = \frac{n^2}{2}\left(5n^2 + 1 - 3l(l+1)\right)\frac{a_0{}^2}{Z^2}, \quad (3.33)$$

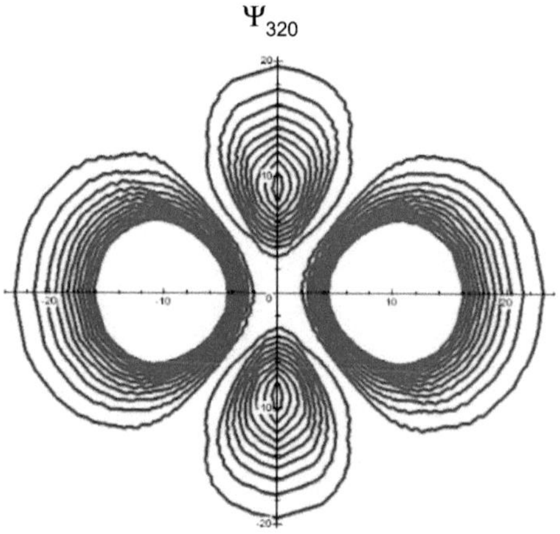

Figure 3.3 Cross section of the electron density distribution in the Ψ_{320} state.

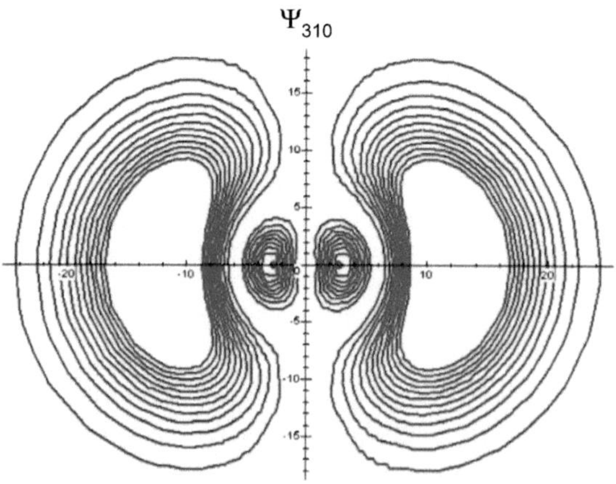

Figure 3.4 Cross section of the electron density distribution in the Ψ_{310} state.

$$\left\langle r^3 \right\rangle = \int \Psi_{nlm}{}^{*} r^3 \Psi_{nlm} r^2 dr$$
$$= \frac{n^2}{8}\left(35n^2(n^2-1) - 30n^2(l+2)(l-1) + 3(l+2)(l+1)l(l-1)\right)\frac{a_0^3}{Z^3}, \quad (3.34)$$

$$\left\langle r^{-1} \right\rangle = \int \Psi_{nlm}{}^{*} r^{-1} \Psi_{nlm} r^2 dr = \frac{1}{n^2}\frac{Z}{a_0}, \quad (3.35)$$

$$\left\langle r^{-2} \right\rangle = \int \Psi_{nlm}{}^{*} r^{-2} \Psi_{nlm} r^2 dr = \frac{1}{n^3\left(l+\frac{1}{2}\right)}\frac{Z^2}{a_0^2}, \quad (3.36)$$

$$\left\langle r^{-3} \right\rangle = \int \Psi_{nlm}{}^{*} r^{-3} \Psi_{nlm} r^2 dr = \frac{1}{n^3(l+1)\left(l+\frac{1}{2}\right)l}\frac{Z^3}{a_0^3}. \quad (3.37)$$

3.4 Selection Rules for the Electric Dipole Transitions

Interaction between electrons in atoms and electromagnetic field of an incident radiation can lead to a formation of the electric dipole moment in the atom (an external electromagnetic field causes displacements of the centers of the charge density for the positive and negative charges in the atom, which no longer coincide in this case). This interaction can also produce transitions between the atomic stationary states. The probability of such a transition from a state Ψ_j into a state Ψ_i under action of the electric dipole operator $\hat{\mathbf{p}} = -e\hat{\mathbf{r}}$ is proportional to the square of the absolute value of the following integral called the matrix element of the transition operator:

$$\int \Psi_i{}^{*} \hat{\mathbf{p}} \Psi_j d\tau . \quad (3.38)$$

If this integral is zero, then such a transition is said to be not allowed or forbidden. It should be pointed out that a direct integration in Eq. (3.38) sometimes can be a quite tedious job. However, there are simple rules, which help in distinguishing between the "zeroth" and "non-zeroth" cases without direct calculation of the above-given integral. These rules are called the selection rules, and now we shall see how one can derive them in the case of the electric dipole operator.

Basically, the dipole operator is nothing else, but the coordinate operator multiplied by the electron's charge $-e$. It will be easier to deal with the x, y, z components of the radius-vector $\hat{\mathbf{r}}$ in the following integrals:

$$x_{ij} = \int \Psi_i^* x \Psi_j d\tau , \tag{3.39}$$

$$y_{ij} = \int \Psi_i^* y \Psi_j d\tau , \tag{3.40}$$

$$z_{ij} = \int \Psi_i^* z \Psi_j d\tau . \tag{3.41}$$

For the next transformations of these integrals, it will be better to write the wave functions in terms of the radial part $R_{nl}(r)$ and associated Legendre polynomials (see Eqs. (3.28)–(3.30)):

$$\Psi_i(r,\theta,\varphi) \cong R_{nl}(r) P_l^m(\cos\theta) e^{im\phi} . \tag{3.42}$$

In the last equation—for the sake of brevity—the $(-1)^m \sqrt{\frac{2l+1}{4\pi}\frac{(l-m)!}{(l+m)!}}$ constant is omitted. Since the spherical system of coordinates is used, $x = r\sin\theta\cos\varphi$, $y = r\sin\theta\sin\varphi$, $z = r\cos\theta$, $d\tau = r^2\sin\theta\, dr\, d\theta\, d\varphi$. It will be also helpful to replace $\cos\varphi$ and $\sin\varphi$ by their exponential expressions: $\cos\varphi = \frac{e^{i\varphi}+e^{-i\varphi}}{2}$, $\sin\varphi = \frac{e^{i\varphi}-e^{-i\varphi}}{2i}$.

The main game will be played around the integrals over φ:

$$\begin{cases} x_{ij} = \frac{1}{2}\int_0^\infty R_{n'l'}R_{nl}r^3 dr \int_0^{2\pi}\left(e^{i(m'-m-1)} + e^{i(m'-m+1)}\right) d\varphi \int_0^\pi P_l^m P_{l'}^{m'} \sin^2\theta d\theta, \\ y_{ij} = \frac{1}{2i}\int_0^\infty R_{n'l'}R_{nl}r^3 dr \int_0^{2\pi}\left(e^{i(m'-m-1)} - e^{i(m'-m+1)}\right) d\varphi \int_0^\pi P_l^m P_{l'}^{m'} \sin^2\theta d\theta, \\ z_{ij} = \int_0^\infty R_{n'l'}R_{nl}r^3 dr \int_0^{2\pi} e^{i(m'-m)} d\varphi \int_0^\pi P_l^m P_{l'}^{m'} \cos\theta\sin\theta d\theta. \end{cases} \tag{3.43}$$

The integrals over φ are not zero, if and only if $m' = m \pm 1$ for x_{ij} and y_{ij} and $m' = m$ for z_{ij}. So, the selection rule for the quantum number m is obtained: An electric dipole transition between two states can take place if and only if the quantum numbers m in these states are equal or differ by unity: $\Delta m = 0, \pm 1$, otherwise the integral over φ is zero.

To find a selection rule for the quantum number l, the integral over θ for z_{ij} is considered. In this case, as we have just seen, $m' = m$. Then the integral over θ is

$$z_{ij} \sim \int_0^{\pi} P_l^m(\cos\theta) P_{l'}^m(\cos\theta)\cos\theta\sin\theta\, d\theta$$

$$= -\int_0^{\pi} P_l^m(\cos\theta) P_{l'}^m(\cos\theta)\cos\theta\, d(\cos\theta).$$

With introduction of a new variable $t = \cos\theta$, we have

$$z_{ij} \sim \int_{-1}^{1} tP_l^m(t)P_{l'}^m(t)dt\,. \tag{3.44}$$

Taking into account the recurrence relation for the associated Legendre polynomials $tP_l^m = \frac{l+m}{2l+1}P_{l-1}^m(t) + \frac{l-m+1}{2l+1}P_{l+1}^m(t)$ and their orthogonal properties $\int_{-1}^{1} P_k^m P_l^m dt = \frac{2(l+m)!}{(2l+1)(l-m)!}\delta_{kl}$, we immediately see that the integral in Eq. (3.44) is not zero if $l' = l \pm 1$. This is the selection rule for the orbital quantum number l: An electric dipole transition between two states can be realized if and only if the orbital quantum numbers l in these states differ by unity: $\Delta l = \pm 1$.

Both selection rules—for the m and l quantum numbers—should be fulfilled simultaneously. If any of them is violated, an electric dipole transition between these states is forbidden. There are no selection rules for the principal quantum number n.

It is also instructive to consider the polarization rules in these cases. If $\Delta m = 0$, both x_{ij} and y_{ij} are zero, and only z_{ij} is not zero. Therefore, in this case emission is linearly polarized along the z axis. If $\Delta m = \pm 1$, then z_{ij} is zero, but both x_{ij} and y_{ij} are not zero, differing by a phase multiplier $\pm i$. In these cases, emission has a circular polarization in the xy plane.

Another formulation of the selection rule for l can be given by using the concept of parity of states. The wave functions Ψ_{nlm} behave in two different ways, if the inversion transformation of the coordinates is considered. In the Cartesian coordinates, this transformation consists in inverting the signs of all coordinates, i.e. in replacing x by $-x$, y by $-y$, z by $-z$. In the spherical coordinates this transformation is as follows: $r \to r$, $\theta \to \pi - \theta$, $\varphi \to \pi + \varphi$. The radial part does not change its sign in this transformation, but the angular part acquires a multiplier $(-1)^l$. Therefore, two situations should be distinguished: If l is an even number, the wave function does not change its sign and such a wave function is said to be an even function. If l is an odd number, the wave function changes its sign, and such a wave function is referred to as an odd function.

The electric dipole operator is an odd function of coordinate. Since the integral in Eq. (3.38) is to be calculated by integrating over the whole space, it will be different from zero if the integrand is an even function, and it will be identically zero otherwise. Therefore, the product of the two wave functions of the initial and final states of the transition should be an odd function (then the whole integrand will be an even function), and this means that the electric dipole transitions can connect only the states with different parity. This selection rule formulated with involving the concept of parity of states is called the Laporte rule.

The parity-based selection rules depend on the symmetry properties of the operator considered as a reason for the transition. If, for example, an electric quadrupole operator (which is an even function of the coordinates) is considered, then $\Delta l = 0, \pm 2$. However, intensities of the electric quadrupole transitions are much lower than the intensities of the electric dipole ones.

3.5 Fine Structure of the Energy Levels and the Spin of an Electron

The electrons moving around a nucleus in an atom have rather high velocities—up to 10^6 - 10^7 m/s. This circumstance makes relativistic effects (dependence of mass on velocity) to be quite noticeable. In this paragraph we shall see the extent to which the energy levels from Eq. (3.23) of a hydrogen-like atom are affected by the relativistic effects.

A total energy of a particle with mass m in a potential field U is: $E_{\text{tot}} = U + \sqrt{c^2p^2 + m^2c^4}$, where p stands for the momentum of the particle. This energy can be expanded into series (with respect to the powers of v/c, where v is the velocity of a particle, and c is the speed of light):

$$\begin{aligned} E_{\text{tot}} &= U + \sqrt{c^2p^2 + m^2c^4} \\ &= U + mc^2\sqrt{1 + \left(\frac{v}{c}\right)^2} \cong mc^2 + U + \frac{p^2}{2m} - \frac{1}{8}\frac{p^4}{m^3c^2} + \cdots \end{aligned} \tag{3.45}$$

The first term in the right-hand side is the rest energy (an additive constant), the two others represent the total non-relativistic energy, and the last one is a small perturbation produced by dependence of mass on velocity. We have already solved the Schrödinger equation in the non-relativistic case—its eigenvalues are given by Eq. (3.23).

As known from quantum mechanics, a correction to the energy levels of the unperturbed system, which is caused by the action of a small perturbation, is given by a diagonal matrix element of the perturbation operator calculated on the wave functions of the corresponding unperturbed level. So, in the considered case the perturbation operator is $V = -\frac{1}{8}\frac{p^4}{m^3c^2}$, and we shall have to evaluate the integrals of the form $\int \Psi_{nlm}{}^* V \Psi_{nlm} d\tau$. First, we can re-write Eq. (3.45):

$$E = E_{\text{tot}} - mc^2 = U + \frac{p^2}{2m} - \frac{1}{8}\frac{p^4}{m^3c^2} = E^{(0)} - \frac{1}{8}\frac{p^4}{m^3c^2},$$

where $E^{(0)} = U + \frac{p^2}{2m}$ stands for the eigenvalues of the non-relativistic Hamiltonian, and transform the perturbation operator to the following form:

$$V = -\frac{1}{8}\frac{p^4}{m^3c^2} = -\frac{1}{2mc^2}(E^{(0)} - U)^2. \tag{3.46}$$

An idea of such a transformation is as follows: Instead of differentiation (which is implied by the momentum operator p = –

$i\hbar\nabla$, with $\Delta = \frac{\partial}{\partial x}\vec{i} + \frac{\partial}{\partial x}\vec{j} + \frac{\partial}{\partial x}\vec{k}$), we get a constant $E^{(0)} = -\frac{mZ^2e^4}{2\hbar^2 n^2}$ and the potential energy as a coordinate function $U = -\frac{Ze^2}{r}$; both are much easier to deal with. We get for the V operator:

$$V = -\frac{1}{2mc^2}\left(\frac{m^2Z^4e^8}{4\hbar^4 n^4} - \frac{mZ^3e^6}{\hbar^2 n^2}\frac{1}{r} + \frac{Z^2e^4}{r^2}\right). \quad (3.47)$$

Then finding a correction produced by the operator V is reduced to finding the averaged values of the $\frac{1}{r}$ and $\frac{1}{r^2}$ operators, which are given by Eqs. (3.35) and (3.36), respectively. Putting these two equations and Eq. (3.47) together, we get that

$$\Delta E_{nl} = -\frac{e^4}{\hbar^2c^2}\left(\frac{1}{l+\frac{1}{2}} - \frac{3}{4n}\right)\frac{Z^4}{n^3} = -\alpha^2\left(\frac{1}{l+\frac{1}{2}} - \frac{3}{4n}\right)\frac{Z^4}{n^3}\mathrm{Ry}, \quad (3.48)$$

where $\alpha = \frac{e^2}{\hbar c} = \frac{1}{137}$ is the fine structure constant and $\mathrm{Ry} = \frac{me^4}{2\hbar^2}$ = 13.535 eV is the unit of energy called Rydberg. What can be immediately seen from Eq. (3.48) is that the correction to the energy levels produced by dependence of mass of an electron on its velocity is proportional to α^2(and is a very fine correction, indeed). On the other hand, it is proportional to Z^4 and is much more important for heavy nuclei from the second half of the periodic table with large values of the atomic numbers Z.

One additional relativistic correction is related to the so-called spin-orbit (SO) interaction. Before calculating its value, we shall need, first of all, to introduce the concept of spin of an electron.

By spin in quantum mechanics is meant an angular momentum, which is an intrinsic property of an object (it should be distinguished from the orbital angular momentum related to the motion of its center of mass about some point which does not coincide with the center of mass). The concept of the electron spin was introduced in 1925 by two Dutch physicists S. A. Goudsmit* and G. E. Uhlenbeck.

*Interestingly, S. Goudsmit was not only a prominent physicist. He made also a considerable contribution to Egyptology.

Figure 3.5 shows a schematic picture (though not completely physically correct!), which may help to illustrate the concept of spin. If a particle simultaneously with the rotation about a point O rotates about its own axis (as shown in the figure), then the angular momentum associated with such a rotation can be thought of as some analogy of spin in quantum mechanics.

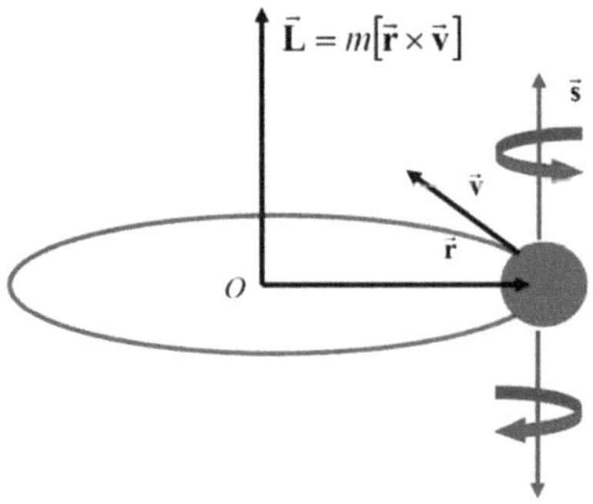

Figure 3.5 To the illustration of an idea of the spin momentum.

The value of spin momentum for an electron (in units of $\hbar$) is 1/2 and its projection onto a certain direction can be ±1/2. This spin momentum can interact with the orbital angular momentum $\vec{\mathbf{l}}$ to form a total angular momentum $\vec{\mathbf{j}} = \vec{\mathbf{l}} + \vec{\mathbf{s}}$, and the values of j can be $l + 1/2$ and $l - 1/2$. So, due to the presence of the spin of an electron, each level with the same orbital quantum number l gives rise to two states with two different values of j. The state with $l = 0$ is an exception from this rule, when only one value of $j = 1/2$ is allowed.

To determine the correction produced by this SO interaction, we recall first that an electron has a magnetic moment $\boldsymbol{\mu}$, which is proportional to its spin: $\boldsymbol{\mu} = -\frac{e\hbar}{mc}\mathbf{s}$. An electron moving in the electric field produced by a nucleus creates a magnetic field **H.** The magnitude of this field is determined by the Biot–Savart law and is $\mathbf{H} = -\frac{[\mathbf{E} \times \mathbf{v}]}{c}$. Since $\mathbf{E} = \frac{\partial U}{\partial r}\frac{\mathbf{r}}{r}$ and $m[\mathbf{r} \times \mathbf{v}] = \hbar\mathbf{l}$, the energy $-\boldsymbol{\mu}\cdot\mathbf{H}$ of this additional interaction is proportional to $\mathbf{l} \cdot \mathbf{s}$. The final result for the operator of the SO interaction V_{so} is

$$\boldsymbol{\mu} \cdot \mathbf{H} = \frac{\hbar^2}{2m^2c^2} \frac{\partial U}{\partial r} \frac{1}{r} \mathbf{l} \cdot \mathbf{s}. \tag{3.49}$$

A scalar product of operators $\mathbf{l} \cdot \mathbf{s}$ can be written as $\mathbf{l} \cdot \mathbf{s} = \frac{1}{2}(\mathbf{j}^2 - \mathbf{l}^2 - \mathbf{s}^2)$ (it can be obtained by squaring the expression $\vec{\mathbf{j}} = \vec{\mathbf{l}} + \vec{\mathbf{s}}$ and solving it with respect to $\mathbf{l} \cdot \mathbf{s}$). Finally, we arrive at

$$V_{SO} = \frac{Ze^2\hbar^2}{2m^2c^2}\frac{1}{r^3}\frac{1}{2}(\mathbf{j}^2 - \mathbf{l}^2 - \mathbf{s}^2). \tag{3.50}$$

Using Eq. (3.37) and the fact that the eigenvalues of the operators $\mathbf{j}^2$, $\mathbf{l}^2$, $\mathbf{s}^2$ are equal to $j(j + 1)$, $l(l + 1)$, $s(s + 1)$, correspondingly, it is not difficult to obtain the expression for the correction to the energy levels produced by the SO interaction operator:

$$\Delta E_{\rm SO} = \alpha^2 \frac{j(j+1) - l(l+1) - s(s+1)}{2l(l+1)\left(l + \frac{1}{2}\right)} \frac{Z^4}{n^3} \mathrm{Ry}. \tag{3.51}$$

As shown in Eq. (3.48), this correction is proportional to Z^4 (heavy ions are mostly affected by the SO interaction) and inversely proportional to n^3 (this correction is more pronounced for the lower energy levels). Now we can find the total correction produced by both considered phenomena. In fact, it will be a simple sum of Eqs. (3.48) and (3.51). Since for one electron $s = 1/2$, $j = l \pm 1/2$. It is interesting that for both values of j the total correction is given by the same expression, which is as follows:

$$\Delta E_{\rm tot} = \alpha^2 \left(\frac{3}{4n} - \frac{1}{j + 1/2} \right) \frac{Z^4}{n^3} \mathrm{Ry}. \tag{3.52}$$

Equation (3.52) describes the fine structure of the energy levels of the hydrogen-like atom.

Figure 3.6 shows the general pattern of the fine-structure of the lowest energy levels for the atom of hydrogen and two hydrogen-like ions He^+ and Li^{2+}. As seen from this figure, the relativistic effects lower down the j energy levels (with respect to the unperturbed energy levels shown from the left side of the figure). The states with the same n and j have the same energy, as follows from Eq. (3.52). Very small splittings for the atom of hydrogen become significantly greater if we move to heavier ions. Nevertheless, when comparing to the energy intervals between the unperturbed levels, these splittings are really small, and this justifies fully the term "fine structure." For a comparison, in the left side of the figure we give the difference E_2–E_1, which can be obtained without relativistic effects. These differences for the considered ions are much greater than their fine-structure splittings. However, even such small splittings of the energy levels, which will produce the splitting of the absorption and emission lines in the experimental spectra, are detectable.

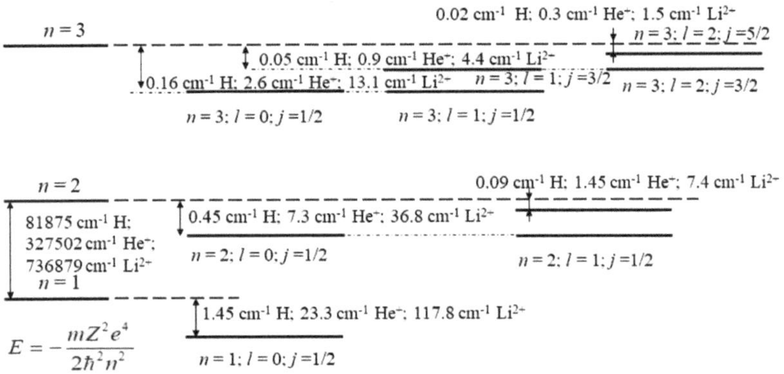

Figure 3.6 Fine structure of the lowest levels for the atom of hydrogen H and two first hydrogen-like ions He^+ and Li^{2+} (not to scale).

3.6 Four Quantum Numbers and Spectral Notation of the Hydrogen-Like Atoms States

Summarizing the results of the previous paragraphs, we can conclude that four quantum numbers are needed to describe the energetic states of the hydrogen-like atoms, including their fine structure. These numbers are as follows: the principal quantum number n (which takes integer values 1, 2, 3, ..., without an upper limit), the orbital quantum number l (which for a given n takes all integer values from 0 to n–1), the magnetic quantum number m (which takes all integer values from $-l$ to $+l$), and the projection of the spin momentum of electron m_S (which can take two values +1/2 and –1/2). Historically, the states of an electron with definite values of l are denoted by the following small letters of the Latin alphabet:

l	0	1	2	3	4	5	6	7	8	9	10	11	12	13	14
	s	p	d	f	g	h	i	k	l	m	n	o	q	r	t

Introduction of an additional quantum number s changes the degree of degeneracy of the energy levels for a single electron configuration: $\sum_{l=0}^{n-1} 2(2l+1) = 2n^2$, because two spin orientations—up and down—become possible.

Various states of atoms are denoted by the capital letters: $^{2s+1}L_j$, e.g. $^2P_{1/2}$, $^2S_{1/2}$, $^2P_{3/2}$, $^2D_{5/2}$, etc. The left superscript is called a multiplicity; it is equal to $2s + 1$ and it denotes the number of projections of the spin momentum. The right subscript shows the value of the total angular momentum j. So, the notation $^2D_{5/2}$ denotes a state with the angular momentum L equal to 2, spin equal to 1/2, and the total angular momentum j equal to 5/2. The group of the $^{2s+1}L_j$ states is called a term.

Bibliography

1. Cohen-Tannoudji C, Diu B, Laloe F (1977). *Quantum Mechanics.* John-Wiley, Paris.
2. Edmonds AR (1957). *Angular Momentum in Quantum Mechanics.* Princeton University Press.
3. Griffiths DJ (1995). *Introduction to Quantum Mechanics.* Prentice-Hall, New Jersey.
4. Henderson B, Bartram RH (2000). *Crystal-Field Engineering of Solid-State Laser Materials*, Cambridge University Press.
5. Landau LD, Lifshitz EM (1977). *Quantum Mechanics.* Pergamon Press, Oxford.
6. Nielsen CW, Koster GF (1963). *Spectroscopic Coefficients for the* p^n, d^n, *and* f^n *Configurations.* MIT Press, Cambridge, Massachusetts.
7. Powell RC (1998). *Physics of Solid-State Laser Materials.* Springer.
8. Schiff LI (1968). *Quantum Mechanics.* McGraw-Hill.
9. Sobelman II (1992). *Atomic Spectra and Radiative Transitions.* Springer.
10. Wang FY (2005). *Physics with Maple.* Wiley-VCH.

Chapter 4

Multielectron Atoms

4.1 Introduction

All results in the previous chapter were obtained assuming that there is only one electron in the hydrogen-like atom or ion. However, in most cases there is more than one electron in the atom. Presence of two or more electrons makes the whole picture of physical interactions in such an atom to be more complicated, because the Coulomb interaction among those electrons has to be considered and added to the total Hamiltonian. The inter-electron Coulomb repulsion does not have a spherical symmetry, like the interaction between the nucleus and single electron in the hydrogen atom. The distribution of electrons through allowed states obeys the Pauli exclusion principle, preventing them from having identical sets of all quantum numbers. Systematics of the multielectron states is the focus of the present chapter.

4.2 Schrödinger Equation for the Multielectron Atoms: What Is Different from the Hydrogen-Like Atom Case?

The Schrödinger equation for the multielectron atom has the following form:

Theoretical Spectroscopy of Transition Metal and Rare Earth Ions: From Free State to Crystal Field
Mikhail G. Brik and Chong-Geng Ma

ISBN 978-981-4800-56-3 (Hardcover), 978-0-429-27875-4 (eBook)
www.jennystanford.com

$$\left\{\sum_{i=1}^{n}\left(-\frac{\hbar^2}{2m}\Delta_i-\frac{Ze^2}{r_i}\right)+\frac{1}{2}\sum_{i,j=1}^{n}\frac{e^2}{\left|\vec{\mathbf{r}}_i-\vec{\mathbf{r}}_j\right|}\right\}\Psi=E\Psi. \tag{4.1}$$

Here n is the total number of electrons in the atom. The first term is just a sum of one-electron Hamiltonians, and the second term describes the Coulomb interaction between electrons. Pre-multiplier 1/2 is needed to avoid counting interaction between the electrons with indices i and j twice, when running a summation over both indices. This term removes the spherical symmetry of the one-electron Hamiltonians, which represents the main mathematical difficulty in this case. It is not possible to find an exact solution of the multielectron Schrödinger equation, and some essential approximations should be made to simplify the problem.

Without going into all details, we just mention that it is possible to introduce for an individual electron an effective potential $U_{\text{add}}(r)$ possessing spherical symmetry, which models the Coulomb potential created by all other $(n - 1)$ electrons. Then the potential energy of this selected electron in the field of the nucleus and the rest of the electrons will be

$$U(r)=-\frac{Ze^2}{r}+U_{\text{add}}(r). \tag{4.2}$$

These $(n - 1)$ electrons can be treated as a screen placed between the selected electron and the nucleus. Then this effective potential can be chosen, for example, as $U_{\text{add}}(r)=\frac{\sigma(r)e^2}{r}$, where $\sigma(r)$ is a function of distance, which accounts for the screening effect produced by the remaining $(n - 1)$ electrons. Then

$$U(r)=-\frac{(Z-\sigma(r))e^2}{r}=-\frac{Z_{\text{eff}}(r)e^2}{r}, \tag{4.3}$$

and $Z_{\text{eff}}(r)$ is an effective charge (in units of the electron charge), which acts upon the selected electron at the distance r from the nucleus. Introduction of such an effective field is an approximation only, which, nevertheless, works perfectly well for the outer electrons (namely these electrons from the most outer electron shells are responsible for the optical spectra). Then it turns out that, like in the hydrogen atom, there are also four quantum numbers n, l, m, m_S, which have the same meaning and the same limits of variation.

We can also go further and introduce the concept of parity of states, which in the case of the multielectron atoms will be defined as a sum of the orbital quantum numbers l of all electrons $\sum_{i=1}^{n} l_i$ in the considered electron configuration.

When dealing with the multielectron atoms, the Pauli exclusion principle should always be considered. It was formulated by W. Pauli in 1925 and states that no two identical fermions (particles with half-integer spin; electrons also belong to this group of particles) may be found simultaneously in the same quantum state. A more rigorous statement of this principle is that, for the system of identical fermions, the total wave function is anti-symmetric with respect to the interchange of two fermions. When applied to the electrons in a single atom, this principle means that no two electrons can have the same four quantum numbers, that is, if n, l, and m are the same for both electrons, their projections of the spin momenta m_S must be different. This implies that these electrons should have opposite spins. If the spins of both chosen electrons are parallel, then at least one of the remaining three quantum numbers n, l, and m should be different.

The electrons with the same principal quantum number n form the electron shell; the electrons with the same (n, l) numbers belong to the same subshell, and the set of the quantum numbers (n, l, m) determines an electron orbital. An electronic configuration is defined as a group (or several groups) of electrons with different values of the principal quantum number n and orbital quantum number l. It is denoted as nl^k with k staying for the number of such electrons. For example, $3d^3$ denotes a configuration consisting of 3 d-electrons with the principal quantum number $n = 3$. The electron configuration of hydrogen atom in its ground state, which has only one s-electron, is $1s^1$. The electron configuration of the next element in the periodic table, helium, is $1s^2$, etc.

The following letters are used to distinguish between different shells: K shell $n = 1$ (only $1s$ states are allowed); L shell $n = 2$ ($2s$, $2p$ states); M shell $n = 3$ ($3s$, $3p$, $3d$ states); N shell $n = 4$ ($4s$, $4p$, $4d$, $4f$ states); O shell $n = 5$ ($5s$, $5p$, $5d$, $5f$, $5g$ states); P shell $n = 6$ ($6s$, $6p$, $6d$, $6f$, $6g$, $6h$ states). The g and h electron states are not filled in atoms under the normal conditions and are listed here just for the sake of completeness.

4.3 Basic Properties of the Operator of Angular Momentum and Its Main Difference from the Classical Momentum

Interactions between electrons in a multielectron atom lead to the formation of a number of states with different energy. The number of these states depends on the number of electrons in the atom, and their energies depend on the arrangement of electrons in the available orbitals as well as on the mutual orientation of the angular and spin momenta of all electrons. In this paragraph, we shall consider the general properties of the angular momenta in quantum mechanics and the rules of their addition—these subjects are very important for a clear understanding of the multielectron atoms spectra.

The operator $\widehat{\mathbf{L}}$ of angular momentum in quantum mechanics is defined as follows:

$$\widehat{\mathbf{L}} = [\hat{\mathbf{r}} \times \hat{\mathbf{p}}] = -i\hbar[\hat{\mathbf{r}} \times \widehat{\nabla}] \cdot \tag{4.4}$$

As it can be seen from Eq. (4.4), this definition is analogous to the angular momentum in classical mechanics; just the position vector $\vec{\mathbf{r}}$ and momentum $\vec{\mathbf{p}}$ are replaced by their corresponding quantum-mechanical operators.

The Cartesian components of the $\widehat{\mathbf{L}}$ operator can be easily obtained from its definition:

$$\hat{L}_x = -i\hbar\left(y\frac{\partial}{\partial z} - z\frac{\partial}{\partial y}\right),$$

$$\hat{L}_y = -i\hbar\left(z\frac{\partial}{\partial x} - x\frac{\partial}{\partial z}\right) \hat{L}_z = -i\hbar\left(x\frac{\partial}{\partial y} - y\frac{\partial}{\partial z}\right). \tag{4.5}$$

It is a common practice to use the spherical coordinates when working with the angular momentum operator. The standard treatment of the angular momentum can be found in any textbook on quantum mechanics, and here we just give a short summary of the main results.

The salient feature of the operator $\widehat{\mathbf{L}}$ is that it is not a vector, like in classical physics. It is possible to show that the quantum-mechanical commutators $[\hat{L}_i, \hat{L}_j] \equiv \hat{L}_i\hat{L}_j - \hat{L}_j\hat{L}_i$, $i, j = x, y, z$ are not zero. A general expression for these commutators is as follows:

$$[\hat{L}_i,\hat{L}_j] \equiv i\hbar\hat{L}_k, i,j,k = x,y,z. \quad (4.6)$$

This means that all three components cannot have definite values simultaneously. On the other hand, each component of the operator $\widehat{\mathbf{L}}$ commutes with the operator $\widehat{\mathbf{L}}^2$, which indicates that the orbital momentum and one of its components (only one out of three) can be known. It should also be pointed out that the eigenvalues of the $\widehat{\mathbf{L}}^2$ operator are $L(L + 1)$; therefore, the "length" of the angular momentum in quantum mechanics is defined as $\sqrt{L(L+1)}$. The maximum value of the component of the angular momentum along a chosen axis is equal to L. The eigenfunctions of the $\widehat{\mathbf{L}}^2$ operator are the spherical functions Y_{lm}.

Since the length of the vector representing the angular momentum in quantum mechanics is $\sqrt{L(L+1)}$, and its projection M onto the z axis takes all integer values from $-L$ to $+L$, the angular momentum is never directed along the z axis, but always makes with it an angle $\theta = \arccos\left(\frac{M}{\sqrt{L(L+1)}}\right)$ (Fig. 4.1). Basically, the vector L makes a precession around the z axis, forming a cone with the angle 2θ and height M. The two remaining components L_x,L_y can take any values to satisfy the condition $L_x{}^2 + L_y{}^2 = L(L+1) - M^2$. The only exception from this rule is the case $L = 0$. Then, obviously, $L_x = L_y = L_z = 0$, and all three components of the angular momentum are known. However, this is the only one exceptional case; in all others, only one component of the angular momentum can be defined.

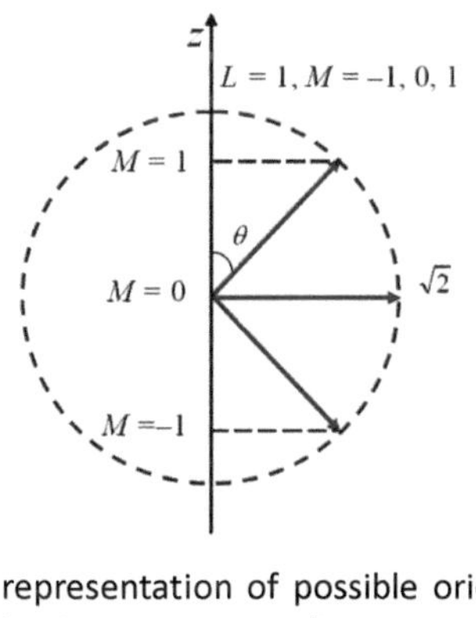

Figure 4.1 Schematic representation of possible orientation of the angular momentum and its projections corresponding to $L = 1$. The "length" of the representing vector is $\sqrt{L(L+1)} = \sqrt{2}$.

4.4 The LS and JJ Coupling Schemes, or Social/Individual Behavior of Electrons

When discussing the fine structure of the hydrogen atom spectra, we introduced the total angular momentum as a sum of the orbital and spin momenta. Depending on the order and magnitude of the inter-electron interaction, two types of coupling can be distinguished: (i) *LS*-coupling (or Russell–Saunders coupling) and (ii) *JJ*- coupling.

(i) The *LS*-coupling is realized when the electrostatic interaction between electrons in the same electron shell is stronger than the SO interaction. This means that the orbital momenta of all electrons are summed up first to get the total orbital momentum L, then their spin momenta are added together to form the total spin momenta S. Finally, these total orbital and spin momenta L and S are combined to form a total angular momentum J of a considered configuration:

$$\sum_{i=1}^{k} \vec{\mathbf{l}}_i = \vec{\mathbf{L}}, \; \sum_{i=1}^{k} \vec{\mathbf{s}}_i = \vec{\mathbf{S}}, \; \vec{\mathbf{L}} + \vec{\mathbf{S}} = \vec{\mathbf{J}}.$$

In this case of the *LS*-coupling, the electrons behave more like "social particles": They "unite" their efforts to make the total orbital and spin momenta without considering first individual SO coupling for each electron.

(ii) The *JJ*-coupling is realized in the opposite situation: when the SO interaction for an individual electron is stronger than the Coulomb interaction between electrons. This is especially the case of heavy atoms, since the SO interaction energy is proportional to the fourth power of the atomic number and increases drastically for the chemical elements in the second half of the periodic table. In this case, the orbital and spin momenta of each electron are summed up to give the total angular momentum j of an individual electron, and then these individual j momenta are added together to form a total angular momentum J of the considered electron configuration:

$$\vec{\mathbf{l}}_i + \vec{\mathbf{s}}_i = \vec{\mathbf{j}}_i \sum_{i=1}^{k} \vec{\mathbf{j}}_i = \vec{\mathbf{J}}.$$

In this case of the *JJ*-coupling the individual properties of the electrons (SO coupling) dominate over the tendency to form the total orbital and spin momenta, and the electrons behave in a more "individual" way, if compared to the above-described case of the *LS*-coupling.

Depending on the type of the angular momenta coupling, different sets of quantum numbers should be used for a description of the electronic states. In the first case, the four quantum numbers L, M_L, S, M_S give a unique description of the multielectron states. In the second case the *nljm* set should be used to characterize the multielectron states.

Finally, it should be pointed out that the orbital quantum number L can take only integer values. The quantum number of the total angular momentum J can take either integer or half-integer values, depending on the value of S, which in the case of the electron configurations with odd number of electrons can be half-integer.

4.5 Addition of Two Angular Momenta or Why Two Plus Two Is Not Always Four

In this paragraph, we shall consider addition of two angular momenta J_1 and J_2 with quantum numbers $-J_1 \leq M_1 \leq J_1$, $-J_2 \leq M_2 \leq J_2$. The main question to be answered is: Which states can be formed after such a summation? All obtained results are valid and can be readily applied to the addition of the orbital angular momenta L_1 and L_2, with the only difference that the L values are always integer, whereas the J values can be both integer and half-integer, depending on the spin value S.

Addition of two vector momenta in classical physics is easy: The corresponding components along the x, y, and z axes should be summed up. In quantum mechanics such a simple approach is not valid, since only one out of three components of the angular momentum can be known.

Therefore, the solution of the problem of addition of two momenta is to sum up their z-components determined by the quantum numbers M_1 and M_2. Since there exist $2J_1 + 1$ values of M_1

and $2J_2 + 1$ values of M_2, there are in total $(2J_1 + 1) \times (2J_2 + 1)$ ways to add these two projections. Let us assume—for convenience only—that $J_1 > J_2$. Then we may build up a table (shown below), the first row of which contains the values of M_1, the first column of which contains the values of M_2, and all entries in the middle part are the sums of two projections in the corresponding rows and columns.

If each value of J determines a set of $2J + 1$ M quantum numbers varying from $-J$ to J, then the reversed statement saying that the set of the numbers $-M, -M + 1, -M + 2, ..., M - 2, M - 1, M$ corresponds to the total angular momentum $J = M$ is also correct.

We see from Table 4.1 that it is possible to select specific sets of the $M = M_1 + M_2$ values corresponding to the different values of the total angular momentum J. For example, the second row and the first column from the right contain the values of M from $J_1 + J_2$ to $-J_1 - J_2$ with the step of one. A value of $J = J_1 + J_2$ corresponds to such a set. The third row and the second column from the right contain the values of M from $J_1 + J_2 - 1$ to $-J_1 - J_2 + 1$, which is the whole range of the allowed M values for the value of $J = J_1 + J_2 - 1$. If we continue in such a way, we end up with a part of the last but one row (highlighted in gray), which contains the values of M from $J_1 - J_2$ to $-J_1 + J_2$, to which the value of $J = J_1 - J_2$ corresponds. The determined values of J are shown in the last row of the table, in the cells corresponding to a particular set of the M values. We see that the values of J, which can be obtained if we add two momenta J_1 and J_2, take all values from $J_1 + J_2$ to $J_1 - J_2$ with a step of one. We can also assume $J_2 > J_1$; then we shall obtain the values of J from the interval $J_1 + J_2$ to $J_2 - J_1$.

Therefore, a general rule of quantum-mechanical addition of two angular momenta J_1 and J_2 with a completely arbitrary relation between both of them can be formulated as follows: The resulting moment J can take any values from the interval $J_1 + J_2$ to $J_2 - J_1$ with the step of one:

$$|J_1 - J_2| \leq J \leq J_1 + J_2.$$

For example, if two angular momenta $J_1 = 2$ and $J_2 = 3$ are added, the values of the resulting J can be 1, 2, 3, 4, 5 (see Table 4.2).

Table 4.1 Addition of the projections of two angular momenta

M_2 \ M_1	J_1	$J_1 - 1$	...	$-J_1 + 2J_2$	$-J_1 + 2J_2 - 1$	...	$-J_1 + 1$	$-J_1$
J_2	$J_1 + J_2$	$J_1 + J_2 - 1$	...	$-J_1 + 3J_2$	$-J_1 + 3J_2 - 1$	...	$-J_1 + J_2 + 1$	$-J_1 + J_2$
$J_2 - 1$	$J_1 + J_2 - 1$	$J_1 + J_2 - 2$	...	$-J_1 + 3J_2 - 1$	$-J_1 + 3J_2 - 2$	...	$-J_1 + J_2$	$-J_1 + J_2 - 1$
...	...	...	...	...	...	...	...	...
$-J_2 + 1$	$J_1 - J_2 + 1$	$J_1 - J_2$	...	$-J_1 + J_2 + 1$	$-J_1 + J_2$	...	$-J_1 - J_2 + 2$	$-J_1 - J_2 + 1$
$-J_2$	$J_1 - J_2$	$J_1 - J_2 - 1$	...	$-J_1 + J_2$	$-J_1 + J_2 - 1$	...	$-J_1 - J_2 + 1$	$-J_1 - J_2$
Set	$J_1 - J_2$				$J_1 - J_2 + 1$	...	$J_1 + J_2 - 1$	$J_1 + J_2$

Table 4.2 Addition of the projections of two angular momenta $J_1 = 3$ and $J_2 = 2$

M_2 \ M_1	**2**	**1**	**0**	**−1**	**-2**
3	5	4	3	2	1
2	4	3	2	1	0
1	3	2	1	0	-1
0	2	1	0	− 1	-2
−1	1	0	− 1	− 2	-3
−2	0	−1	−2	−3	−4
−3	−1	−2	−3	−4	-5
Set	1	2	3	4	5

Hence, because of impossibility to determine simultaneously all three components of the angular momentum in quantum mechanics, the sum of two momenta is not determined unambiguously. Therefore, in the quantum world two plus two is not always four, but can be zero, one, two, three, or four, following the general rule of addition of two momenta.

4.6 The Wave Functions of the Resulting Angular Momentum, Clebsch–Gordan Coefficients and Wigner 3j-Symbols

So, as we have seen, the states with quantum numbers (J_1, M_1) and (J_2, M_2) can give rise to a state with $M = M_1 + M_2$ and any J from the interval $|J_1 - J_2| \leq J \leq J_1 + J_2$, so that $|M| \leq J$. We know the wave functions, which correspond to the states (J_1, M_1) and (J_2, M_2). Since the given value of M can be obtained from different combinations of $M_1 + M_2$, the resulting wave function of the (J, M) state will be written as a linear combination of the products of the wave functions of the (J_1, M_1) and (J_2, M_2) states with the $M = M_1 + M_2$ condition fulfilled:

$$\Psi_{JM} = \sum_{M_1+M_2=M} C^{JM}_{J_1M_1J_2M_2} \Psi_{J_1M_1} \Psi_{J_2M_2}. \tag{4.7}$$

The sum is running over M_1 and M_2 with the above-mentioned condition. The coefficients $C^{JM}_{J_1M_1J_2M_2}$ are called the Clebsch–Gordan coefficients. They have a simple physical meaning: The square of the absolute value of the $C^{JM}_{J_1M_1J_2M_2}$ coefficient is equal to the probability that two states (J_1, M_1) and (J_2, M_2) after addition produce the state (J, M). If $M \neq M_1 + M_2$, the corresponding Clebsch–Gordan coefficient is zero. There are various notations for the Clebsch–Gordan coefficients in the literature, for example: $C^{JM}_{J_1M_1J_2M_2}$, $C^{J}_{M_1M_2}$, $\left(J_1J_2M_1M_2\middle|J_1J_2JM\right)$.

In some applications, it is more convenient to use the Wigner 3*j*-symbols $\begin{pmatrix} j_1 & j_2 & j \\ m_1 & m_2 & -m \end{pmatrix}$, defined by the following expression:

$$C^{jm}_{j_1m_1j_2m_2} = (-1)^{j_1-j_2+m}\sqrt{2j+1}\begin{pmatrix} j_1 & j_2 & j \\ m_1 & m_2 & -m \end{pmatrix}. \tag{4.8}$$

The general expression for calculating the Wigner 3*j*-symbols is as follows:

$$\begin{pmatrix} j_1 & j_2 & j_3 \\ m_1 & m_2 & m_3 \end{pmatrix} = \sqrt{\left[\frac{(j_1+j_2-j_3)!(j_1-j_2+j_3)!(-j_1+j_2+j_3)!}{(j_1+j_2+j_3+1)!}\right]}$$
$$\times\sqrt{(j_1+m_1)!(j_1-m_1)!(j_2+m_2)!(j_2-m_2)!(j_3+m_3)!(j_3-m_3)!}$$
$$\times\sum_z \frac{(-1)^{z+j_1-j_2-m_3}}{z!(j_1+j_2-j_3-z)!(j_1-m_1-z)!(j_2+m_2-z)!D} \tag{4.9}$$

where $\quad D = (j_3-j_2+m_1+z)!(j_3-j_1-m_2+z)!$

The index of summation *z* runs through all integer numbers, starting from zero. Since the factorial of a negative number is not defined, the number of terms in the right-hand side is always finite. It is not difficult to write a computer program for calculating the Wigner 3*j*-symbols, or the Clebsch–Gordan coefficients. The convenience of the Wigner 3*j*-symbols is that they possess more pronounced symmetry properties with respect to the interchange of different indices than the Clebsch–Gordan coefficients. For example,

$$\begin{pmatrix} j_2 & j_1 & j_3 \\ m_2 & m_1 & m_3 \end{pmatrix} = (-1)^{j_1+j_2+j_3}\begin{pmatrix} j_1 & j_2 & j_3 \\ m_1 & m_2 & m_3 \end{pmatrix}. \tag{4.10}$$

The even transposition of the columns does not change the value of the 3*j*-symbol; the odd transposition introduces a multiplier

$(-1)^{j_1+j_2+j_3}$, which may or may not change the sign of the 3*j*-symbol depending on the sum of numbers from the first row. The same multiplier appears if we simultaneously change the signs of all entries in the second row of the 3*j*-symbol:

$$\begin{pmatrix} j_1 & j_2 & j_3 \\ -m_1 & -m_2 & -m_3 \end{pmatrix} = (-1)^{j_1+j_2+j_3} \begin{pmatrix} j_1 & j_2 & j_3 \\ m_1 & m_2 & m_3 \end{pmatrix}. \tag{4.11}$$

The Clebsch–Gordan coefficients or 3*j*-symbols are very useful for calculations of the matrix elements of the so-called tensor operators. In particular, the integrals from a product of three spherical functions over all space can be expressed in terms of the product of two 3*j*-symbols:

$$\int_0^{\pi}\int_0^{2\pi} Y_{l_1m_1} Y_{l_2m_2} Y_{l_3m_3} \sin\theta\, d\phi\, d\theta = \sqrt{\frac{(2l_1+1)(2l_2+1)(2l_3+1)}{4\pi}} \begin{pmatrix} l_1 & l_2 & l_3 \\ 0 & 0 & 0 \end{pmatrix} \begin{pmatrix} l_1 & l_2 & l_3 \\ m_1 & m_2 & m_3 \end{pmatrix} , \tag{4.12}$$

$$\begin{aligned} \int_0^{\pi}\int_0^{2\pi} Y_{l_1m_1}{}^{*} Y_{l_2m_2} Y_{l_3m_3} \sin\theta\, d\phi\, d\theta &= (-1)^{m_1} \int_0^{\pi}\int_0^{2\pi} Y_{l_1-m_1} Y_{l_2m_2} Y_{l_3m_3} \sin\theta\, d\phi\, d\theta \\ &= (-1)^{m_1} \sqrt{\frac{(2l_1+1)(2l_2+1)(2l_3+1)}{4\pi}} \begin{pmatrix} l_1 & l_2 & l_3 \\ 0 & 0 & 0 \end{pmatrix} \begin{pmatrix} l_1 & l_2 & l_3 \\ -m_1 & m_2 & m_3 \end{pmatrix}. \end{aligned} \tag{4.13}$$

It is also worthwhile to note that the $\begin{pmatrix} l_1 & l_2 & l_3 \\ 0 & 0 & 0 \end{pmatrix}$ 3*j*-symbol is not zero if and only if the $l_1 + l_2 + l_3$ sum is an even number. In addition, the following conditions should be also fulfilled in this case $l_1 + l_2 \geq l_3$, $l_2 + l_3 \geq l_1$, $l_1 + l_3 \geq l_3$. Equations (4.12) and (4.13) are widely used in the crystal field theory.

Let us consider here at least one example: addition of two momenta $J_1 = 2$, $J_2 = 1$. The allowed values of J will be 1, 2, 3. If we take the states with the $J = 3$, there will be seven possible states differing by the M values in the range from −3 to +3 with the step of one. There is only one possibility to obtain the value of $M = 3$ with the values of $M_1 = 2$, $M_2 = 1$. The wave function of the state with $J = 3$, $M = 3$ will be as follows (in what follows we use the $|JM\rangle$ notation for the wave functions; we also write the wave functions in the general form, using Eq. (4.7) and calculate then the numerical values of the corresponding Clebsch–Gordan coefficients):

$$|33\rangle = C^{33}_{2211}|22\rangle|11\rangle = |22\rangle|11\rangle.$$

The value of $M = 2$ can be obtained if $M_1 = 2$, $M_2 = 0$ or $M_1 = 1$, $M_2 = 1$. Therefore, the wave functions of the $|32\rangle$ state will have two terms:

$$|32\rangle = C^{32}_{2111}|21\rangle|11\rangle + C^{32}_{2210}|22\rangle|10\rangle = \frac{\sqrt{6}}{3}|21\rangle|11\rangle + \frac{\sqrt{3}}{3}|22\rangle|10\rangle,$$

The value of $M = 1$ can be obtained by three possible combinations: $M_1 = 2$, $M_2 = -1$; $M_1 = 1$, $M_2 = 0$; $M_1 = 0$, $M_2 = 1$. As a result, the wave function of the $|31\rangle$ state will look as follows:

$$\begin{aligned}|31\rangle &= C^{31}_{221-1}|22\rangle|1-1\rangle + C^{31}_{2110}|21\rangle|10\rangle + C^{31}_{2011}|20\rangle|11\rangle \\ &= \frac{\sqrt{15}}{15}|22\rangle|1-1\rangle + \frac{2\sqrt{30}}{15}|21\rangle|10\rangle + \frac{\sqrt{10}}{5}|20\rangle|11\rangle.\end{aligned}$$

Appendix I and Appendix II list the orbital parts of all wave functions of the d^2 and f^2 electron configurations, respectively.

As it should be (since the wave functions should be normalized to unity), the sum of squares of the Clebsch–Gordan coefficients for any function represented by Eq. (4.7) is equal to 1. This is a manifestation of the orthonormalization property of the Clebsch–Gordan coefficients:

$$\begin{aligned}&\sum_{m_1,m_2} C^{jm}_{j_1m_1j_2m_2}{}^{*} C^{j'm'}_{j_1m_1j_2m_2} \\ &= \sum_{m_1,m_2} \sqrt{(2j+1)(2j'+1)} \begin{pmatrix} j_1 & j_2 & j \\ m_1 & m_2 & -m \end{pmatrix} \begin{pmatrix} j_1 & j_2 & j' \\ m_1 & m_2 & -m' \end{pmatrix}. \\ &= \delta_{jj'}\delta_{mm'}\end{aligned} \quad (4.14)$$

4.7 Addition of the Spin Momenta

The total value of the spin momentum for a particular electron configuration plays a very important role in the systematic description of the multielectron states. Since the spin of one electron is 1/2 (in units of ħ), then for a system of two electrons the total spin can be either 1, if both spins are parallel, or 0, if both spins are antiparallel. For the configuration of three electrons, we may have the total values of spin as 3/2 and 1/2. These numbers can be obtained if we add the spin of an electron to the values of total spin of the preceding two-electron configuration. The value of 3/2 is obtained if all three spins are parallel, and the value of 1/2 is for the configuration when one of the spins has an opposite direction with two parallel spins. If we continue in this way—by adding the spin of one electron to the obtained values of spin for the preceding configuration, we shall get the values shown in Table 4.3. The values of the orbital momenta of these electrons are assumed to be different, to avoid possible violation of the Pauli principle for the configurations with large number of electrons.

Table 4.3 Addition of spin momenta for multielectron configurations

Number of electrons	Possible total spin values									
1		1/2								
2	0		1							
3		1/2		3/2						
4	0		1		2					
5		1/2		3/2		5/2				
6	0		1		2		3			
7		1/2		3/2		5/2		7/2		
8	0		1		2		3		4	
9		1/2		3/2		5/2		7/2		5/2
10	0		1		2		3		4		5

The states with various values of S are given specific names. For example, the states with $S = 0$ are called the spin-singlets; $S = 1/2$ – spin-doublets; $S = 1$ – spin-triplets; $S = 3/2$ – spin-quartets; $S = 2$ – spin-quintets; $S = 5/2$ – spin-sextets; $S = 3$ – spin-septets, etc. The meaning of these names becomes clear if we recall that the value $2S + 1$ is called a multiplicity, which determines a number of levels in the fine structure of a term, or a number of the projections of a spin momentum onto a given axis. So, for $S = 0$ we have the multiplicity 1 (the spin-singlet state), for $S = 1/2$ we have the multiplicity 2 (the spin-doublet state), etc.

4.8 Terms of Configurations of Inequivalent Electrons

After we have derived the rules of the angular and spin momenta addition, we are in a right position to determine the possible states or terms of electronic configurations consisting of inequivalent electrons. By those inequivalent electrons we understand the electrons with different values of either the principal quantum number n or the orbital quantum number l. In this case we do not have to take care of the Pauli exclusion principle, which forbids two electrons to have identical sets of all quantum numbers.

Determination of the allowed ^{2S+1}L terms for such configurations is very easy and straightforward: We have to add the angular momenta first, then we should add the spin momenta to find all possible multiplicity values and write down all obtained results in the ^{2S+1}L form.

Let us illustrate this by an example of the *nfn′d* configuration, one *f*-electron and one *d*-electron with different principal quantum numbers. In this case, the Pauli principle imposes no limitations on the quantum numbers of both electrons. Since we have two electrons, the total values of spin can be 0 or 1 with multiplicity $2S + 1$ equal to 1 and 3, respectively. Since we add two momenta $l_1 = 3$, $l_2 = 2$, the possible values of the total momentum can be 1, 2, 3, 4, 5, according to the general rule of the angular momenta addition. So, the possible terms will be written as ^{1}P, ^{1}D, ^{1}F, ^{1}G, ^{1}H, ^{3}P, ^{3}D, ^{3}F, ^{3}G, ^{3}H. We have 10 terms in this configuration: five spin-singlets and five spin-triplets. Each term has a degree of degeneracy equal to the $(2S + 1) \times (2L + 1)$ product. Thus, the ^{3}G term contains 27 states, the ^{3}F term has 21 states, etc. The total number of states for the *nfn′d* configuration is 140—this number can be obtained by adding degeneracies of each of the above-listed terms. If the fine structure of these terms is neglected, the number of energy levels can be equal to the number of the terms or less than it (some terms may accidentally have the same energy).

4.9 Terms of Configurations of Equivalent Electrons

Another situation is when an electron configuration is formed by the so-called equivalent electrons—those having identical principal and orbital quantum numbers. Since the principal and orbital quantum numbers for such electrons are already fixed and identical, the Pauli exclusion principle should be carefully applied to eliminate the states with identical sets of all quantum numbers.

Let us consider in all details an example of the nd^2 configuration, and both electrons have the same principal and orbital quantum numbers. Tables 4.4 and 4.5 contain all the possible values of sums of the projections of both angular momenta and spin momenta, respectively. The analysis of the obtained states should be performed with utmost care and some possibilities will be dismissed as violating the Pauli principle.

Table 4.4 Addition of the projections of two angular momenta l_1 = 2 and l_2 = 2.

M_2 \ M_1	**2**	**1**	**0**	**–1**	**–2**
2	4	3	2	1	0
1	3	~~2~~	1	0	–1
0	2	1	~~0~~	–1	–2
–1	1	0	–1	~~–2~~	–3
–2	0	–1	–2	–3	~~–4~~

Table 4.5 Addition of the projections of two spin momenta $S_1 = S_2 = 1/2$

M_{S2} \ M_{S1}	**1/2**	**–1/2**
1/2	1	0
–1/2	0	–1

If both spins are simultaneously up or down, which corresponds to the diagonal cells of Table 4.5 with the total spin ±1, we have to eliminate the states, which correspond to the diagonal cells of

Table 4.4. Otherwise, as is easy to see, the projections of the orbital and spin momenta of both electrons would be identical for both electrons, which is a clear violation of the Pauli principle.

These eliminated states are marked in Table 4.4 with a diagonal cross. Analysis of the remaining values of the projection of the total orbital momenta reveals two complete sets, corresponding to the values of L = 3 and L = 1 (these sets in Table 4.4 are shadowed), or F and P terms, respectively. The cells below the main diagonal give identical sets for the total value of L and thus are not considered. If one spin is up and the other one is down, the total spin is zero, and all values of M from Table 4.4 are allowed (since the spins are different), thus giving the values of total L as 0, 1, 2, 3, 4, which correspond to the S, P, D, F, G terms. These results are summarized in Table 4.6.

Table 4.6 Allowed states of the nd^2 configuration

	Allowed states of L				
$M_S = 1$		P		F	
$M_S = 0$	S	P	D	F	G
$M_S = -1$		P		F	

We see now, that the P and F terms are allowed for all possible values of M_S = 0, ±1. This set of the M_S values corresponds to the spin-triplet state with S = 1. So, we have then the terms ^{3}P and ^{3}F. The remaining spin states are for M_S = 0 and they form the spin-singlets ^{1}S, ^{1}D, and ^{1}G. As can be easily observed, the number of the LS terms for the configurations of the inequivalent electrons is always greater than that for the configurations consisting of the equivalent ones. The difference in the number of states is because some states are not allowed in the latter case.

The states of the nd^3 configuration can be obtained by adding one d-electron to the allowed LS terms of the nd^2 configuration. Again, the Pauli principle should be taken into account at each step. Adding in this way the electrons, one by one, to the LS terms of the preceding electron configuration, it is possible to get all allowed states for all configurations of the p, d, f-electrons (all elements of the periodic table do not have other electrons—with greater numbers of the orbital momentum l—under normal conditions in the ground and lower excited states).

It should be pointed out, however, that there is no need to consider all possible configurations. The configuration having $4l + 2 - k$ electrons with the orbital quantum number l has the same *LS* terms as the configuration with k electrons. This is because the configuration with $4l + 2 - k$ electrons can be considered as having k positive holes (or electron vacancies) in their complete electron shell. Table 4.7 shows all allowed *LS* terms for the configurations of the *s*, *p*, *d*, *f*-electrons, with the number of the *LS* terms and total number of states for each of these configurations.

Since all states with the same L and S form one term and have the same energy (the fine structure of the terms produced by the spin-orbit interaction is neglected at this point), the total number of states of one term can be found as $(2S + 1) \times (2L + 1)$. Hence, all terms except for ^1S are the highly degenerated states. The total number of states for a particular electron configuration can be found by summing up the numbers of states for individual terms:

$$N = \sum_k (2S_k + 1)(2L_k + 1), \tag{4.15}$$

where the "k" subscript runs through all terms. There is also another way of finding the number of states of a particular electron configuration. It is equal to the number of possible ways of distributing n electrons through the q allowed orbitals:

$$N = \frac{q!}{n!(q-n)!}. \tag{4.16}$$

Table 4.7 clearly shows that the number of the *LS* terms and the total number of the allowed states increase drastically when moving from the *s*- to the *f*-electron configurations.

There is an empirical rule (called Hund's rule, after a German physicist F. Hund), which allows to determine the ground term of any electron configuration. This rule is as follows: (i) the term with the lowest energy is one of those having the highest multiplicity, or the highest spin; (ii) among the terms with the highest multiplicity, the term with the greatest value of L has the lowest energy. So, the term with the lowest energy (the ground state) has the maximum multiplicity and maximum orbital momentum among all terms of a considered electron configuration. The ground terms of all configurations from Table 4.7 are listed as the last entries in the corresponding lists of terms.

Table 4.7 *LS* terms of configurations of equivalent electrons

Configurations	*LS*-terms	Number of *LS* terms	Number of states
	Configurations of *s*-electrons		
s	^{2}S	1	2
s^2	^{1}S	1	1
	Configurations of *p*-electrons		
p^1 and p^5	^{2}P	1	6
p^2 and p^4	^{1}SD, ^{3}P	3	15
p^3	^{2}PD, ^{4}S	3	20
	Configurations of *d*-electrons		
d^1 and d^9	^{2}D	1	10
d^2 and d^8	^{1}SDG, ^{3}PF	5	45
d^3 and d^7	^{2}PD$_{(2)}$FGH, ^{4}PF	8	120
d^4 and d^6	^{1}S$_{(2)}$D$_{(2)}$FG$_{(2)}$I, ^{3}P$_{(2)}$DF$_{(2)}$GH, ^{5}D	16	210
d^5	^{2}SPD$_{(3)}$F$_{(2)}$G$_{(2)}$HI, ^{4}PDFG, ^{6}S	16	252
	Configurations of *f*-electrons		
f^1 and f^{13}	^{2}F	1	14
f^2 and f^{12}	^{1}SDGI, ^{3}PFH	7	91
f^3 and f^{11}	^{2}PD$_{(2)}$F$_{(2)}$G$_{(2)}$H$_{(2)}$IKL, ^{4}SDFGI	17	364
f^4 and f^{10}	^{1}S$_{(2)}$D$_{(4)}$FG$_{(4)}$H$_{(2)}$I$_{(3)}$KL$_{(2)}$N, ^{3}P$_{(3)}$ D$_{(2)}$F$_{(4)}$G$_{(3)}$H$_{(4)}$I$_{(2)}$K$_{(2)}$LM, ^{5}SDFGI	47	1001
f^5 and f^9	^{2}P$_{(4)}$D$_{(5)}$F$_{(7)}$G$_{(6)}$H$_{(7)}$I$_{(5)}$K$_{(5)}$L$_{(3)}$M$_{(2)}$ NO, ^{4}SP$_{(2)}$D$_{(3)}$F$_{(4)}$G$_{(4)}$H$_{(3)}$I$_{(3)}$K$_{(2)}$LM, ^{6}PFH	73	2002
f^6 and f^8	^{1}S$_{(4)}$PD$_{(6)}$F$_{(4)}$G$_{(8)}$H$_{(4)}$I$_{(7)}$K$_{(3)}$L$_{(4)}$M$_{(2)}$ N$_{(2)}$Q, ^{3}P$_{(6)}$D$_{(5)}$F$_{(9)}$G$_{(7)}$H$_{(9)}$I$_{(6)}$K$_{(6)}$ L$_{(3)}$M$_{(3)}$NO, ^{5}SPD$_{(3)}$F$_{(2)}$G$_{(3)}$H$_{(2)}$I$_{(2)}$ KL, ^{7}F	119	3003
f^7	^{2}S$_{(2)}$P$_{(5)}$D$_{(7)}$F$_{(10)}$G$_{(10)}$H$_{(9)}$I$_{(9)}$K$_{(7)}$L$_{(5)}$ M$_{(4)}$N$_{(2)}$OQ, ^{4}S$_{(2)}$P$_{(2)}$D$_{(6)}$F$_{(5)}$G$_{(7)}$H$_{(5)}$ I$_{(5)}$K$_{(3)}$L$_{(3)}$MN, ^{6}PDFGHI, ^{8}S	119	3432

Note: The subscript means the number of different terms with the same values of L and S in a particular configuration. For the sake of brevity, the terms multiplicity is given only once for the term listed in the first position of the considered group, i.g. ^{1}SD denotes two terms ^{1}S, ^{1}D, etc.

If the transitions between the terms with different multiplicity are considered, it becomes clear that they should be strictly forbidden, since the spin parts of their wave functions are different (and orthogonal!). Therefore, the matrix element of any operator with the Cartesian coordinates between such states would be identically zero. This rule is known as the spin-selection rule. However, the so-called spin-forbidden transitions can be detected in the experimental spectra, although they have a low intensity. This is because various interactions (SO interaction, first of all) can mix up the states with different multiplicity but with the same total momenta J. For example, the wave function of the spin-singlet ^{1}D from the above-considered example of the d^2 configuration may contain a minor contribution from the ^{3}F ground term. Because of such admixture the transitions become "slightly allowed," such mechanism is sometimes referred to as the "intensity stealing" or "intensity borrowing" from spin-allowed transitions.

The wave functions of the configurations of equivalent electrons can be obtained by using Eq. (4.7) for the configurations of two electrons as a first step. Then one has to add one l-electron to a particular L-term and again write down an analogue of the Eq. (4.7) keeping in mind that one multiplier in the product of the wave functions is the one-electron wave function of the added l-electron, and another one is in fact a two-electron wave function obtained at the previous step. This procedure is not a trivial one and is time-consuming, especially because of the large number of terms with the same L and S quantum numbers. In this situation, it is very important to track from which term of a preceding configuration a given LS term arises. It requires one additional quantum number, the so-called seniority number, along with the coefficients of the fractional parentage (showing the contributions of the wave functions of the l^{n-1} configuration to the functions of the l^n configuration) to be introduced. So, the wave functions of the two-electron configurations listed in Appendix I and Appendix II should be used to get the wave functions of the three-electron configuration, etc. A more detailed description of this procedure can be found in the books recommended at the end of this chapter.

4.10 Energies of the LS Terms of Configurations of Equivalent Electrons, Slater Integrals and Racah Parameters

Once the wave functions of the *LS* terms of a particular electron configuration are known, it becomes possible to calculate the energies of all terms. This is a necessary step to get a complete picture of the energy levels location and relate the calculated energy levels to the experimental spectra. The main interaction, which produces the difference in the *LS*-states energies, is the Coulomb interaction between electrons. From the quantum mechanical point of view, calculating the energy of a particular *LS* terms would mean finding a diagonal value of the $\frac{e^2}{r_{12}} = \frac{e^2}{\left|\vec{\mathbf{r}}_1 - \vec{\mathbf{r}}_2\right|}$ operator ($\vec{\mathbf{r}}_1$ and $\vec{\mathbf{r}}_2$ are the position vectors of two electrons) on the wave functions of the corresponding term:

$$E = \int \Psi^* \frac{e^2}{r_{12}} \Psi dV_1 dV_2 . \tag{4.17}$$

The integral is taken over all coordinates of both electrons and is, in fact, a six-dimensional integral. We already know that the wave function Ψ finally is expressed as a linear combination of the products of one-electron wave functions, each of which can be separated into the radial and angular parts. Consequently, the integral (4.17) can be represented as a combination of the following integrals:

$$E = \int\!\!\int\!\!\int\!\!\int\!\!\int\!\!\int Y_{lm}^*(\theta_2,\phi_2) Y_{lm}^*(\theta_1,\phi_1) R_{nl}(r_2) R_{nl}(r_1) \frac{e^2}{r_{12}} Y_{lm}(\theta_2,\phi_2) Y_{lm}(\theta_1,\phi_1) R_{nl}(r_2) R_{nl}(r_1) dV_1 dV_2 , \tag{4.18}$$

with $dV_1 = r_1^2 \sin\theta_1 dr_1 d\theta_1 d\phi_1$ and $dV_1 = r_2^2 \sin\theta_2 dr_2 d\theta_2 d\phi_2$ being the volume elements in the spherical coordinates. The value of this integral does not depend on the magnetic quantum number *m*. The integral (4.18) can be simplified considerably by using the following expansion:

$$\frac{1}{r_{12}} = \sum_{k=0}^{\infty} \frac{4\pi}{2k+1} \frac{r_<^k}{r_>^{k+1}} \sum_{t=-k}^{k} Y_{kt}(\theta_1,\phi_1) Y_{kt}^*(\theta_2,\phi_2), \tag{4.19}$$

where $r_<$ and $r_>$ are the smaller and the greater values of the r_1 and r_2 distances. Such a notation will become clear if we consider a system of coordinates (r_1, r_2) shown in Fig. 4.2. The straight line $r_1 = r_2$ divides the area of integration into two parts, and in each of these parts determination of the smallest and greatest values of r is straightforward. Then the integration over the radial coordinates in Eq. (4.18) will be well determined:

$$\iint \frac{r_<^k}{r_>^{k+1}} r_1^2 r_2^2 dr_1 dr_2 = \int_0^\infty r_1^2 dr_1 \left(\int_0^{r_1} \frac{r_2^k}{r_1^{k+1}} r_2^2 dr_2 + \int_{r_1}^\infty \frac{r_1^k}{r_2^{k+1}} r_2^2 dr_2 \right). \quad (4.20)$$

After substituting Eq. (4.19) into Eq. (4.18), we arrive at the following long expression:

$$E = \sum_{k=0}^{\infty} \frac{4\pi}{2k+1} \iint R_{nl}^2(r_2) \frac{r_<^k}{r_>^{k+1}} R_{nl}^2(r_1) r_1^2 r_2^2 dr_1 dr_2 \times$$
$$\times \sum_{t=-k}^{k} \iint Y_{lm}^*(\theta_1,\phi_1) Y_{kt}(\theta_1,\phi_1) Y_{lm}(\theta_1,\phi_1) \sin\theta_1 d\theta_1 d\phi_1 \,. \quad (4.21)$$
$$\times \iint Y_{lm}(\theta_2,\phi_2) Y_{lm}^*(\theta_2,\phi_2) Y_{kt}^*(\theta_2,\phi_2) \sin\theta_2 d\theta_2 d\phi_2$$

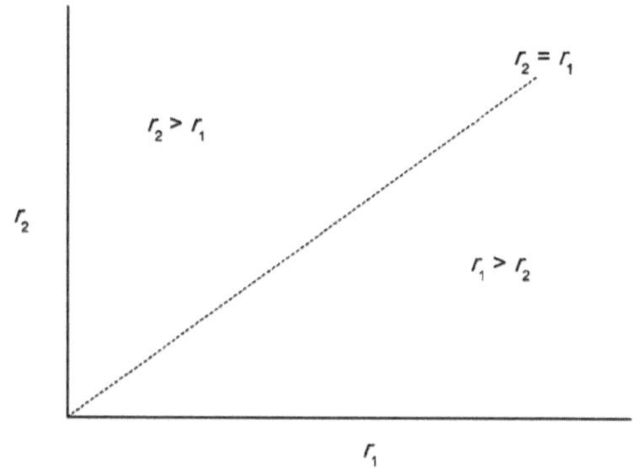

Figure 4.2 To the definition of the smaller and the greater values of the r_1 and r_2 distances in Eq. (4.19).

A further significant simplification is possible by applying the following identities for the integrals containing the product of three spherical functions:

$$\int_0^{\pi}\int_0^{2\pi} Y_{l_1m_1}Y_{l_2m_2}Y_{l_3m_3}\sin\theta\, d\phi\, d\theta = \sqrt{\frac{(2l_1+1)(2l_2+1)(2l_3+1)}{4\pi}}$$

$$\begin{pmatrix} l_1 & l_2 & l_3 \\ 0 & 0 & 0 \end{pmatrix}\begin{pmatrix} l_1 & l_2 & l_3 \\ m_1 & m_2 & m_3 \end{pmatrix},$$

$$\int_0^{\pi}\int_0^{2\pi} Y_{l_1m_1}{}^{*}Y_{l_2m_2}Y_{l_3m_3}\sin\theta\, d\phi\, d\theta = (-1)^{m_1}\int_0^{\pi}\int_0^{2\pi} Y_{l_1-m_1}Y_{l_2m_2}Y_{l_3m_3}\sin\theta\, d\phi\, d\theta$$

$$= (-1)^{m_1}\sqrt{\frac{(2l_1+1)(2l_2+1)(2l_3+1)}{4\pi}}$$

$$\begin{pmatrix} l_1 & l_2 & l_3 \\ 0 & 0 & 0 \end{pmatrix}\begin{pmatrix} l_1 & l_2 & l_3 \\ -m_1 & m_2 & m_3 \end{pmatrix}. \tag{4.22}$$

Since the $\begin{pmatrix} l_1 & l_2 & l_3 \\ 0 & 0 & 0 \end{pmatrix}$ 3*j*-symbol is not zero if and only if $l_1 + l_2 + l_3$ is an even number, the number of terms in the sum over k becomes, in fact, quite small. The index k can take only the values 0, 2, and 4 in the case of the *d*-electrons and 0, 2, 4, and 6 in the case of the *f*-electrons. So, finally the energy of any *LS* term will be expressed as a linear combination of the following so-called Slater integrals F^k having dimension of energy:

$$F^k = e^2 \int\int \frac{r_<^{\,k}}{r_>^{\,k+1}} R_1(r_1)^2 R_2(r_2)^2 r_1^{\,2} r_2^{\,2} dr_1 dr_2, \tag{4.23}$$

which can be calculated numerically, since the radial parts of the corresponding wave functions can be easily found in the literature.

In the case of the 3*d*-ions it is a common practice to use the so-called Racah parameters *A*, *B*, and *C* instead of the Slater integrals. The relationship between them is as follows:

$$A = F^0 - \frac{1}{9}F^4; B = \frac{1}{49}F^2 - \frac{5}{441}F^4; C = \frac{5}{63}F^4. \tag{4.24}$$

The matrix of the electrostatic interaction is diagonal in the basis spanned by all functions of the LS terms of the given electronic configurations (unless there are the terms with the same *L* and *S* quantum numbers; then the non-diagonal matrix elements will appear). Table 4.8 below gives the energies of the d^n configurations. These energies are widely cited in the literature and can be obtained by using the corresponding matrix elements from the below-mentioned book by C. W. Nielsen and G. F. Koster listed at the end of this chapter.

Table 4.8 Energies of the *LS* terms of the d^n configurations

	d^2, d^8		d^5
^{3}F	0	^{6}S	0
^{3}P	$15B$	^{4}P	$7B + 7C$
^{1}D	$5B + 2C$	^{4}D	$17B + 5C$
^{1}G	$12B + 2C$	^{4}F	$22B + 7C$
^{1}S	$22B + 7C$	^{4}G	$10B + 5C$
	d^3, d^7	^{2}S	$32B + 8C$
^{4}F	0	^{2}P	$55B + 10C$
^{4}P	$15B$	^{2}D$_{(1)}$	$31B + 10C$
^{2}P	$9B + 3C$	^{2}D$_{(2)}$, ^{2}D$_{(3)}$	$32B+11C \pm 3\sqrt{57B^2+2BC+C^2}$
^{2}D	$20B+5C \pm\sqrt{193B^2+8BC+4C^2}$	^{2}F	$26B + 8C$
^{2}F	$24B + 3C$	^{2}F'	$10B + 10C$
^{2}G	$4B + 3C$	^{2}G	$22B + 8C$
^{2}H	$9B + 3C$	^{2}G'	$38B + 10C$
	d^4, d^6	^{2}H	$13B + 10C$
^{5}D	0	^{2}I	$11B + 8C$
^{3}P$_{(1),(2)}$	$16B+\frac{11}{2}C \pm\frac{1}{2}\sqrt{912B^2-24BC+9C^2}$		
^{3}D	$16B + 4C$		
^{3}F	$16B+\frac{11}{2}C \pm\frac{3}{2}\sqrt{68B^2+4BC+C^2}$		
^{3}G	$9B + 4C$		
^{3}H	$4B + 4C$		

	d^4, d^6 (continued)
$^1S_{(1),(2)}$	$31B+10C \pm 2\sqrt{193B^2+8BC+4C^2}$
$^1D_{(1),(2)}$	$30B+\frac{15}{2}C \pm \frac{3}{2}\sqrt{144B^2+8BC+C^2}$
^1F	$27B$
$^1G_{(1),(2)}$	$16B+\frac{15}{2}C \pm \frac{1}{2}\sqrt{708B^2-12BC+9C^2}$
^1I	$6B + 6C$

Note: In all cases, the energy of the ground term is taken as zero.

It should be pointed out here that the energies of most terms of the f^n configurations (due to many repeating terms) cannot be expressed in such a simple way. The reader is advised to use the electrostatic matrices for the corresponding configurations from the above-mentioned book.

Table 4.9 contains typical values of the Racah parameters *B* and *C* for some 3*d*-ions. These parameters were estimated for free ions, and it should be kept in mind that the values of the Racah parameters for the ions embedded into a crystal and/or a glass are considerably reduced with respect to those for free ions. This is a manifestation of the so-called nephelauxetic effect, it is due to the formation of the covalent bonds between impurity ions and crystal lattice ions and will be addressed later.

It is instructive to consider some trends in the values of the Racah parameters. Table 4.9 shows that practically always increasing the electrical charge of the same ion is accompanied by the increase of both parameters *B* and *C*. If the ions with the same electronic configurations or the so-called isoelectronic ions are considered, then again increase of the electrical charge is followed by increase of the Racah parameters. Figure 4.3 shows the energy levels of four ions (V^{2+}, Cr^{3+}, Mn^{4+}, Fe^{5+}), all having the $3d^3$ configuration, calculated with the values given in Table 4.9. There is a clear trend—the shift of all terms to the higher energies when going from V^{2+} to Fe^{5+}.

Table 4.9 Typical values (in cm^{-1}) of the Racah parameters B and C for a number of the 3d ions

Ion	B	C
Ti^{2+}	718	2629
V^{2+}	766	2855
V^{3+}	861	4165
Cr^{3+}	918	3850
Cr^{4+}	1039	4238
Mn^{2+}	960	3325
Mn^{3+}	965	3675
Mn^{4+}	1160	4303
Fe^{3+}	1015	4800
Fe^{4+}	1144	4459
Fe^{5+}	1210	5066
Co^{2+}	971	4366
Ni^{2+}	1041	4831
Cu^{2+}	1238	4659

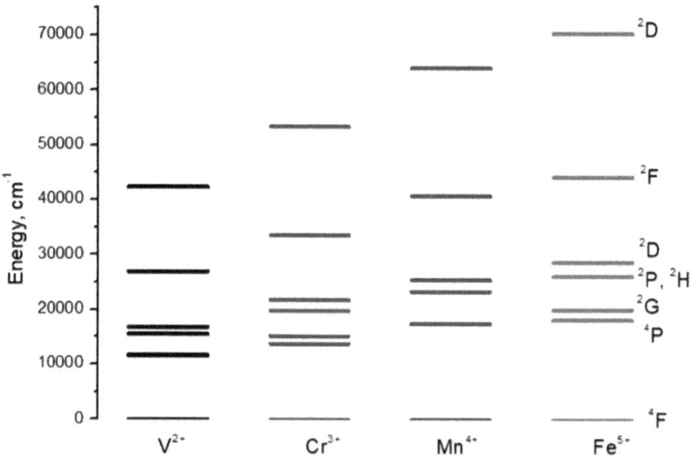

Figure 4.3 Energy levels of free V^{2+}, Cr^{3+}, Mn^{4+}, Fe^{5+} ions. Energy of the ground ^4F term is taken as zero. For Mn^{4+} and V^{2+} the energies of the ^4P and ^2G terms are very close and seem to be almost indistinguishable in the figure. The energies of the ^2P and ^2H terms coincide in the 3d^3-configuration (see Table 4.8).

4.11 Fine-Structure of the LS Terms of Multielectron Configurations

As in the case of the hydrogen atom, the relativistic effects (primarily, the SO interaction) produce the fine structure of the *LS* terms of the multielectron atoms as well. Interaction between the *S* and *L* momenta gives rise to the states with the total angular momentum *J*, $|L-S| \leq J \leq L + S$. If $L \geq S$, then $2S + 1$ values of *J* are possible (the term splits into $2S + 1$ levels). If $L < S$, then $2L + 1$ levels for *J* are allowed. The group of the *J* states arising from a given *LS* term is called a multiplet, or a *J*-manifold. The states with different *J* have different energies. The energy intervals between the states with different *J* follow the empirical Lande's interval rule: The energy separation between the states *J* and $J - 1$ is proportional to *J*:

$$E_J - E_{J-1} = A(LS)J, \tag{4.25}$$

where *A* is the multiplet splitting constant. It depends on the values of *L* and *S* and can have different signs. When *A* is positive, the multiplet with the smallest possible value of *J* for a given *LS* term has the lowest energy, and the multiplet with the greatest possible value of *J* has the highest energy. These multiplets are called the normal multiplets. Schematically formation of the normal manifold from the *LS* term with the $J = 0, 1, 2, 3$ is shown in Fig. 4.4a.

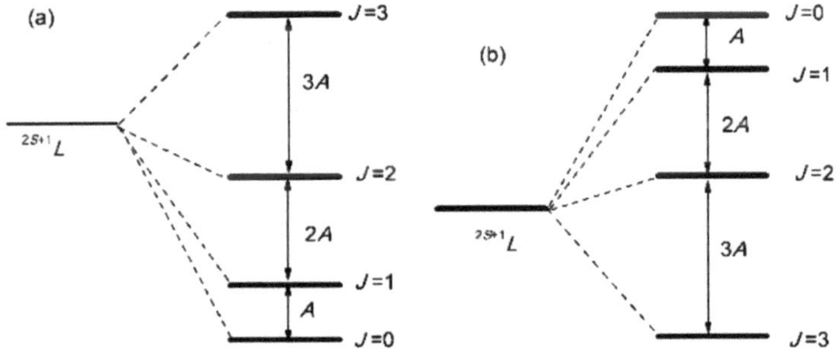

Figure 4.4 Formation of the normal (a) and inverted (b) manifolds from an *LS* term in the case with $J = 0, 1, 2, 3$.

It should be noticed that the energy levels intervals between the adjacent levels increase with increasing the value of *J*.

When *A* is negative, the multiplet with the smallest possible value of *J* for a given *LS* term has the highest energy, and the multiplet with the greatest possible value of J has the lowest energy. These multiplets are called the inverted manifolds (Fig. 4.4b).

Empirically it has been established that the normal multiplets correspond to the configurations of the equivalent electrons having less than half of the maximum number of electrons for a particular shell (1, 2 for the p-electrons, 1, 2, 3, 4 for the d-electrons, 1, 2, 3, 4, 5, 6 for the f-electrons; all these configurations are referred to as the less-than-half-filled electron shells). The configurations with more than half from the possible number of electrons (4, 5 for the p-electrons, 6, 7, 8, 9 for the d-electrons, 8, 9, 10, 11, 12, 13 for the f-electrons; they are called the more-than-half-filled electron shells) have the inverted multiplets. Finally, when the electron shell is exactly half-filled (3 p-electrons, 5 d-electrons, 7 f-electrons), the A constant is zero, and there is no fine structure in the first order of the perturbation theory.

The value of A also can be calculated numerically using the following expression:

$$A = \xi_{nl} \sqrt{\frac{l(l+1)(2l+1)}{S(S+1)(2S+1)L(L+1)(2L+1)}} \left(l^n \gamma SL \middle\| V^{11} \middle\| l^n \gamma SL \right), \tag{4.26}$$

where ξ_{nl} is the spin-orbit constant, and the matrix elements $\left(l^n \gamma SL \middle\| V^{11} \middle\| l^n \gamma SL \right)$ can be found in the book by Nielsen and Koster listed at the end of this chapter.

Finally, we add a couple of observations regarding the properties of the half-filled and completely filled electron shells. For the ground term of a half-filled shell the orbital momentum is zero, and the spin momentum is $N/4$ (where N is the total number of electrons, which a given electron shell may have). For a completely filled shell, both spin and orbital momenta are zero, and there is only one allowed non-degenerate term ^1S, which cannot be split in any external field.

4.12 Periodic System of Elements Explained by the Combinations of Quantum Numbers

In this paragraph we shall consider the order of the chemical elements in the Periodic Table and associated with it sequence of the electron shells filling.

Electrons in the atoms occupy the states with the lowest energy, on one hand, and allowed by the Pauli exclusion principle, on the other hand. The simplest atom—atom of hydrogen—has only one electron in the $1s$ state. When moving from one atom to another in the Periodic table, both the nucleus charge and the number of electrons in the atom increase by one. When considering the chemical elements in the Periodic table (Table 4.10), one should keep in mind that for each value of n we may have n values of l in the range from 0 to $n - 1$. Then it is clear that after hydrogen H comes helium He with its $1s^2$ configuration, and both electrons in the $1s$ shell have opposite spins. Then two metals Li and Be follow with their $1s^2 2s$ and $1s^2 2s^2$ configurations, respectively. The $2p$ shell is filling from B to Ne, where we end up with the completely filled $2p$ shell with 6 electrons (the inert or noble gas Ne, $1s^2 2s^2 2p^6$ configuration).

In the same order, we have the filling of the electrons in the $3s$ and $3p$ shells, from Na to Kr. Then, instead of the allowed $3d$ shell, the $4s$ shell is filling for K and Ca, and only then, for the elements from Sc to Zn, the $3d$ shell is gradually filled with the electrons. Such a seeming irregularity occurs since for the principal quantum number $n \geq 3$ the states with the lowest energies are those for which the sum $(n + l)$ is smaller. The value $(n + l)$ is 5 for the $3d$ shell, and only 4 for the $4s$ shell, which explains this first seeming irregularity of the periodic table.

Then—after the $3d$ shell is finished (there are a few more irregularities within this $3d$ group, since there is an energetical competition between the $3d$ and $4s$ states, as seen from Table 4.10)—the $4p$ shell starts to be filled from Ga to Kr. It should be pointed out that for both $3d$ and $4p$ shells the $(n + l)$ sum is 5, but the energy of the states with a smaller n quantum number is lower, if the $(n + l)$ sum is the same for several shells. Rb and Sr come with their $5s$ and $5s^2$ outer configurations, respectively. Then we have the second row of the d-elements with the $4d$ shell being gradually filled from Y to Cd. After that, from In to Xe the $5p$ shell is completed. It should be noted here, that the $4f$ shell is not started yet to fill. Only after Cs and Ba with their $6s$ and $6s^2$ outer configurations the $4f$ shell is completed from La to Yb (the so-called lanthanide elements). Only after the $4f$ shell, the $5d$ states start to be occupied from Lu to Hg, and then $6p$ states are completed from Tl to Rn. The $7s$ shell is filled in Fr and Ra, and only then the $5f$ shell is completed in the elements from Ac to Md (the so-called actinide series). The remaining elements follow the $n + l$ rule as well, with the $6d$ and $7p$ shells filled.

Table 4.10 Periodic system of elements and electron shells. P and S stand for the "period" and "shell," respectively.

P	S																							
		1	2																					
1	1*s*	H	He																					
		s	s^2																					
		3	4																					
	2*s*	Li	Be																					
2		*s*	s^2	5	6	7	8	9	10															
	2*p*			B	C	N	O	P	Ne															
				p	p^2	p^3	p^4	p^5	p^6															
		11	12																					
	3*s*	Na	Mg																					
3		*s*	s^2	13	14	15	16	17	18															
	3*p*			Al	Si	P	S	Cl	Ar															
				p	p^2	p^3	p^4	p^5	p^6															
		19	20																					
	4*s*	K	Ca																					
		s	s^2	21						22	23	24	25	26	27	28	29	30						
4	3*d*			Sc						Ti	V	Cr	Mn	Fe	Co	Ni	Cu	Zn						
				ds^2						d^2s^2	d^3s^2	d^5s	d^5s^2	d^6s^2	d^7s^2	d^8s^2	$d^{10}s$	$d^{10}s^2$	31	32	33	34	35	36
	4*p*																		Ga	Ge	As	Se	Br	Kr
																			p	p^2	p^3	p^4	p^5	p^6

		37	38																														
	5*s*	Rb	Sr																														
		s	s^2	39															40	41	42	43	44	45	46	47	48						
5	4*d*			Y															Zr	Nb	Mo	Tc	Ru	Rh	Pd	Ag	Cd						
				ds^2															d^2s^2	d^4s	d^5s	d^5s^2	d^7s	d^8s	d^{10}	$d^{10}s$	$d^{10}s^2$	49	50	51	52	53	54
	5*p*																											In	Sn	Sb	Te	I	Xe
																												p	p^2	p^3	p^4	p^5	p^6
		55	56																														
	6*s*	Cs	Ba																														
		s	s^2	57	58	59	60	61	62	63	64	65	66	67	68	69	70	71															
	4*f*			La	Ce	Pr	Nd	Pm	Sm	Eu	Gd	Tb	Dy	Ho	Er	Tm	Yb	Lu															
6				ds^2	f^2s^2	f^3s^2	f^4s^2	f^5s^2	f^6s^2	f^7s^2	f^7ds^2	f^8ds^2	$f^{10}s^2$	$f^{11}s^2$	$f^{12}s^2$	$f^{13}s^2$	$f^{14}s^2$	ds^2	72	73	74	75	76	77	78	79	80						
	5*d*																		Hf	Ta	W	Re	Os	Ir	Pt	Au	Hg						
																			d^2s^2	d^3s^2	d^4s^2	d^5s^2	d^6s^2	d^7s^2	d^9s	$d^{10}s$	$d^{10}s^2$	81	82	83	84	85	86
	6*p*																											Tl	Pb	Bi	Po	At	Rn
																												p	p^2	p^3	p^4	p^5	p^6
		87	88																														
	7*s*	Fr	Ra																														
		s	s^2	89	90	91	92	93	94	95	96	97	98	99	100	101	102																
	5*f*			Ac	Th	Pa	U	Np	Pu	Am	Cm	Bk	Cf	Es	Fm	Md	No																
7				ds^2	d^2s^2	f^2ds^2	f^3ds^2	f^4ds^2	f^6s^2	f^7s^2	f^7ds^2	f^8ds^2	$f^{10}s^2$	$f^{11}s^2$	$f^{12}s^2$	$f^{13}s^2$	$f^{14}s^2$	103	104	105	106	107	108	109	110	111	112						
	6*d*																	Lr	Rf	Db	Sg	Bh	Hs	Mt	Ds	Rg	Cn						
																		p	d^2	d^3	d^4	d^5	d^6	d^7	d^8	d^9	d^{10}	113	114	115	116	117	118
	7*p*																											Nh	Fl	Mc	Lv	Ts	Og
																												p	p^2	p^3	p^4	p^5	p^6

So, the sequence of the elements in the Periodic table can be explained from the quantum mechanical point of view by gradual filling of the electron shells with electrons, occupying the states of the lowest energy, allowed by the Pauli exclusion principle. The shells are filled not in the increasing order of the principal quantum number n, but rather in the increasing order of the sum of the principal quantum number n and the orbital quantum number l. This sum, obviously, can have the same values for various combinations of n and l. Then the state with smaller n is more energetically favorable. With taking this rule into account, the seeming irregularities of the periodic table can be readily explained and understood.

The elements with unfilled d- and f-shells are most interesting from the spectroscopic point of view, since they have large numbers of the allowed states (see Table 4.7), which makes their spectra to be attractive for various applications, some of which will be considered in the next chapters.

Bibliography

1. Cohen-Tannoudji C, Diu B, Laloe F (1977). *Quantum Mechanics.* John-Wiley, Paris.
2. Cowan RD (1981). *The Theory of Atomic Structure and Spectra.* University of California Press, Berkeley.
3. Edmonds AR (1957). *Angular Momentum in Quantum Mechanics.* Princeton University Press.
4. Griffiths DJ (1995). *Introduction to Quantum Mechanics.* Prentice-Hall, New Jersey.
5. Henderson B, Bartram RH (2000). *Crystal-Field Engineering of Solid-State Laser Materials.* Cambridge University Press.
6. Landau LD, Lifshitz EM (1977). *Quantum Mechanics.* Pergamon Press, Oxford.
7. Nielsen CW, Koster GF (1963). *Spectroscopic Coefficients for the p^n, d^n, and f^n Configurations.* MIT Press, Cambridge, Massachusetts.
8. Powell RC (1998). *Physics of Solid-State Laser Materials.* Springer.
9. Schiff LI (1968). *Quantum Mechanics.* McGraw-Hill.
10. Sobelman II (1992). *Atomic Spectra and Radiative Transitions.* Springer.

Chapter 5

Basic Spectroscopic Properties of the Ions with Unfilled *d* Electron Shell*

5.1 Introduction

In this chapter, we shall see how the results obtained in the previous chapters can be applied to get a consistent systematic description of the spectroscopic properties and free ion Hamiltonian parameters of the ions with unfilled $3d$, $4d$, $5d$ electronic shells. We shall see how these properties vary with increasing atomic number. Several groups of isovalent ions, such as having the electrical charges +2, +3, +4, +5, +6, +7 (which cover all important practical cases) will be considered in detail. The chapter is in a certain sense a bridge from the basic theoretical spectroscopy of free atoms to a real practical research.

*Some parts of this chapter are reproduced with permission from: Brik MG and Srivastava AM (2013). Systematic analysis of the spectroscopic characteristics of $3d$ ions in a free state and some cubic crystals, *Opt Mater* **35**(10), (2013) 1776–1782 and Ma CG and Brik MG (2014). Systematic analysis of spectroscopic characteristics of heavy transition metal ions with $4d^N$ and $5d^N$ (N = 1... 10) electronic configurations in a free state, *J Lumin* **145** (2014) 402–409.

Theoretical Spectroscopy of Transition Metal and Rare Earth Ions: From Free State to Crystal Field
Mikhail G. Brik and Chong-Geng Ma

ISBN 978-981-4800-56-3 (Hardcover), 978-0-429-27875-4 (eBook)
www.jennystanford.com

5.2 Possible Electron Configurations of the 3*d*, 4*d*, 5*d* Ions

We shall focus on several groups of chemical elements, all of which have one common feature: one of their *d* electron shells is unfilled. The number of *d* electrons in such a shell is increasing by one when moving across the periodic table rows from left to right until the *d* shell becomes completed with 10 electrons. These elements are called the "transition metals." The first such group with unfilled 3*d* shell is located in the fourth period and includes the chemical elements from Sc (atomic number $Z = 21$) to Zn ($Z = 30$); the second group with unfilled 4*d* shell includes the elements from Y ($Z = 39$) to Cd ($Z = 48$) in the fifth period, and the third group of transition metals is that one from the sixth period with the unfilled 5*d* shell, from Lu ($Z = 71$) to Hg ($Z = 80$). The groups of chemical elements with the unfilled 4*f* and 5*f* shells will be considered in the next chapter.

These elements can easily form positively charged ions with several electrons in their unfilled open *d*-electron shells and varying oxidation states. Being placed into suitable crystals and/or glasses, these ions can be a reason for noticeable changes of the optical properties of these host materials. As we have seen in a previous chapter, the unfilled electron configurations have many energetic terms, arising from the Coulomb interaction between the electrons of those unfilled shells. These terms are highly degenerated, and these levels can be split further by a crystal field, to form rich energy level schemes with numerous absorption/emission bands in a wide spectral region, from infrared to ultraviolet, depending on the considered ion. Electronic transitions between those split states are the reason for different colors of many crystals and water solutions of the transition metal ions salts. Numerous applications of the transition metal ions in different fields (e.g., laser materials, phosphors, infrared to visible up-conversion materials, solar energy conversion, new photovoltaic devices, phosphorescent sensors, electroluminescent devices, etc.[1-6] and references therein) are well-known. Although the $3d^N$ ions are the main research objects in all these applications, it is worth mentioning that recently co-doping of the $4d^N$ ions as the sensitizers together with lanthanide ions into crystalline hosts has become a new method for designing the phosphor materials for the tunable white luminescence.[7] In addition, the research interests on the electron paramagnetic resonance spectra have also moved from the light $3d^N$ ions to the heavy $4d^N$ and $5d^N$ ions because the latter are more sensitive to the local environment due to much stronger couplings of the $4d/5d$ electrons with the host lattice[8,9], which allows their use as reliable probes of the local structure.

In this chapter, we consider systematically these ions with the unfilled $3d$, $4d$ and $5d$ electron shells in several oxidation states.[10,11] High oxidation state, such as +5, +6, +7, which are not uncommon for many transition metal ions, were also included into the consideration. Although for some ions such high oxidation states are hard to be realized, we keep all the obtained data for the sake of completeness of our systematic study to follow full variation of the spectroscopic parameters from the d^1 electron configuration to d^{10}.

We start from the analysis of the parameters of electrostatic interaction (Racah parameters A, B and C), SO interaction constant ζ and relations between all those parameters and atomic number Z. The averaged values of the second and fourth power of the d electrons radial coordinate $<r^2>$ and $<r^4>$ were also calculated, since these quantities are necessary to calculate the crystal field parameters. The third, fourth and fifth ionization energies for all considered ions were calculated and compared with the available experimental data. Only for some of those ions the corresponding experimental data were found; agreement between those experimental and our calculated values is very good.

Table 5.1 lists all considered ions and their corresponding electron configurations of the d-shells. The energy levels of free ions with the electronic configurations from Table 5.1 can be expressed as the eigenvalues of the following free-ion Hamiltonian H_{FI}.[12]

$$H_{\mathrm{FI}} = \sum_{k=0,2,4} F^{(k)} h_k + \zeta \sum_{i=1}^{N} \mathbf{l}_i \cdot \mathbf{s}_i + \alpha L(L+1), \tag{5.1}$$

where $F^{(k)}(k = 0, 2, 4)$ are the Slater parameters describing the Coulomb interaction between electrons of the unfilled electron shell (for obvious reasons, these parameters can be defined only for the ions having more than one electron or one positively charged hole in the electron configuration. As "it takes two to tango," it also takes two charges for the Coulomb interaction), ζ is the SO interaction constant,

Table 5.1 Atomic numbers and electron configurations for the considered polyvalent 3*d*, 4*d* and 5*d* ions

Atomic number Z	Symbol	Electron configuration $3d^N$					
		Divalent	Trivalent	Tetravalent	Pentavalent	Hexavalent	Heptavalent
21	Sc	$3d^1$	$3d^0$	—	—	—	—
22	Ti	$3d^2$	$3d^1$	$3d^0$	—	—	—
23	V	$3d^3$	$3d^2$	$3d^1$	$3d^0$	—	—
24	Cr	$3d^4$	$3d^3$	$3d^2$	$3d^1$	$3d^0$	—
25	Mn	$3d^5$	$3d^4$	$3d^3$	$3d^2$	$3d^1$	$3d^0$
26	Fe	$3d^6$	$3d^5$	$3d^4$	$3d^3$	$3d^2$	$3d^1$
27	Co	$3d^7$	$3d^6$	$3d^5$	$3d^4$	$3d^3$	$3d^2$
28	Ni	$3d^8$	$3d^7$	$3d^6$	$3d^5$	$3d^4$	$3d^3$
29	Cu	$3d^9$	$3d^8$	$3d^7$	$3d^6$	$3d^5$	$3d^4$
30	Zn	$3d^{10}$	$3d^9$	$3d^8$	$3d^7$	$3d^6$	$3d^5$
31	Ga	—	$3d^{10}$	$3d^9$	$3d^8$	$3d^7$	$3d^6$
32	Ge	—	—	$3d^{10}$	$3d^9$	$3d^8$	$3d^7$
33	As	—	—	—	$3d^{10}$	$3d^9$	$3d^8$
34	Se	—	—	—	—	$3d^{10}$	$3d^9$
35	Br	—	—	—	—	—	$3d^{10}$

		Electron configuration $4d^N$					
		Divalent	Trivalent	Tetravalent	Pentavalent	Hexavalent	Heptavalent
39	Y	$4d^1$	$4d^0$	—	—	—	—
40	Zr	$4d^2$	$4d^1$	$4d^0$	—	—	—
41	Nb	$4d^3$	$4d^2$	$4d^1$	$4d^0$	—	—
42	Mo	$4d^4$	$4d^3$	$4d^2$	$4d^1$	$4d^0$	—
43	Tc	$4d^5$	$4d^4$	$4d^3$	$4d^2$	$4d^1$	$4d^0$
44	Ru	$4d^6$	$4d^5$	$4d^4$	$4d^3$	$4d^2$	$4d^1$
45	Rh	$4d^7$	$4d^6$	$4d^5$	$4d^4$	$4d^3$	$4d^2$
46	Pd	$4d^8$	$4d^7$	$4d^6$	$4d^5$	$4d^4$	$4d^3$
47	Ag	$4d^9$	$4d^8$	$4d^7$	$4d^6$	$4d^5$	$4d^4$
48	Cd	$4d^{10}$	$4d^9$	$4d^8$	$4d^7$	$4d^6$	$4d^5$
49	In	—	$4d^{10}$	$4d^9$	$4d^8$	$4d^7$	$4d^6$
50	Sn	—	—	$4d^{10}$	$4d^9$	$4d^8$	$4d^7$
51	Sb	—	—	—	$4d^{10}$	$4d^9$	$4d^8$
52	Te	—	—	—	—	$4d^{10}$	$4d^9$
53	I	—	—	—	—	—	$4d^{10}$

(*Continued*)

Table 5.1 *(Continued)*

		Electron configuration $5d^N$					
		Divalent	**Trivalent**	**Tetravalent**	**Pentavalent**	**Hexavalent**	**Heptavalent**
71	Lu	$5d^1$	$5d^0$	—	—	—	—
72	Hf	$5d^2$	$5d^1$	$5d^0$	—	—	—
73	Ta	$5d^3$	$5d^2$	$5d^1$	$5d^0$	—	—
74	W	$5d^4$	$5d^3$	$5d^2$	$5d^1$	$5d^0$	—
75	Re	$5d^5$	$5d^4$	$5d^3$	$5d^2$	$5d^1$	$5d^0$
76	Os	$5d^6$	$5d^5$	$5d^4$	$5d^3$	$5d^2$	$5d^1$
77	Ir	$5d^7$	$5d^6$	$5d^5$	$5d^4$	$5d^3$	$5d^2$
78	Pt	$5d^8$	$5d^7$	$5d^6$	$5d^5$	$5d^4$	$5d^3$
79	Au	$5d^9$	$5d^8$	$5d^7$	$5d^6$	$5d^5$	$5d^4$
80	Hg	$5d^{10}$	$5d^9$	$5d^8$	$5d^7$	$5d^6$	$5d^5$
81	Tl	—	$5d^{10}$	$5d^9$	$5d^8$	$5d^7$	$5d^6$
82	Pb	—	—	$5d^{10}$	$5d^9$	$5d^8$	$5d^7$
83	Bi	—	—	—	$5d^{10}$	$5d^9$	$5d^8$
84	Po	—	—	—	—	$5d^{10}$	$5d^9$
85	At	—	—	—	—	—	$5d^{10}$

$\mathbf{l}_i \cdot \mathbf{s}_i$ stands for the SO interaction operator with the first and second terms being the operators of the angular and spin momenta of an individual electron with index i, and α is the Trees correction[13] to the energy of a term with the total orbital momentum L. The h_k operators can be calculated for a given electron configuration with N electrons as[12].

$$h_k = \frac{1}{2}\sum_{i,j=1}^{N} \frac{4\pi}{2k+1} \sum_{q=-k}^{k} Y_{kq}^{*}(\theta_i, \phi_i) Y_{kq}(\theta_j, \phi_j) , \tag{5.2}$$

with Y_{kq} being the spherical functions of the d-electrons' coordinates θ, ϕ. In the case of the d-ions, it is a common practice to express the energies of the calculated terms using the Racah parameters A, B and C, which are the linear functions of $F^{(0)}$, $F^{(2)}$, $F^{(4)}$ (see Eq. (4.24)). We shall restrict our consideration by the first and second terms of Hamiltonian (5.1), since their contribution to the formation of the energy levels is dominant.

5.3 Results of Calculations: Relations between Main Spectroscopic Parameters for Free 3*d*/4*d*/5*d* Ions and Atomic Number Z

The numerical values of the Racah parameters A, B, C and SO interaction constant ζ were calculated by using the suite of the Cowan programs.[14] It should be noted here that there exists certain discrepancy between the experimentally deduced parameters of the Hamiltonian (5.1) as reported in various literature sources; the difference can be about 7–10%[12] and can be explained by uncertainty of detection of some optical transitions. As a rule, the Hartree–Fock calculated Slater parameters (and thus, the Racah parameters as well) are somewhat overestimated with respect to those ones deduced from the experimental spectra; the overall observation is that $F^{(k)}{}_{\text{exp}} \approx 0.8F_{\text{HF}}$, $\zeta_{\text{exp}} \approx 0.9\zeta_{\text{HF}}$ [12]. These scaling coefficients can be introduced, if necessary; they do not question the validity of the obtained results, and we did not use them in our analysis. All parameters calculated in the present work are collected in Table 5.2.

Table 5.2 Calculated energy parameters (in cm^{-1}, apart from the non-dimensional C/B ratio) and radial integrals (in $Å^k$) of the d^N configurations for the considered $3d$, $4d$ and $5d$ ions with the charges +2, +3, +4, +5, +6, +7 (see text for more details)

Ion	Sc^{2+}	Ti^{2+}	V^{2+}	Cr^{2+}	Mn^{2+}	Fe^{2+}	Co^{2+}	Ni^{2+}	Cu^{2+}	Zn^{2+}	
$3d^N$ N	1	2	3	4	5	6	7	8	9	10	
A	—	139887	151916	163479	174732	185755	196599	207299	217881	228364	
B	—	892.10	967.75	1040.04	1110.12	1178.59	1245.80	1312.00	1377.37	1442.04	
C	—	3296.12	3573.18	3837.10	4092.50	4341.71	4586.15	4826.77	5064.25	5299.10	
C/B	—	3.69481	3.69225	3.68937	3.68652	3.68382	3.68130	3.67895	3.67676	3.67472	
E_{avg}	—	141063	459569	989073	1761882	1123791	594684	209012	—	—	
E_I	24.58	25.95	28.08	30.09	32.00	33.83	35.59	37.29	38.94	40.55	
ζ_d	90.31	131.66	184.47	249.84	329.57	425.51	539.63	673.99	830.74	1012.18	
r^2	0.80788	0.69447	0.59226	0.51469	0.45365	0.40380	0.36264	0.32819	0.29879	0.27364	
r^4	1.41855	1.08136	0.79906	0.61251	0.48257	0.38832	0.31790	0.26403	0.22192	0.18851	
Ion		Ti^{3+}	V^{3+}	Cr^{3+}	Mn^{3+}	Fe^{3+}	Co^{3+}	Ni^{3+}	Cu^{3+}	Zn^{3+}	Ga^{3+}
$3d^N$ N		1	2	3	4	5	6	7	8	9	10
A		—	167896	178875	189677	200342	210896	221357	231739	242052	252307
B		—	1094.59	1162.95	1230.01	1296.08	1361.35	1425.94	1489.97	1553.51	1616.62
C		—	4095.57	4343.87	4587.13	4826.63	5063.09	5297.03	5528.84	5758.82	5987.22

C/B	—	3.74165	3.73522	3.72934	3.72402	3.71917	3.71475	3.71070	3.70698	3.70355	
E_{avg}	—	169379	541334	1147989	2020799	1276298	669776	233721	—	—	
E_I	43.39	44.98	47.61	50.13	52.57	54.94	57.23	59.48	61.67	63.81	
ζ_d	164.18	222.18	293.69	380.19	483.56	605.77	748.92	915.19	1106.89	1326.44	
r^2	0.51805	0.46345	0.41136	0.36824	0.33211	0.30159	0.27552	0.25289	0.23312	0.21576	
r^4	0.54499	0.44783	0.35781	0.29108	0.24042	0.20106	0.17001	0.14515	0.12500	0.10837	
Ion		**V^{4+}**	**Cr^{4+}**	**Mn^{4+}**	**Fe^{4+}**	**Co^{4+}**	**Ni^{4+}**	**Cu^{4+}**	**Zn^{4+}**	**Ga^{4+}**	**Ge^{4+}**
$3d^N$ N		1	2	3	4	5	6	7	8	9	10
A		—	192491	203059	213531	223922	234246	244510	254723	264892	275019
B		—	1271.99	1337.65	1402.56	1466.87	1530.65	1593.99	1656.94	1719.54	1781.83
C		—	4798.07	5035.67	5270.43	5502.87	5733.34	5962.13	6189.48	6415.55	6640.46
C/B		—	3.77210	3.76457	3.75772	3.75144	3.74568	3.74038	3.73549	3.73097	3.72676
E_{avg}		—	194244	614685	1292691	2259202	1417946	740003	256960	—	—
E_I		65.64	67.58	70.75	73.83	76.82	79.74	82.60	85.40	88.14	90.83
ζ_d		261.76	339.72	433.44	544.67	675.44	827.86	1004.14	1206.63	1437.77	1700.09
r^2		0.37860	0.34500	0.31195	0.28395	0.25978	0.23878	0.22041	0.20420	0.18986	0.17703
r^4		0.27955	0.23737	0.19682	0.16522	0.14013	0.11998	0.10351	0.08994	0.07865	0.06923

(*Continued*)

Table 5.2 *(Continued)*

Ion	Y^{2+}	Zr^{2+}	Nb^{2+}	Mo^{2+}	Tc^{2+}	Ru^{2+}	Rh^{2+}	Pd^{2+}	Ag^{2+}	Cd^{2+}	
$4d^N$	1	2	3	4	5	6	7	8	9	10	
A	—	95564.6	103884.5	111790.2	119392.5	126765.3	133967.5	141016.6	147953.6	154792.0	
B	—	651.04	706.76	759.50	810.08	859.01	906.69	953.33	999.15	1044.25	
C	—	2573.03	2800.39	3015.44	3221.48	3420.63	3614.50	3804.06	3990.10	4173.09	
C/B	—	3.95218	3.96229	3.97030	3.97674	3.98206	3.98648	3.99029	3.99349	3.99626	
E_{avg}	—	96554	314891	677723	1206385	768536	406107	142493	—		
E_I	19.84855	22.057	24.167	26.218	28.230	30.217	32.191	34.154	36.115	38.074	
ζ_d	302.68	418.78	550.14	700.42	871.11	1064.26	1282.10	1525.15	1796.48	2098.22	
r^2	1.6200	1.3335	1.1299	0.9770	0.8577	0.7617	0.6827	0.6166	0.5606	0.5127	
r^4	4.7081	3.2127	2.3203	1.7432	1.3495	1.0680	0.8610	0.7054	0.5851	0.4908	
Ion	Y^{3+}	Zr^{3+}	Nb^{3+}	Mo^{3+}	Tc^{3+}	Ru^{3+}	Rh^{3+}	Pd^{3+}	Ag^{3+}	Cd^{3+}	In^{3+}
$4d^N$	0	1	2	3	4	5	6	7	8	9	10
A	—	—	111353.0	118861.6	126157.0	133285.1	140290.0	147171.8	153968.9	160687.8	167330.6
B	—	—	767.15	816.81	865.01	912.05	958.16	1003.52	1048.24	1092.39	1136.04
C	—	—	3085.26	3287.14	3482.94	3673.87	3860.87	4044.79	4225.96	4404.74	4581.40

C/B	—	—	4.02172	4.02436	4.02647	4.02815	4.02946	4.03060	4.03148	4.03220	4.03278
E_{avg}	—	—	112559	360444	765122	1347237	850815	446271	155625	—	—
E_I	—	33.694	36.362	38.972	41.540	44.079	46.599	49.105	51.603	54.096	56.586
ζ_d	—	501.67	640.79	799.03	978.18	1180.34	1408.39	1661.47	1944.05	2257.68	2603.84
r^2	—	1.1070	0.9602	0.8446	0.7513	0.6743	0.6096	0.5547	0.5077	0.4668	0.4310
r^4	—	2.0765	1.5730	1.2249	0.9747	0.7897	0.6485	0.5396	0.4537	0.3852	0.3297
Ion		**Zr^{4+}**	**Nb^{4+}**	**Mo^{4+}**	**Tc^{4+}**	**Ru^{4+}**	**Rh^{4+}**	**Pd^{4+}**	**Ag^{4+}**	**Cd^{4+}**	**In^{4+}**
$4d^N$		0	1	2	3	4	5	6	7	8	9
A		—	—	125153.5	132244.5	139198.5	146053.1	152820.4	159514.4	166144.2	172706.8
B		—	—	866.44	913.10	958.88	1003.91	1048.42	1092.38	1135.88	1178.99
C		—	—	3525.51	3714.56	3899.98	4082.25	4262.34	4440.10	4615.93	4790.18
C/B		—	—	4.06896	4.06808	4.06722	4.06635	4.06549	4.06461	4.06375	4.06295
E_{avg}		—	—	126548	401141	844444	1476664	927026	483804	167967	
E_I		—	49.787	52.944	56.060	59.143	62.204	65.250	68.284	71.310	74.329
ζ_d		—	732.87	899.91	1088.35	1299.71	1537.60	1801.63	2095.79	2421.67	2779.66
r^2		—	0.8435	0.7505	0.6738	0.6093	0.5545	0.5074	0.4665	0.4307	0.3993
r^4		—	1.1637	0.9277	0.7521	0.6188	0.5154	0.4337	0.3684	0.3155	0.2722

(Continued)

Table 5.2 *(Continued)*

Ion	Lu^{2+}	Hf^{2+}	Ta^{2+}	W^{2+}	Re^{2+}	Os^{2+}	Ir^{2+}	Pt^{2+}	Au^{2+}	Hg^{2+}	
$5d^N$	1	2	3	4	5	6	7	8	9	10	
A	—	90631.4	96881.6	102774.3	108395.7	113810.3	119063.7	124177.1	129185.2	134091.4	
B	—	628.23	673.05	715.05	754.86	793.04	829.90	865.67	900.58	934.68	
C	—	2499.48	2691.71	2872.02	3042.94	3206.83	3364.94	3518.33	3667.93	3813.84	
C/B	—	3.97861	3.99927	4.01653	4.03113	4.04372	4.05463	4.06429	4.07285	4.08037	
E_{avg}	—	91598	293786	623373	1095885	690424	361171	125567	—	—	
E_I	19.276	21.163	22.986	24.771	26.532	28.279	30.021	31.761	33.506	35.253	
ζ_{5d}	1352.89	1722.32	2122.68	2556.93	3029.06	3540.98	4094.41	4692.35	5333.47	6022.08	
r^2	1.7191	1.4679	1.2814	1.1364	1.0199	0.9238	0.8429	0.7740	0.7144	0.6626	
r^4	5.2562	3.8220	2.9045	2.2780	1.8302	1.4977	1.2445	1.0476	0.8908	0.7650	
Ion	Lu^{3+}	Hf^{3+}	Ta^{3+}	W^{3+}	Re^{3+}	Os^{3+}	Ir^{3+}	Pt^{3+}	Au^{3+}	Hg^{3+}	Tl^{3+}
$5d^N$	0	1	2	3	4	5	6	7	8	9	10
A	—	—	103562.7	108936.7	114156.0	119252.4	124241.8	129137.9	133959.9	138703.9	143389.7
B	—	—	729.78	767.54	804.03	839.53	874.16	908.09	941.41	974.11	1006.39
C	—	—	2965.43	3127.47	3283.89	3436.04	3584.29	3729.44	3871.86	4011.49	4149.28

C/B	—	—	4.06346	4.07467	4.08429	4.09281	4.10027	4.10691	4.11283	4.11811	4.12293
E_{avg}	—	—	104734	330527	692757	1206185	754020	391878	135507	—	—
E_I	—	32.287	34.473	36.639	38.789	40.931	43.068	45.204	47.344	49.484	51.633
ζ_d	—	2060.08	2466.94	2909.88	3393.33	3918.78	4486.87	5100.58	5758.90	6465.64	7219.63
r^2	—	1.2234	1.0952	0.9893	0.9006	0.8250	0.7597	0.7032	0.6533	0.6093	0.5701
r^4	—	2.4858	1.9918	1.6256	1.3464	1.1300	0.9590	0.8210	0.7087	0.6165	0.5397
Ion		**Hf^{4+}**	**Ta^{4+}**	**W^{4+}**	**Re^{4+}**	**Os^{4+}**	**Ir^{4+}**	**Pt^{4+}**	**Au^{4+}**	**Hg^{4+}**	**Tl^{4+}**
$5d^N$		0	1	2	3	4	5	6	7	8	9
A		—	—	114316.6	119253.0	124116.5	128909.8	133639.1	138317.4	142935.6	147511.0
B		—	—	812.39	846.56	880.14	913.16	945.70	977.81	1009.47	1040.82
C		—	—	3349.31	3495.47	3638.98	3779.96	3918.79	4055.71	4190.59	4324.10
C/B		—	—	4.12279	4.12903	4.13455	4.13943	4.14380	4.14775	4.15128	4.15451
E_{avg}	—	—	—	115658	361964	753469	1304295	811295	419851	144625	—
E_I		—	47.060	49.579	52.092	54.605	57.117	59.631	62.151	64.671	67.199
z		—	2812.34	3266.26	3762.24	4302.43	4885.5	5797.56	6192.08	6916.33	7689.66
r^2		—	0.9683	0.8855	0.8138	0.7516	0.6967	0.6486	0.6057	0.5673	0.5329
r^4		—	1.4923	1.2492	1.0563	0.9018	0.7763	0.6731	0.5875	0.5159	0.4554

(*Continued*)

Table 5.2 (*Continued*)

Ion	Cr^{5+}	Mn^{5+}	Fe^{5+}	Co^{5+}	Ni^{5+}	Cu^{5+}	Zn^{5+}	Ga^{5+}	Ge^{5+}	As^{5+}	
$3d^N$ N	1	2	3	4	5	6	7	8	9	10	
A	—	215334	225704	236008	246256	256456	266614	276734	286819	296875	
B	—	1436.37	1500.84	1564.76	1628.23	1691.29	1754.00	1816.41	1878.54	1940.42	
C	—	5449.78	5682.72	5913.57	6142.63	6370.18	6596.41	6821.48	7045.51	7268.61	
C/B	—	3.79413	3.78637	3.77922	3.77259	3.76646	3.76077	3.75547	3.75052	3.74589	
ζ_d	388.27	489.64	609.14	748.87	910.99	1097.75	1311.50	1554.72	1829.98	2139.99	
E_{avg}	—	217338	683368	1429040	2485008	1552678	807048	279214	—	—	
r^2	0.29487	0.27185	0.24886	0.22890	0.21139	0.19596	0.18227	0.17003	0.15906	0.14920	
r^4	0.16507	0.14295	0.12139	0.10398	0.08979	0.07805	0.06830	0.06012	0.05320	0.04730	
E_I	91.19	93.53	97.29	100.95	104.52	108.03	111.46	114.83	118.15	121.41	
Ion		Mn^{6+}	Fe^{6+}	Co^{6+}	Ni^{6+}	Cu^{6+}	Zn^{6+}	Ga^{6+}	Ge^{6+}	As^{6+}	Se^{6+}
$3d^N$ N		1	2	3	4	5	6	7	8	9	10
A		—	237126	247388	257599	267766	277896	287992	298059	308098	318114
B		—	1592.91	1656.8	1720.25	1783.30	1846.01	1908.43	1970.58	2032.49	2094.18
C		—	6070.72	6301.51	6530.48	6757.93	6984.06	7209.03	7432.98	7656.02	7878.24

C/B	—	3.81110	3.80341	3.79625	3.78957	3.78332	3.77747	3.77198	3.76682	3.76197	
ζ_d	548.85	677.31	826.40	998.64	1196.29	1421.76	1677.53	1966.23	2290.58	2653.41	
E_{avg}	—	239370	749136	1560014	2702481	1682739	871891	300775	—	—	
r^2	0.23872	0.22192	0.20495	0.19000	0.17673	0.16488	0.15424	0.14469	0.13601	0.12817	
r^4	0.10618	0.09324	0.08045	0.06992	0.06117	0.05382	0.04762	0.04233	0.03780	0.03389	
E_I	119.96	122.75	127.11	131.37	135.55	139.65	143.68	147.64	151.54	155.39	
Ion		**Fe^{7+}**	**Co^{7+}**	**Ni^{7+}**	**Cu^{7+}**	**Zn^{7+}**	**Ga^{7+}**	**Ge^{7+}**	**As^{7+}**	**Se^{7+}**	**Br^{7+}**
$3d^N$ N		1	2	3	4	5	6	7	8	9	10
A		—	258222	268421	278577	288698	298787	308848	318885	328899	338892
B		—	1744.24	1807.86	1871.08	1933.95	1996.51	2058.8	2120.85	2182.68	2244.31
C		—	6671.25	6901.05	7129.17	7355.90	7581.4	7805.83	8029.31	8251.95	8473.83
C/B		—	3.82472	3.81724	3.81019	3.80357	3.79733	3.79145	3.78590	3.78065	3.77569
ζ_d		748.87	908.37	1091.12	1300.10	1537.73	1806.52	2109.14	2448.33	2826.97	3248.03
E_{avg}		—	260697	812929	1687268	2914109	1809468	935150	321831	—	—
r^2		0.19854	0.18571	0.17269	0.16107	0.15066	0.14130	0.13282	0.12512	0.11812	0.11170
r^4		0.07232	0.06419	0.05611	0.04932	0.04361	0.03874	0.03457	0.03097	0.02787	0.02516
E_I		151.92	155.17	160.15	165.03	169.83	174.54	179.18	183.74	188.25	192.69

(Continued)

Table 5.2 (*Continued*)

Ion	Mo^{5+}	Tc^{5+}	Ru^{5+}	Rh^{5+}	Pd^{5+}	Ag^{5+}	Cd^{5+}	In^{5+}	Sn^{5+}	Sb^{5+}	
$4d^N$ N	1	2	3	4	5	6	7	8	9	10	
A	—	136989	143619	150199	156735	163226	169677	176089	182464	188806	
B	—	940.28	983.84	1026.98	1069.77	1112.22	1154.39	1196.27	1237.90	1279.30	
C	—	3854.86	4029.46	4202.21	4373.55	4543.52	4712.32	4879.98	5046.57	5212.20	
C/B	—	4.09968	4.09563	4.09183	4.08832	4.08509	4.08209	4.07932	4.07671	4.07426	
ζ_d	1095.28	1298.15	1522.34	1773.32	2052.90	2362.85	2705.07	3081.59	3494.45	3945.80	
E_{avg}	—	138525	435668	911219	1584726	990178	514639	178024	—	—	
r^2	0.68551	0.62362	0.56846	0.52085	0.47913	0.44273	0.41052	0.38168	0.35620	0.33323	
r^4	0.75005	0.62780	0.52429	0.44234	0.37648	0.32292	0.27900	0.24254	0.21219	0.18655	
E_I	67.43	70.11	73.79	77.4	80.95	84.45	87.91	91.32	94.71	98.07	
Ion		Tc^{6+}	Ru^{6+}	Rh^{6+}	Pd^{6+}	Ag^{6+}	Cd^{6+}	In^{6+}	Sn^{6+}	Sb^{6+}	Te^{6+}
$4d^N$ N		1	2	3	4	5	6	7	8	9	10
A		—	149179	155647	162087	168497	174879	181232	187556	193852	200124
B		—	1024.18	1066.80	1109.16	1151.26	1193.12	1234.77	1276.19	1317.41	1358.45
C		—	4224.83	4395.61	4565.24	4733.79	4901.41	5068.09	5233.87	5398.82	5562.98

C/B	—	4.12509	4.12035	4.11594	4.11185	4.10805	4.10450	4.10117	4.09804	4.09509	
ζ_d	1428.85	1666.69	1927.96	2218.58	2540.32	2895.09	3284.89	3711.79	4177.91	4685.55	
E_{avg}	—	150872	472219	983474	1703880	1061011	549759	189642	—	—	
r^2	0.56930	0.52309	0.48137	0.44469	0.41220	0.38336	0.35760	0.33435	0.31363	0.29459	
r^4	0.50939	0.43427	0.36957	0.31696	0.27367	0.23784	0.20796	0.18271	0.16138	0.14319	
E_I	87.54	90.64	94.82	98.93	102.99	106.99	110.95	114.87	118.76	122.62	
Ion		**Ru^{7+}**	**Rh^{7+}**	**Pd^{7+}**	**Ag^{7+}**	**Cd^{7+}**	**In^{7+}**	**Sn^{7+}**	**Sb^{7+}**	**Te^{7+}**	**I^{7+}**
$4d^N$ N		1	2	3	4	5	6	7	8	9	10
A		—	160853	167208	173545	179866	186165	192443	198701	204937	211154
B		—	1103.86	1145.90	1187.73	1229.38	1270.84	1312.13	1353.24	1394.19	1434.99
C		—	4575.69	4744.18	4911.79	5078.64	5244.71	5410.04	5574.62	5738.51	5901.73
C/B		—	4.14516	4.14015	4.13545	4.13106	4.12696	4.12309	4.11945	4.11600	4.11275
ζ_d		1810.74	2086.76	2388.51	2722.18	3089.64	3492.91	3934.04	4415.16	4938.57	5506.53
E_{avg}		—	162695	507346	1053106	1819037	1129604	583829	200932	—	—
r^2		0.48417	0.44833	0.41556	0.38644	0.36040	0.33687	0.31587	0.29655	0.27927	0.26345
r^4		0.36416	0.31476	0.27171	0.23603	0.20616	0.18106	0.15973	0.14162	0.12609	0.11268
E_I		109.61	113.14	117.82	122.44	127.00	131.51	135.97	140.40	144.80	149.17

(*Continued*)

Table 5.2 (*Continued*)

Ion	W^{5+}	Re^{5+}	Os^{5+}	Ir^{5+}	Pt^{5+}	Au^{5+}	Hg^{5+}	Tl^{5+}	Pb^{5+}	Bi^{5+}	
$5d^N$ N	1	2	3	4	5	6	7	8	9	10	
A	—	122023	126443	130873	135306	139734	144143	148535	152905	157252	
B	—	867.79	897.82	927.82	957.80	987.71	1017.46	1047.08	1076.54	1105.84	
C	—	3616.64	3743.06	3869.24	3995.24	4120.94	4245.98	4370.40	4494.11	4617.08	
C/B	—	4.16766	4.16907	4.17023	4.17128	4.17224	4.17312	4.17389	4.17458	4.17519	
ζ_d	3737.36	4263.01	4803.75	5394.38	6037.50	6727.26	7467.60	8261.40	9110.06	10015.01	
E_{avg}	—	123486	383873	794635	1369235	848416	437588	150305	—	—	
r^2	0.82720	0.77008	0.71743	0.67011	0.62726	0.58862	0.55362	0.52169	0.49257	0.46597	
r^4	1.05234	0.91825	0.79749	0.69680	0.61172	0.53950	0.47810	0.42541	0.38008	0.34087	
E_I	62.69	64.65	67.62	70.54	73.43	76.28	79.10	81.89	84.67	87.43	
Ion		Re^{6+}	Os^{6+}	Ir^{6+}	Pt^{6+}	Au^{6+}	Hg^{6+}	Tl^{6+}	Pb^{6+}	Bi^{6+}	Po^{6+}
$5d^N$ N		1	2	3	4	5	6	7	8	9	10
A		—	130952	135224	139517	143825	148132	152433	156725	161002	165266
B		—	932.56	961.48	990.49	1019.56	1048.60	1077.60	1106.51	1135.32	1164.04
C		—	3917.99	4039.34	4160.96	4282.90	4404.68	4526.28	4647.50	4768.26	4888.59
C/B		—	4.20135	4.20116	4.20093	4.20072	4.20053	4.20034	4.20013	4.19991	4.19967

ζ_d	4680.04	5267.21	5869.81	6526.57	7231.87	7989.28	8801.51	9669.84	10595.53	11581.98	
E_{avg}	—	132549	410610	847275	1455702	899560	462832	158618	—	—	
r^2	0.70875	0.66423	0.62334	0.58610	0.55194	0.52057	0.49201	0.46597	0.44160	0.41948	
r^4	0.75766	0.66967	0.59055	0.52288	0.46454	0.41435	0.37083	0.33303	0.30010	0.27124	
E_I	80.38	82.63	85.94	89.21	92.44	95.64	98.82	101.97	105.11	108.23	
Ion		**Os^{7+}**	**Ir^{7+}**	**Pt^{7+}**	**Au^{7+}**	**Hg^{7+}**	**Tl^{7+}**	**Pb^{7+}**	**Bi^{7+}**	**Po^{7+}**	**At^{7+}**
$5d^N$ N		1	2	3	4	5	6	7	8	9	10
A		—	139345	143515	147716	151929	156150	160371	164586	168795	172993
B		—	992.51	1020.72	1049.10	1077.54	1106.00	1134.46	1162.87	1191.22	1219.50
C		—	4196.60	4314.71	4433.48	4552.54	4671.77	4790.92	4909.87	5028.57	5146.93
C/B		—	4.22828	4.22711	4.22599	4.22496	4.22401	4.22308	4.22220	4.22135	4.22052
ζ_d		5705.01	6357.95	7027.74	7748.01	8522.12	9352.60	10240.50	11186.95	12195.29	13264.98
E_{avg}		—	141065	435849	897194	1537937	948379	486998	166596	—	—
r^2		0.61998	0.58414	0.55110	0.52057	0.49229	0.46653	0.44273	0.42060	0.40016	0.38112
r^4		0.57118	0.50978	0.45442	0.40635	0.36440	0.32778	0.29571	0.26748	0.24270	0.22074
E_I		99.50	102.04	105.70	109.32	112.90	116.46	119.99	123.50	126.99	130.47

Figures 5.1–5.3 show how the calculated values of *A*, *B* and *C* from Table 5.2 vary with the atomic number *Z* for the doubly, triply and quadruply positively charged 3*d*, 4*d* and 5*d* ions. It is seen that the Racah parameters perfectly behave as the linear functions of *Z*; the fitting equations are given in the figures and (with the calculated uncertainties) in Table 5.3.

Table 5.3 Linear approximations $a_1 + a_2Z$ (a_1 in cm^{-1}; a_2 in cm^{-1}/at.number) of the Racah parameters *A*, *B* and *C* as functions of the atomic number *Z* for the 3*d*, 4*d*, 5*d* ions with the charges +2, +3, +4

			Divalent ions	Trivalent ions	Tetravalent ions
3*d* ions	*A*	a_1	-101466.33 ±2325.43	-73970.08±1165.24	-54628.40 ±743.13
		a_2	11021.83±89.00	10538.57±42.96	10309.72±26.43
	B	a_1	-606.25±16.45	-400.31±8.42	-253.98±5.67
		a_2	68.47±0.63	65.16±0.31	63.68±0.21
	C	a_1	-2157.75±63.06	-1320.13±31.85	-715.51±21.33
		a_2	249.30±2.41	236.09±1.17	230.11±0.76
4*d* ions	*A*	a_1	-198114.35 ±3817.95	-174273.91±2301.17	-159567.48 ±1578.80
		a_2	7369.08±86.62	6981.66±51.05	6786.75±34.65
	B	a_1	-1297.34±27.32	-1114.67±16.01	-1004.75±10.78
		a_2	48.90±0.62	46.00±0.36	44.60±0.24
	C	a_1	-5354.31±114.26	-4546.05±67.27	-4045.31±45.14
		a_2	198.99±2.59	186.57±1.49	180.48±0.99
5*d* ions	*A*	a_1	-297460.83 ±5992.86	-258638.61±3177.19	-236175.12 ±1911.41
		a_2	5403.74±78.81	4968.38±41.24	4739.35±24.65
	B	a_1	-2105.58±47.88	-1784.15±25.29	-1598.90±14.97
		a_2	38.08±0.63	34.49±0.33	32.61±0.19
	C	a_1	-9227.80±207.08	-7792.98±112.09	-6939.60±67.17
		a_2	163.35±2.72	147.61±1.45	139.14±0.87

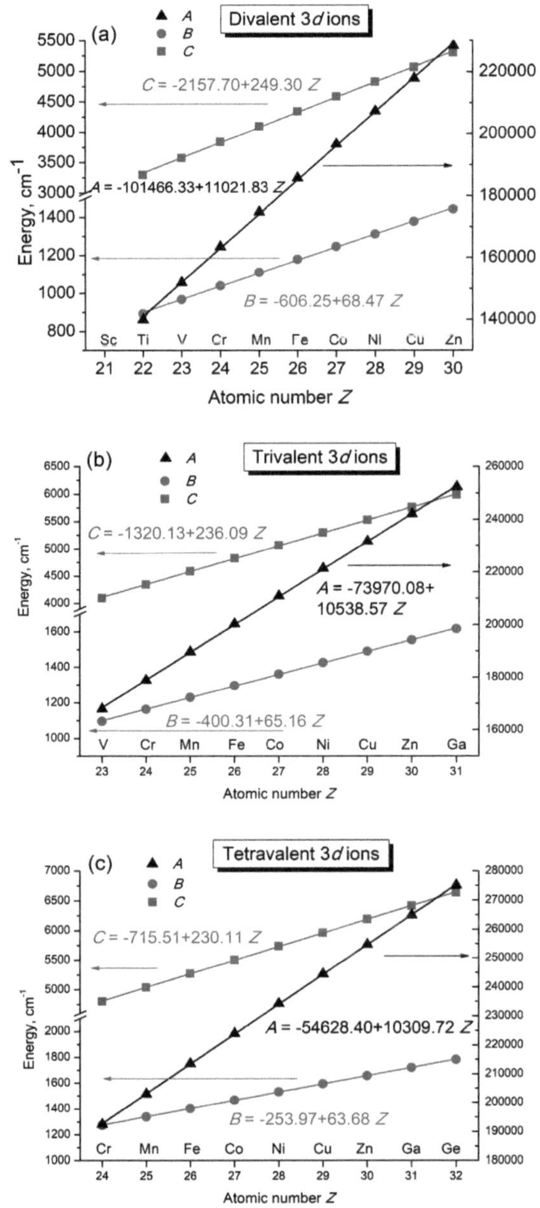

Figure 5.1 Dependence of the *A*, *B*, *C* parameters on the atomic number *Z* for the di- (a), tri- (b) and tetravalent (c) 3*d* ions. The left vertical axis is used for the *B*, *C* parameters, the right axis—for the *A* parameter.

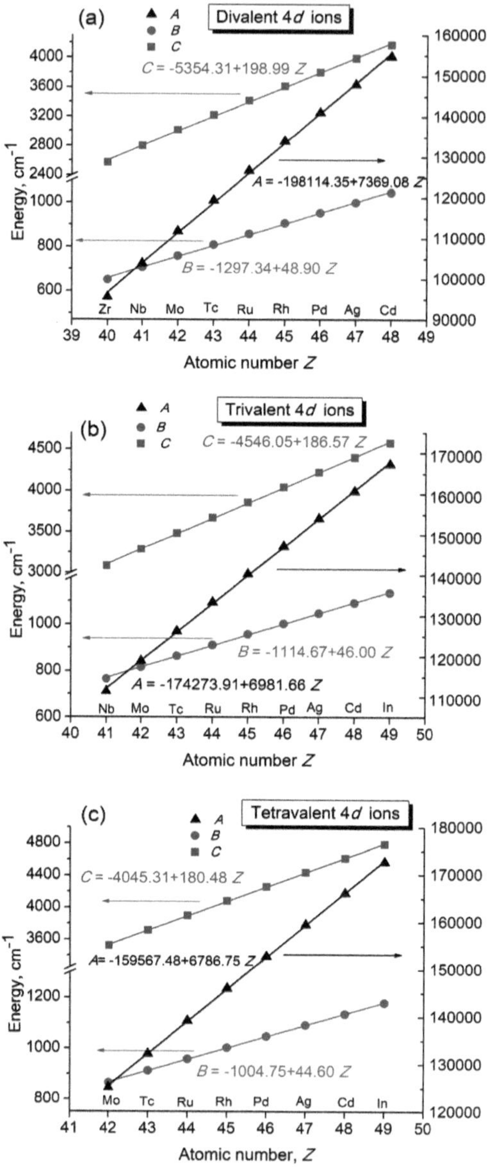

Figure 5.2 Dependence of the *A*, *B*, *C* parameters on the atomic number *Z* for the di- (a), tri- (b) and tetravalent (c) 4*d* ions. The left vertical axis is used for the *B*, *C* parameters, the right axis—for the *A* parameter.

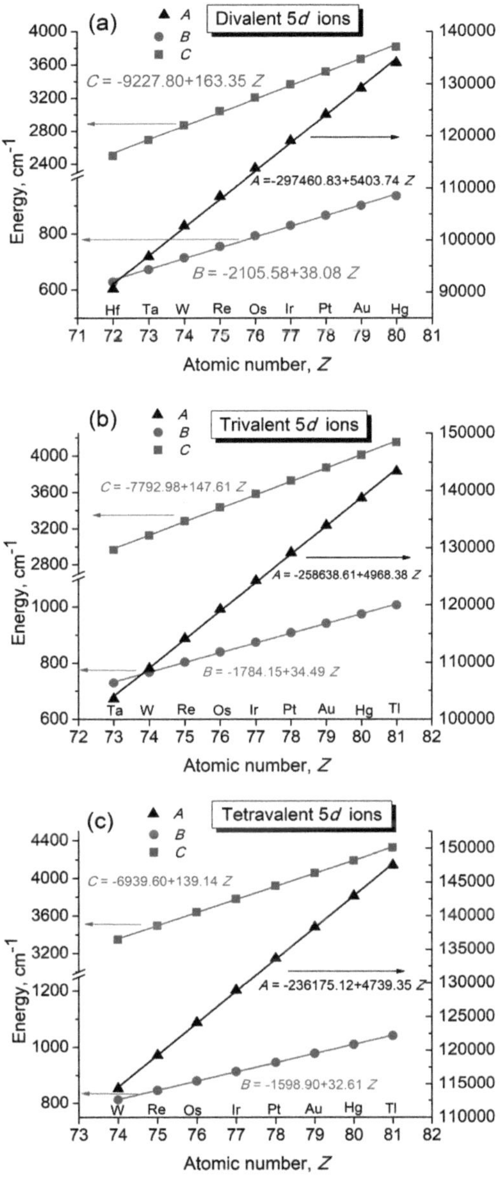

Figure 5.3 Dependence of the *A*, *B*, *C* parameters on the atomic number *Z* for the di- (a), tri- (b) and tetravalent (c) 5*d* ions. The left vertical axis is used for the *B*, *C* parameters, the right axis—for the *A* parameter.

As follows from Table 5.3, the absolute values of the fitting coefficients decrease with increasing charge of all ions; therefore, the steepest fits are for the divalent ions. It can be also noticed that the slopes of all straight lines decrease when moving from 3*d* to 4*d* and 5*d* ions.

An additional interest represents a non-dimensional C/B ratio, since it allows the reduction of the number of electrostatic parameters and is used for a simplified representation of the energy levels splitting of *d*-electrons in cubic crystal fields by the Tanabe-Sugano diagrams.[15] As is shown by Fig. 5.4, this ratio decreases with increasing atomic number for the 3*d* ions (Fig. 5.4a), increases with increasing atomic number (or the number of *d* electrons) for the di- and trivalent 4*d* ions and slightly decreases for the tetravalent ions (Fig. 5.4b). The C/B ratio is an increasing function of the atomic number for the 5*d* ions (Fig. 5.4c).

However, the slopes of all linear approximations in Fig. 5.4 are very small, and on average C/B is about 3.68, 3.72 and 3.75 for the di-, tri- and tetravalent 3*d* ions, 3.98, 4.03 and 4.07 for the di-, tri- and tetravalent 4*d* ions, and slightly greater—4.04, 4.06, and 4.14 for the di-, tri- and tetravalent 5*d* ions.

It is a well-known fact that the SO constant ζ is a linear function of Z^4, with Z being the atomic number of the chemical element under consideration. Therefore, plotting $\zeta^{1/4}$ against Z should yield a linear dependence. Indeed, Fig. 5.5 confirms such an expectation. The derived linear equations are as follows (Z^* denotes an effective nuclear charge):

$$\begin{aligned}&\text{Divalent } 3d \text{ ions: } \zeta^{1/4} = 0.283(Z-10.021) = 0.283\,Z^*\\&\text{Trivalent } 3d \text{ ions: } \zeta^{1/4} = 0.273\,(Z-8.822) = 0.273\,Z^*\\&\text{Tetravalent } 3d \text{ ions: } \zeta^{1/4} = 0.266\,(Z-7.879) = 0.266\,Z^*\end{aligned} \tag{5.3}$$

$$\begin{aligned}&\text{Divalent } 4d \text{ ions: } \zeta^{1/4} = 0.286\,(Z-24.10) = 0.286\,Z^*\\&\text{Trivalent } 4d \text{ ions: } \zeta^{1/4} = 0.267\,(Z-22.06) = 0.267\,Z^*\\&\text{Tetravalent } 4d \text{ ions: } \zeta^{1/4} = 0.257\,(Z-20.60) = 0.257\,Z^*\end{aligned} \tag{5.4}$$

$$\begin{aligned}&\text{Divalent } 5d \text{ ions: } \zeta^{1/4} = 0.302\,(Z-50.60) = 0.302\,Z^*\\&\text{Trivalent } 5d \text{ ions: } \zeta^{1/4} = 0.275\,(Z-47.20) = 0.275\,Z^*\\&\text{Tetravalent } 5d \text{ ions: } \zeta^{1/4} = 0.262\,(Z-45.06) = 0.262\,Z^*\end{aligned} \tag{5.5}$$

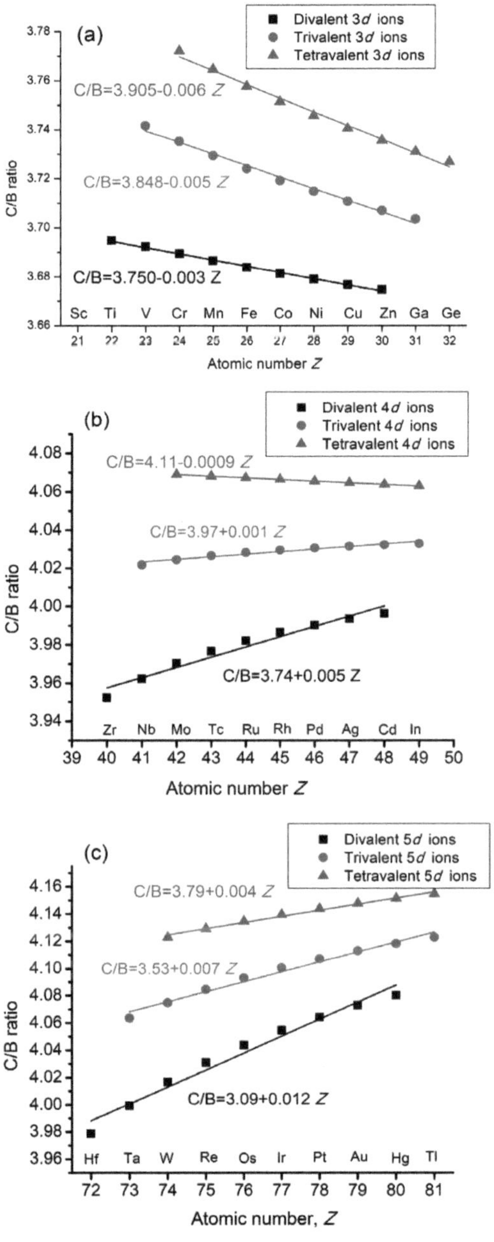

Figure 5.4 Dependence of the *C/B* ratio on the atomic number *Z* for the 3*d* (a), 4*d* (b) and 5*d* (c) ions.

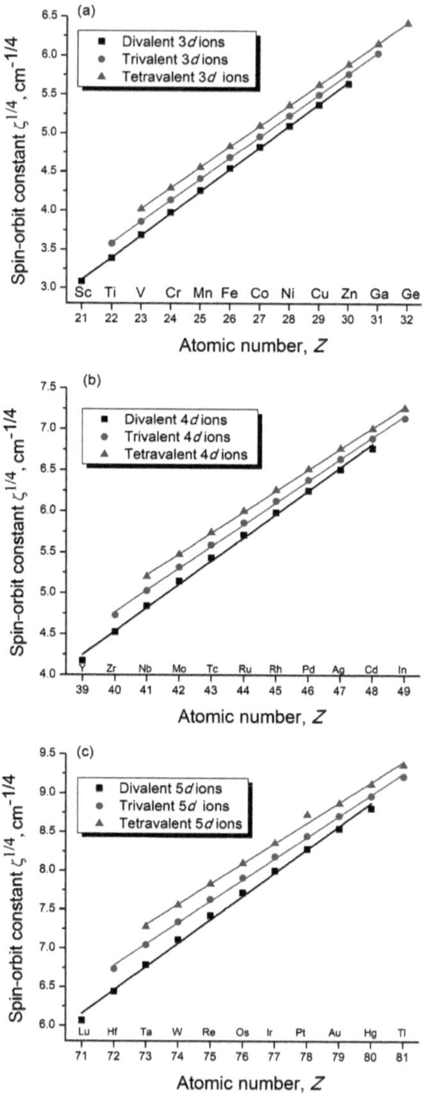

Figure 5.5 Linear dependencies of the $\zeta^{1/4}$ value on the atomic number Z for the di-, tri- and tetravalent 3*d* (a), 4*d* (b) and 5*d* (c) ions. The fitting lines equations are given by Eqs. (5.3)–(5.5).

Since the Racah parameters A, B, and C are also the linear functions of the atomic number Z, one then may expect another linear relation to exist between these parameters and $\zeta^{1/4}$. Figure 5.6 illustrates

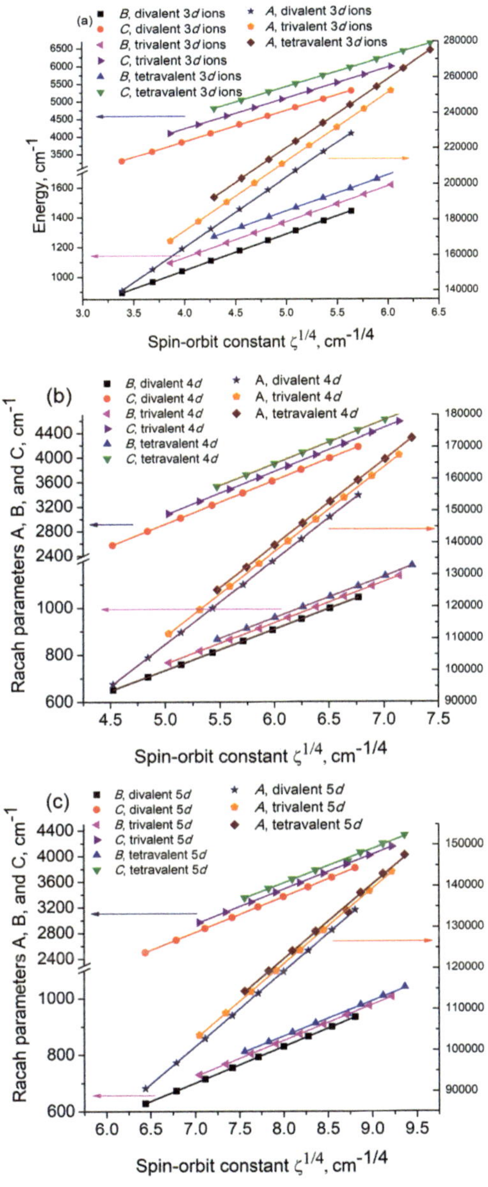

Figure 5.6 Linear dependencies of the Racah parameters *A*, *B*, and *C* on the $\zeta^{1/4}$ value for the di-, tri-, tetravalent 3*d* (a), 4*d* (b) and 5*d* (c) ions. The left vertical axis is used for the *B* and *C* parameters (the upper and lower groups of three lines), the right axis—for the *A* parameter (the central group of three lines). The fitting lines equations are given by Eqs. (5.6)–(5.14).

those linear trends, whose linear equations are given below. These data, taken together, allow for estimating all four parameters A, B, C and ζ if the atomic number Z and oxidation state of a particular ion from the $3d$, $4d$, $5d$ groups are known. This demonstrates that the number of independent parameters in Hamiltonian (5.1) can be reduced enormously, from four to one only.

Divalent $3d$ ions:

$$A = (7380.01 \pm 478.74) + (39227.18 \pm 104.34)\zeta^{1/4}$$
$$B = (69.88 \pm 4.23) + (243.70 \pm 0.92)\zeta^{1/4} \quad (5.6)$$
$$C = (303.99 \pm 17.37) + (887.33 \pm 3.79)\zeta^{1/4}$$

Trivalent $3d$ ions:

$$A = (18233.19 \pm 279.31) + (38811.66 \pm 55.81)\zeta^{1/4}$$
$$B = (169.77 \pm 2.55) + (239.97 \pm 0.51)\zeta^{1/4} \quad (5.7)$$
$$C=(745.36 \pm 10.16) + (869.48 \pm 2.03)\zeta^{1/4}$$

Tetravalent $3d$ ions:

$$A = (26169.79 \pm 248.30) + (38773.67 \pm 45.94)\zeta^{1/4}$$
$$B = (245.09 \pm 2.31) + (239.50 \pm 0.42)\zeta^{1/4} \quad (5.8)$$
$$C = (1087.84 \pm 8.97) + (865.42 \pm 1.66)\zeta^{1/4}$$

Divalent $4d$ ions:

$$A = (-24020 \pm 178.07) + (26409.81 \pm 31.07)\zeta^{1/4}$$
$$B = (-142.08 \pm 0.30) + (175.28 \pm 0.05)\zeta^{1/4} \quad (5.9)$$
$$C = (-653.58 \pm 0.96) + (713.25 \pm 0.17)\zeta^{1/4}$$

Trivalent $4d$ ions:

$$A = (-22124.59 \pm 94.05) + (26518.53 \pm 15.30)\zeta^{1/4}$$
$$B = (-112.15 \pm 0.30) + (174.74 \pm 0.05)\zeta^{1/4} \quad (5.10)$$
$$C = (-480.31 \pm 1.33) + (708.68 \pm 0.22)\zeta^{1/4}$$

Tetravalent $4d$ ions:

$$A = (-20883.23 \pm 34.06) + (26661.23 \pm 5.32)\zeta^{1/4}$$
$$B = (-93.28 \pm 0.28) + (175.23 \pm 0.04)\zeta^{1/4} \quad (5.11)$$
$$C = (-357.41 \pm 1.69) + (709.00 \pm 0.26)\zeta^{1/4}$$

Divalent 5d ions:

$$A = (-27761.42 \pm 243.47) + (18361.59 \pm 31.55)\zeta^{1/4}$$
$$B = (-205.34 \pm 0.32) + (129.42 \pm 0.04)\zeta^{1/4} \quad (5.12)$$
$$C = (-1076.28 \pm 1.75) + (555.17 \pm 0.22)\zeta^{1/4}$$

Trivalent 5d ions:

$$A = (-25844.18 \pm 299.81) + (18347.26 \pm 36.59)\zeta^{1/4}$$
$$B = (-168.15 \pm 0.86) + (127.38 \pm 0.10)\zeta^{1/4} \quad (5.13)$$
$$C = (-877.03 \pm 2.28) + (545.17 \pm 0.28)\zeta^{1/4}$$

Tetravalent 5d ions:

$$A = (-23640.42 \pm 3940.01) + (18226.06 \pm 462.87)\zeta^{1/4}$$
$$B = (-136.66 \pm 26.75) + (125.41 \pm 3.14)\zeta^{1/4} \quad (5.14)$$
$$C = (-700.18 \pm 113.57) + (535.16 \pm 13.34)\zeta^{1/4}$$

Table 5.4 in the present chapter contains the energies of all terms of the d^N configurations; their barycenters can be easily calculated to be as follows: E_{avg} (d^2, d^8) = A −14B/9 + 7C/9; E_{avg} (d^3, d^7) = 3A −14B/3 + 7C/3; E_{avg} (d^4, d^6) = 6A −28B/3 + 14C/3; E_{avg} (d^5) = 10A −140B/9 + 70C/9.

Table 5.4 *LS*-terms and their energies for the d^N electron configurations (including the A parameter)

Configuration	*LS*-terms	Energies
d^1, d^9	^{2}D	
d^2, d^8	^{3}F	$A - 8B$
	^{3}P	$A + 7B$
	^{1}G	$A + 4B + 2C$
	^{1}D	$A - 3B + 2C$
	^{1}S	$A - 14B + 7C$
d^3, d^7	^{4}F	$3A - 15B$
	^{4}P	$3A$
	^{2}H, ^{2}P	$3A - 6B + 3C$
	^{2}G	$3A - 11B + 3C$
	^{2}F	$3A + 9E + 3C$
	^{2}D′, ^{2}D″	$3A+5B+5C\pm\sqrt{193B^2+8BC+4C^2}$

Configuration	*LS*-terms	Energies
d^4, d^6	^{5}D	$6A - 21B$
	^{3}H	$6A - 17B + 4C$
	^{3}G	$6A - 12B + 4C$
	^{3}F′, ^{3}F″	$6A - 5B + 11/2\,C \pm 3/2\sqrt{68B^2 + 4BC + C^2}$
	^{3}D	$6A - 5B + 4C$
	^{3}P′, ^{3}P″	$6A - 5B + 11/2\,C \pm 1/2\sqrt{912B^2 - 24BC + 9C^2}$
	^{1}I	$6A - 15B + 6C$
	^{1}G′, ^{1}G″	$6A - 5B + 15/2\,C \pm 1/2\sqrt{708B^2 - 12BC + 9C^2}$
	^{1}F	$6A + 6C$
	^{1}D′, ^{1}D″	$6A + 9B + 15/2\,C \pm 3/2\sqrt{144B^2 + 8BC + C^2}$
	^{1}S′, ^{1}S″	$6A + 10B + 10C \pm 2\sqrt{193B^2 + 8BC + 4C^2}$
d^5	^{6}S	$10A - 35B$
	^{4}G	$10A - 25B + 5C$
	^{4}F	$10A - 13B + 7C$
	^{4}D	$10A - 18B + 5C$
	^{4}P	$10A - 28B + 7C$
	^{2}I	$10A - 24B + 8C$
	^{2}H	$10A - 22B + 10C$
	^{2}G′	$10A - 13B + 8C$
	^{2}G″	$10A + 3B + 10C$
	^{2}F′	$10A - 9B + 8C$
	^{2}F″	$10A - 25B + 10C$
	^{2}D′, ^{2}D″	$10A - 3B + 11C \pm 3\sqrt{57B^2 + 2BC + C^2}$
	^{2}D″	$10A - 4B + 10C$
	^{2}P	$10A + 20B + 10C$
	^{2}S	$10A - 3B + 8C$

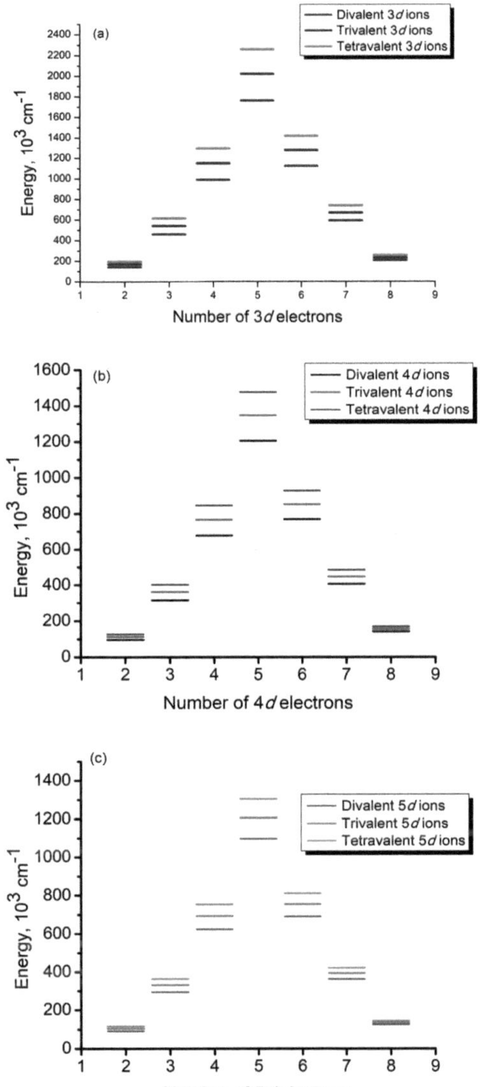

Figure 5.7 Barycenters E_{avg} of the d^N configurations for the di-, tri-, and tetravalent 3*d* (a), 4*d* (b) and 5*d* (c) ions.

The corresponding numerical values are collected in Table 5.2 and the plots of the E_{avg} values as the functions of the number of *d* electrons (Fig. 5.7) look like asymmetric stairs with a maximum at $N = 5$, and right "steps" (or energies of the more than half filled *d* shells) somewhat higher than the left "steps" (energies of the less than half filled d shells). It can be also noticed that the E_{avg} values increase with increasing the oxidation state. The results of calculations for the penta-, hexa- and heptavalent 3d, 4d, 5d ions are shown in Figs. 5.8–5.13, Table 5.5 and Eqs. (5.15)–(5.26).

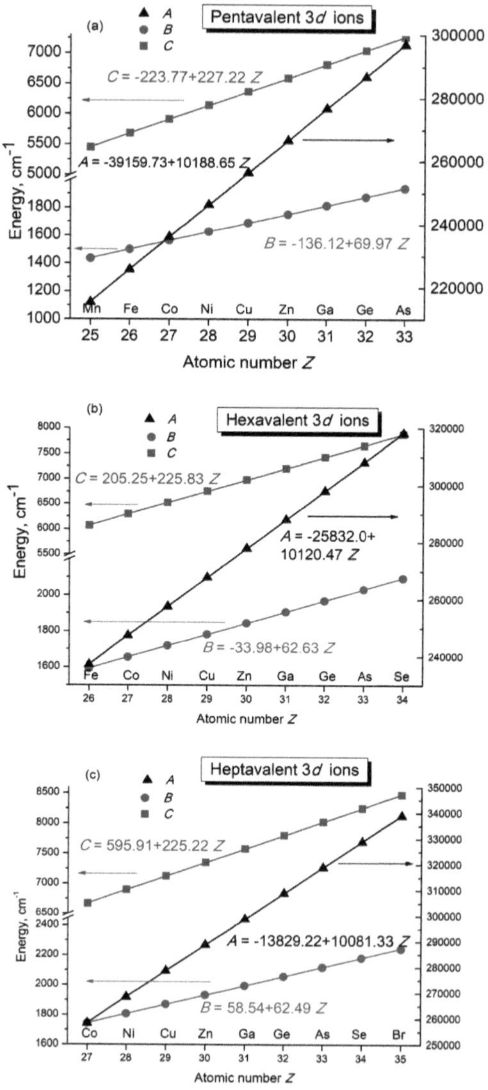

Figure 5.8 Dependence of the A, B, C parameters on the atomic number for the penta- (a), hexa- (b) and heptavalent (c) 3d ions. The left vertical axis is used for the B, C parameters, the right axis—for the A parameter.

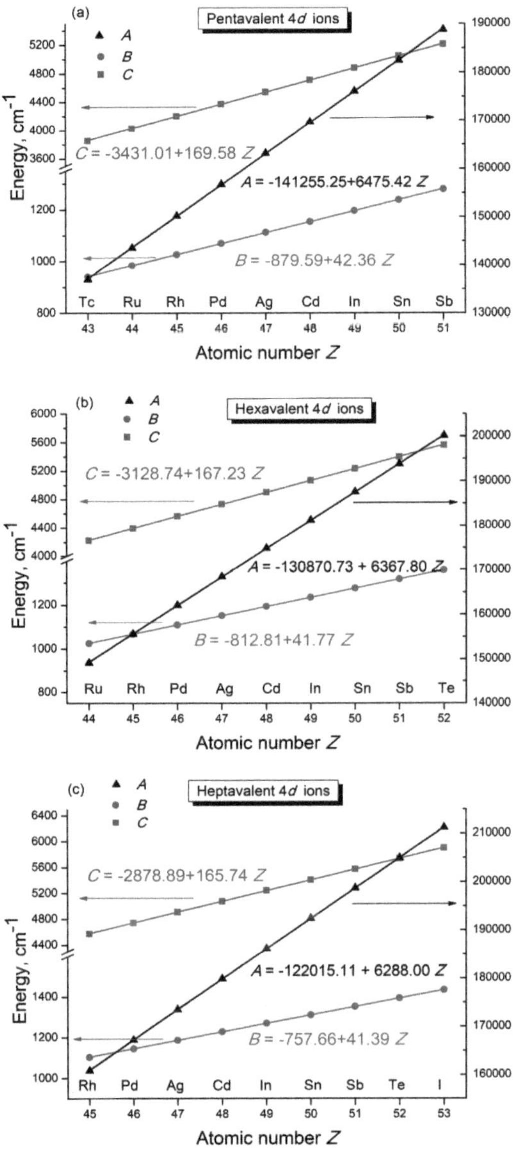

Figure 5.9 Dependence of the *A*, *B*, *C* parameters on the atomic number for the penta- (a), hexa- (b) and heptavalent (c) 4*d* ions. The left vertical axis is used for the *B*, *C* parameters, the right axis—for the *A* parameter.

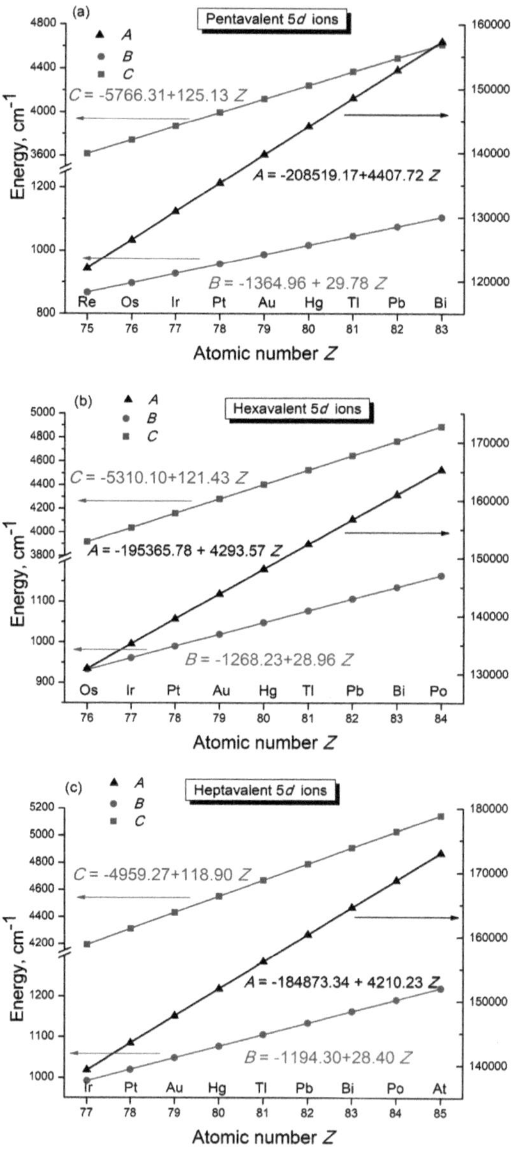

Figure 5.10 Dependence of the *A*, *B*, *C* parameters on the atomic number for the penta- (a), hexa- (b) and heptavalent (c) 5*d* ions. The left vertical axis is used for the *B*, *C* parameters, the right axis—for the *A* parameter.

Table 5.5 Linear approximations $a_1 + a_2Z$ (a_1 in cm^{-1}; a_2 in cm^{-1}/at.number) of the Racah parameters *A*, *B* and *C* as functions of the atomic number *Z* for the 3*d*, 4*d*, 5*d* ions with the charges +5, +6, +7

			Pentavalent ions	Hexavalent ions	Heptavalent ions
3*d* ions	*A*	a_1	-39159.74±550.96	-25832.00±446.83	-13829.22±384.46
		a_2	10188.65±18.92	10120.47±14.84	10081.33±12.36
	B	a_1	-136.12±4.53	-33.98±4.00	58.54±3.74
		a_2	62.97±0.16	62.63±0.13	62.49±0.12
	C	a_1	-223.77±17.22	-205.25±15.52	595.91±14.83
		a_2	227.22±0.59	225.83±0.52	225.22±0.47
4*d* ions	*A*	a_1	-141255.25±829.03	-130870.73±585.44	-122015.11±423.50
		a_2	6475.42±17.61	6367.80±12.18	6288.00±8.63
	B	a_1	-879.59±6.15.62	-812.81±4.66	-757.66±3.72
		a_2	42.35±0.13	41.77±0.10	41.39±0.08
	C	a_1	-3431.01±25.27	-3128.74±19.32	-2878.89±15.76
		a_2	169.58±0.54	167.23±0.40	165.74±0.32
5*d* ions	*A*	a_1	-208519.17±424.38	-195365.78±181.81	-184873.34±172.23
		a_2	4407.72±5.37	4293.57±2.27	4210.23±2.13
	B	a_1	-1364.96±3.78	-1268.23±1.54	-1194.30±0.89
		a_2	29.78±0.05	28.96±0.02	28.40±0.01
	C	a_1	-5766.31±17.25	-5310.10±6.99	-4959.27±3.93
		a_2	125.13±0.21	121.43±0.09	118.90±0.05

Pentavalent 3*d* ions: $\zeta^{1/4} = 0.262(Z-7.071) = 0.262\ Z^*$
Hexavalent 3*d* ions: $\zeta^{1/4} = 0.260\ (Z-7.288) = 0.260\ Z^*$ (5.15)
Heptavalent 3*d* ions: $\zeta^{1/4} = 0.257\ (Z-5.690) = 0.257\ Z^*$

Pentavalent 4*d* ions: $\zeta^{1/4} = 0.241\ (Z-18.116) = 0.241\ Z^*$
Hexavalent 4*d* ions: $\zeta^{1/4} = 0.236\ (Z-16.919) = 0.236\ Z^*$ (5.16)
Heptavalent 4*d* ions: $\zeta^{1/4} = 0.232\ (Z-15.909) = 0.232\ Z^*$

Pentavalent 5*d* ions: $\zeta^{1/4} = 0.242\ (Z-41.653) = 0.242\ Z^*$
Hexavalent 5*d* ions: $\zeta^{1/4} = 0.233\ (Z-39.464) = 0.233\ Z^*$ (5.17)
Heptavalent 5*d* ions: $\zeta^{1/4} = 0.226\ (Z-37.571) = 0.226\ Z^*$

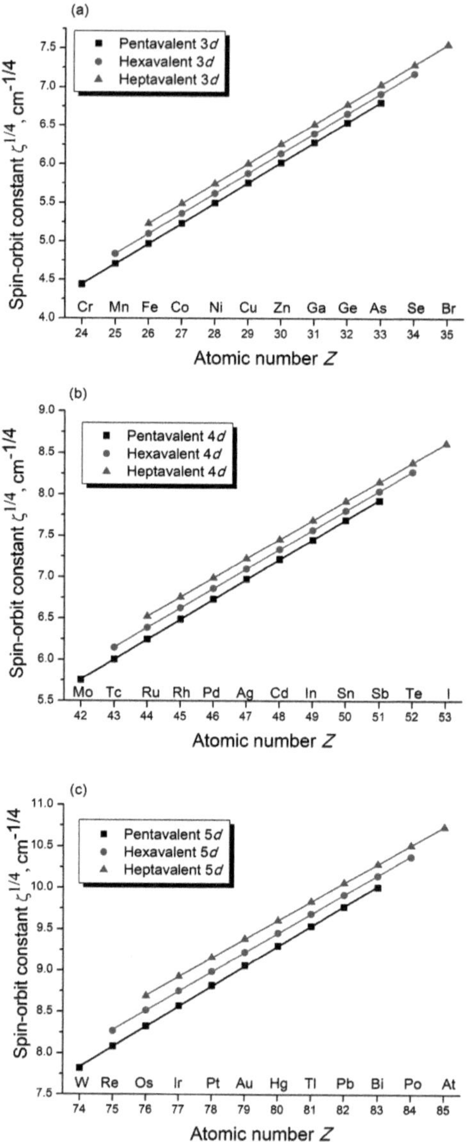

Figure 5.11 Linear dependencies of the $\zeta^{1/4}$ value on the atomic number for the penta-, hexa-, and heptavalent 3*d* (a), 4*d* (b) and 5*d* (c) ions. The fitting lines equations are given by Eqs. (5.15)–(5.17).

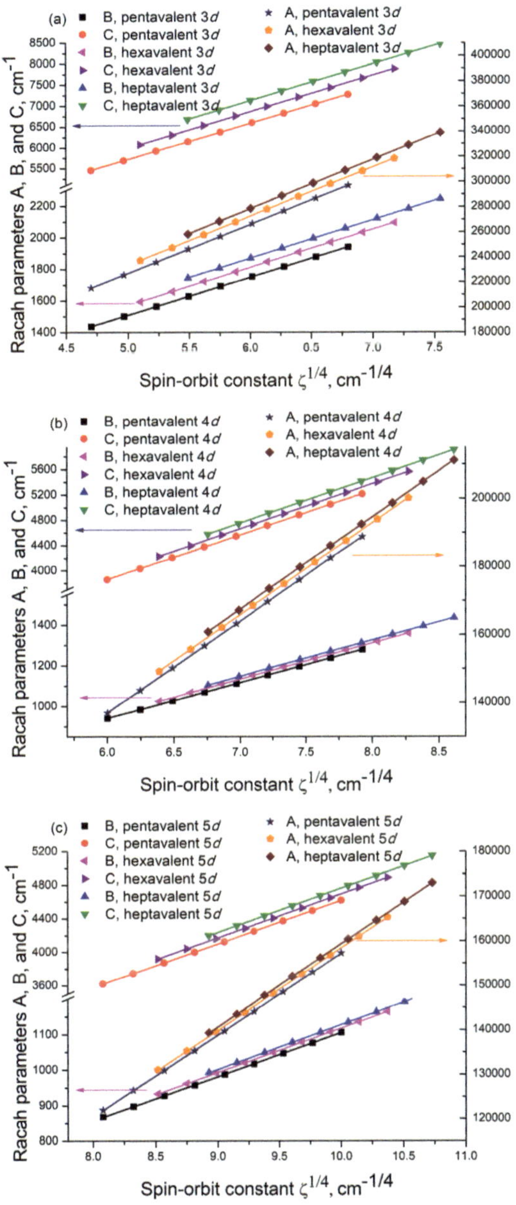

Figure 5.12 Linear dependencies of the Racah parameters *A*, *B*, and *C* on the $\zeta^{1/4}$ value for the highly charged 3*d* (a), 4*d* (b) and 5*d* (c) ions. The left vertical axis is used for the *B* and *C* parameters (the upper and lower groups of three lines), the right axis—for the *A* parameter (the central group of three lines). The fitting lines equations are given by Eqs. (5.18)–(5.26).

Pentavalent 3*d* ions:

$$A = (32630.05 \pm 264.44) + (38869.28 \pm 45.64)\zeta^{1/4}$$
$$B = (307.59 \pm 2.49) + (240.24 \pm 0.43)\zeta^{1/4} \quad (5.18)$$
$$C = (1377.23 \pm 9.64) + (866.84 \pm 1.66)\zeta^{1/4}$$

Hexavalent 3*d* ions:

$$A = (38253.88 \pm 291.52) + (39010.87 \pm 47.20)\zeta^{1/4}$$
$$B = (362.63 \pm 2.78) + (241.43 \pm 0.45)\zeta^{1/4} \quad (5.19)$$
$$C = (1635.26 \pm 10.89) + (870.49 \pm 1.76)\zeta^{1/4}$$

Heptavalent 3*d* ions:

$$A = (43353.11 \pm 321.76) + (39165.34 \pm 49.10)\zeta^{1/4}$$
$$B = (412.96 \pm 3.10) + (242.75 \pm 0.47)\zeta^{1/4} \quad (5.20)$$
$$C = (1873.38 \pm 12.29) + (874.97 \pm 1.87)\zeta^{1/4}$$

Pentavalent 4*d* ions:

$$A = (-24641.04 \pm 179.06) + (26939.21 \pm 25.59)\zeta^{1/4}$$
$$B = (-116.79 \pm 1.63) + (176.22 \pm 0.23)\zeta^{1/4} \quad (5.21)$$
$$C = (-377.05 \pm 6.91) + (705.51 \pm 0.99)\zeta^{1/4}$$

Hexavalent 4*d* ions:

$$A = (-23485.24 \pm 210.56) + (27035.53 \pm 28.61)\zeta^{1/4}$$
$$B = (-108.34 \pm 1.92) + (177.36 \pm 0.26)\zeta^{1/4} \quad (5.22)$$
$$C = (-308.62 \pm 8.11) + (710.00 \pm 1.10)\zeta^{1/4}$$

Heptavalent 4*d* ions:

$$A = (-22263.56 \pm 231.10) + (27104.95 \pm 29.97)\zeta^{1/4}$$
$$B = (-101.12 \pm 2.16) + (178.40 \pm 0.28)\zeta^{1/4} \quad (5.23)$$
$$C = (-249.68 \pm 9.23) + (714.42 \pm 1.20)\zeta^{1/4}$$

Pentavalent 5*d* ions:

$$A = (-26100.60 \pm 341.51) + (18319.43 \pm 37.65)\zeta^{1/4}$$
$$B = (-132.67 \pm 1.82) + (123.76 \pm 0.20)\zeta^{1/4} \quad (5.24)$$
$$C = (-587.59 \pm 7.01) + (520.09 \pm 0.77)\zeta^{1/4}$$

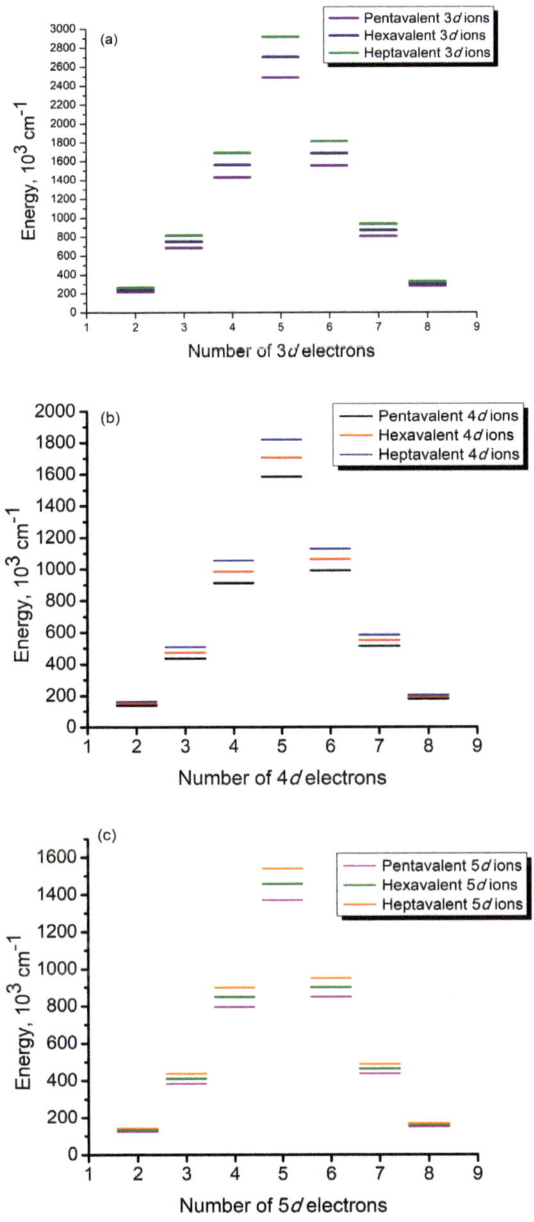

Figure 5.13 Barycenters E_{avg} of the d^N configurations for the penta-, hexa-, and heptavalent 3*d* (a), 4*d* (b) and 5*d* (c) ions.

Hexavalent 5*d* ions:

$$A = (-26806.12 \pm 274.69) + (18508.86 \pm 29.01)\zeta^{1/4}$$

$$B = (-131.36 \pm 1.51) + (124.84 \pm 0.16)\zeta^{1/4} \quad (5.25)$$

$$C = (-543.06 \pm 6.10) + (523.45 \pm 0.64)\zeta^{1/4}$$

Heptavalent 5*d* ions:

$$A = (-27511.32 \pm 195.00) + (18679.20 \pm 19.80)\zeta^{1/4}$$

$$B = (-132.87 \pm 1.01) + (125.99 \pm 0.10)\zeta^{1/4} \quad (5.26)$$

$$C = (-515.22 \pm 4.25) + (527.51 \pm 0.43)\zeta^{1/4}$$

5.4 Energy Level Diagrams for Free 3*d* Ions

The calculated Racah parameters *B* and *C* can be used further to estimate the positions of the free ions *LS* terms. It can be easily done for the 3*d* ions. For the 4*d* and 5*d* ions, due to the increased atomic numbers and enhanced SO coupling, the fine structure of the energy levels should be taken into account and we shall not consider the corresponding energy level schemes. On the other hand, the 3*d* ions are much more often used for optical applications; therefore, the emphasis placed on this group of ions is fully justified.

The energy level schemes of free 3*d* ions shown below give an overall idea of the sequence of the free ion *LS* terms, their relative locations and energy intervals between them. They also can serve as a starting point to approach the Tanabe-Sugano diagram describing the energy level splitting of the *d* ions in a crystal field (these topics will be discussed in Chapter 8).

The ions, whose energy levels are shown, are grouped in such a way, that each diagram contains the energy levels of the d^N and d^{10-N} configurations, which have the same *LS* terms. In all diagrams the ground term energy is taken as zero.

As can be seen from Figs. 5.14–5.17, the energy intervals between the ground state and the first excited term are increasing with the atomic number, if we consider identical electron configurations. This is a consequence of an increase of the Racah parameters with the atomic number. Due to the same reason, the energy range spanned by all terms of the particular electron configuration is widened if the atomic number of an ion with the same configuration is increased. Density of the energy levels is increased for the $3d^4$, $3d^5$ and $3d^6$ ions, which means that if the SO coupling would be taken into account, many states with the same total angular momentum J will be strongly mixed, which may make energy level assignment somewhat ambiguous. This is especially true in the case of the $4f/5f$ ions (Chapter 6).

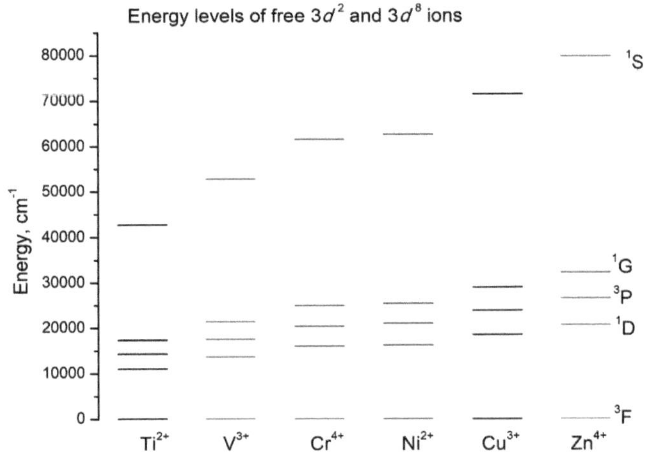

Figure 5.14 Energy level diagram for the $3d^2$ and $3d^8$ ions.

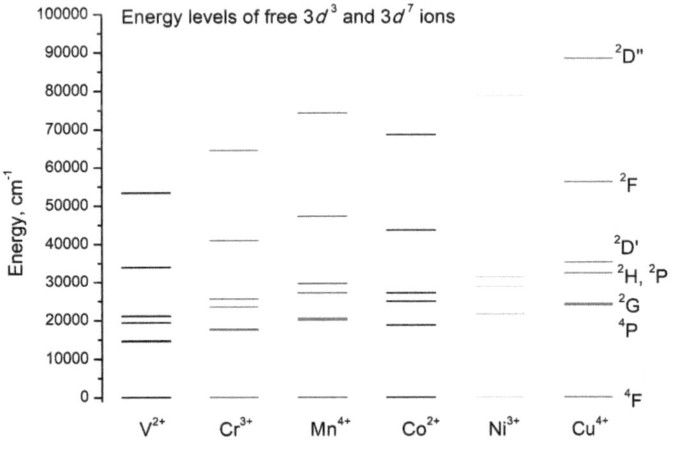

Figure 5.15 Energy level diagram for the $3d^3$ and $3d^7$ ions.

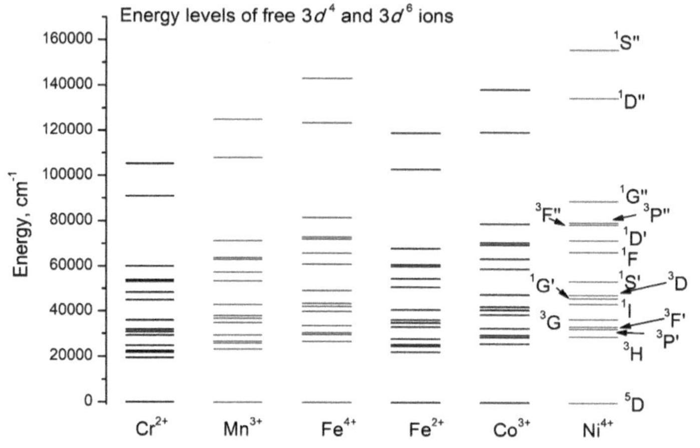

Figure 5.16 Energy level diagram for the $3d^4$ and $3d^6$ ions.

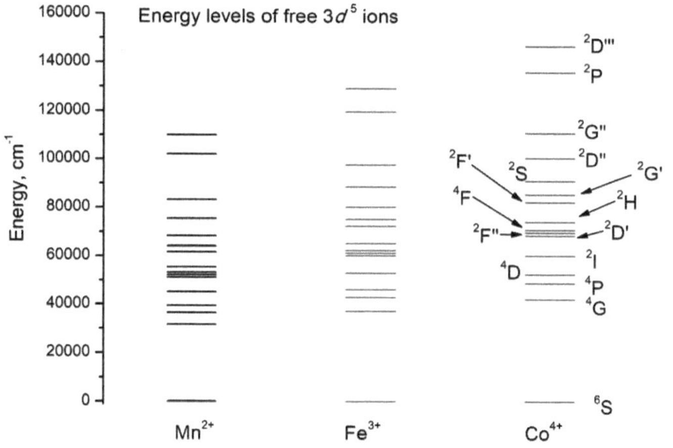

Figure 5.17 Energy level diagram for the $3d^5$ ions.

5.5 Ionization Energies for Free 4*d*, 5*d* Ions

In addition to the spectroscopic quantities, in this section we give the results of calculations of the ionization energies for the free 4*d*, 5*d* ions. The ionization energy is equal to the energy, which is required to remove one electron from a neutral atom or a charged ion. Thus, the first ionization energy is defined as an amount of energy needed to form a singly positively charged ion M^+ from a neutral atom M^0, the second ionization energy is determined as the energy of formation of a doubly positively charged ion M^{2+} from an M^+ ion, etc. Therefore, starting from the divalent ions, we can speak of the third, fourth, and fifth ionization energies for the di-, tri- and tetravalent ions, respectively. The calculated values of these energies are plotted in Fig. 5.18 against the atomic number Z. The ionization energy appears to be a linear function of Z. Its slope increases with

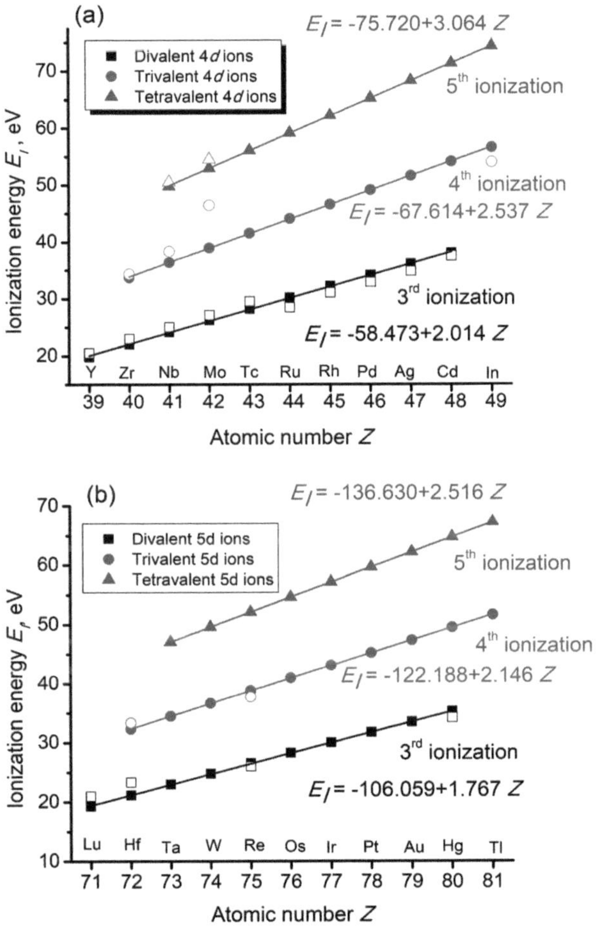

Figure 5.18 Ionization energies for the 4*d* (a) and 5*d* (b) di-, tri- and tetravalent ions. The calculated and experimental (if available) values are shown by the filled and empty symbols, respectively. The experimental values are taken from http://www.webelements.com.

the oxidation state, i.e. with moving to the ionization energies of higher order. It also decreases when moving from the 4*d* to 5*d* ions (comparing the ionization energies of the same order). For some of the ions considered in the present paper, the ionization energies are collected at the well-known website http://www.webelements.com. However, very few experimental data are available on the ionization energies for the 5*d* ions, and high ionization energies for the 4*d* ions. Therefore, our calculated results can serve as first estimations of those characteristic ionization energies; validity of these estimations can be justified by good agreement between the calculated and reference data for the third ionization energy for the 4*d* ions.

5.6 Average Values of the *d* Electrons Radial Coordinate

The non-energetical quantities calculated in the present work are the moments of the electron density of the *d*-electrons for all considered ions. They were found in a standard and a straightforward way

$$\left\langle r^k \right\rangle = \int_0^\infty r^{k+2} \left|R_{nd}(r)\right|^2 dr, \qquad (5.27)$$

where $k = 2, 4$; $R_{nd}(r)$ is the corresponding radial wave function with n being a principal quantum number (3, 4 or 5), and d corresponds to the orbital quantum number $l = 2$ for the *d*-electrons. The data given in Table 5.2 show systematic decrease of $\langle r^k \rangle$ with increase of atomic number and oxidation state. The $\langle r^k \rangle$ values can be used for calculations of the parameters of crystal field acting upon the 3*d*, 4*d* and 5*d* ions in solids. Considering the decreasing dependence of $\langle r^k \rangle$ related to the crystal-field parameters, one can easily conclude that the crystal field effects on the *d* electrons must decrease across the considered *nd* series.

Bibliography

1. Yersin H (ed) (2004). *Transitional Metal and Rare Earth Compounds III*, Springer–Verlag, Berlin, Heidelberg, New York.

2. Svelto O, Longhi S, Della Valle G et al. (2012). Lasers and coherent light sources, in *Springer Handbook of Lasers and Optics*. Träger F, ed., Springer-Verlag, Berlin, Heidelberg, pp. 641–1170.
3. Henderson B, Bartram RH (2000). *Crystal-Field Engineering of Solid-State Laser Materials*. Cambridge University Press.
4. Kück S (2001). Laser-related spectroscopy of ion-doped crystals for tunable solid-state lasers, *Appl Phys B*, **72**(5), 515–562.
5. Mironova NA, Ulmanis UA (1988). *Radiation Defects and Ions of the Iron Group in Oxides* (in Russian). Riga, Zinatne.
6. Avram NM, Brik MG (eds) (2013). *Optical Properties of 3d-Ions in Crystals: Spectroscopy and Crystal Field Analysis*. Springer and Tsinghua University Press.
7. Guo H, Wang XF, Chen JD, Li F (2010). Ultraviolet light induced white light emission in Ag and Eu^{3+} co-doped oxyfluoride glasses. *Opt Express* **18**(18), 18900–18905.
8. Zheng WC, Fang F, Mei Y (2007). Optical spectra and spin Hamiltonian parameters for rhombic Zr^{3+} in $Y_3Al_5O_{12}$. *J Appl Phys* **101**(5), 053911.
9. Zheng WC, Wang F, Qu GQ (2007). A study of spin-Hamiltonian parameters for Hf^{3+} ions in orthophosphate MPO_4 (M = Sc, Lu, Y) crystals, *Solid State Commun* **144**(5–6), 202–205.
10. Brik MG, Srivastava AM (2013). Systematic analysis of the spectroscopic characteristics of 3*d* ions in a free state and some cubic crystals, *Opt Mater* **35**(10), 1776–1782.
11. Ma CG, Brik MG (2014). Systematic analysis of spectroscopic characteristics of heavy transition metal ions with $4d^N$ and $5d^N$ (N = 1... 10) electronic configurations in a free state, *J Lumin* **145** (2014) 402–409.
12. Morrison CA (1992), *Crystal Fields for Transition-Metal Ions in Laser Host Materials*. Springer-Verlag.
13. Trees RE (1964). $4f^3$ and $4f^25d$ configurations of doubly ionized praseodymium (Pr iii). *J Opt Soc Am* **54**(5), (1964) 651–657.
14. Cowan RD (1981). *The Theory of Atomic Structure and Spectra*. University of California Press, Berkeley.
15. Sugano S, Tanabe Y, Kamimura H (1970). *Multiplets of Transition-Metal Ions in Crystals*. New York and London, Academic Press.

Chapter 6

Basic Spectroscopic Properties of the Ions with Unfilled *f* Electron Shell*

6.1 Introduction

The spectra of the lanthanide (especially) and actinide ions with the unfilled 4*f* and 5*f* electron shells, respectively, in the free state will be analyzed in detail in this chapter. The interest to these ions is driven by their numerous optical applications in lighting, lasing, thermometry, etc., and the knowledge and understanding of the free ions energy level schemes is a necessary step to understand their spectra in crystals, glasses, and crystalline powders (phosphor materials). This chapter will be also a good practical application of the basic results from the free ion's spectra theory.

*Some parts of this chapter are reproduced with permission from: Ma C-G, Brik MG, Liu D-X, Feng B, Tian Y, Suchocki A (2016). Energy level schemes of f^N electronic configurations for the di-, tri-, and tetravalent lanthanides and actinides in a free state, *J Lumin* **170** (2016), 369–374; Ma CG, Brik MG, Li QX, Tian Y (2014) Systematic analysis of spectroscopic characteristics of the lanthanide and actinide ions with the $4f^N$ and $5f^N$ (N = 1...14) electronic configurations in a free state. *J Alloys Compds* **599** 93–101; Ma CG, Brik MG, Tian Y, Li QX (2014). Systematic analysis of spectroscopic characteristics of the lanthanide and actinide ions with the $4f^{N-1}5d$ and $5f^{N-1}6d$ electronic configurations in a free state. *J Alloys Compds* **603** 255–267.

Theoretical Spectroscopy of Transition Metal and Rare Earth Ions: From Free State to Crystal Field
Mikhail G. Brik and Chong-Geng Ma

ISBN 978-981-4800-56-3 (Hardcover), 978-0-429-27875-4 (eBook)
www.jennystanford.com

6.2 Possible Electron Configurations of the 4*f*/5*f* Ions

In this chapter, we shall consider two particular groups of chemical elements in the periodic table, which share a certain similarity in their electronic and optical properties. These are the so-called the "lanthanides" and "actinides," which have the 4*f* and 5*f* unfilled electron shells, respectively. These names are derived from the names of the first elements opening the considered groups: lanthanum ($Z = 57$) and actinium ($Z = 89$). Their *f*-electron shells are gradually built up by adding one additional electron when moving across the considered series until the *f* shell becomes completed with fourteen electrons. It has also to be mentioned that the unfilled 4*f* (lanthanides) and 5*f* (actinides) shells are screened by the completely filled $5s^25p^6$ and $6s^26p^6$ shells, respectively. This circumstance explains a strongly decreased influence of environment of these ions on their inner *f* shells; it also has very important consequences, resulting in very sharp absorption and emission lines corresponding to the intrashell *f*-*f* electronic transitions, even when these ions are placed into crystals.

The lanthanide ions—also known as the rare earth ions—are widely used in various optical applications, such as laser materials, infrared to visible up-conversion materials, phosphors, etc.[1-7] Recently, the lanthanide ions have also entered the area of biomedical applications.[8] The actinide ions, although somewhat resembling lanthanides from the point of view of electronic structure, stand apart due to their radioactivity, which determines their specific areas of application in nuclear energetic and military spheres.[9] For example, the photophysics and photochemistry of the actinide ions were developed to address the safety and security issues arising from nuclear fission.[10]

All spectroscopic characteristics of these ions come from the properties of their electron configurations. In Table 6.1 the number of the *J*-manifolds (majority of them are highly degenerated) and the number of states for each considered configuration are given. Those energetic *J*-manifolds arise from the Coulomb interaction between the *f* electrons and the *f* electron's SO coupling. Since almost all those levels are highly degenerated, they can be split up by a crystal field, to form rich energy level schemes with numerous sharp absorption/emission lines in a wide spectral region, from infrared to ultraviolet, depending on the considered ion. Table 6.2 lists the 4*f* and 5*f* ions in the oxidation states +2, +3 and +4, which correspond to these electron configurations.

Table 6.1 Electronic configurations and the number of possible states for the f^N configurations

Electron configuration	Number of states (including possible orientation of spin)	Number of *J* manifolds (number of energy levels of free ions)
f^1	14	2
f^2	91	13
f^3	364	41
f^4	1001	107
f^5	2002	198
f^6	3003	295
f^7	3432	327
f^8	3003	295
f^9	2002	198
f^{10}	1001	107
f^{11}	364	41
f^{12}	91	13
f^{13}	14	2
f^{14}	1	1

The semi-empirical effective Hamiltonian model operating within the f^N electron configurations has been well developed to understand the electronic energy level schemes observed in the *f*-*f* transition spectra of both lanthanide and actinide ions in solids.[11,12] In this model, effective interactions within the f^N configuration can be expressed as the product of the radial fitting parameters (depending on the *f*-electrons radial coordinates only) and the angular operators (acting on the *f*-electrons angular coordinates only). The matrix elements of all operators entering the free ion Hamiltonian acting between the basis wave functions $|f^N \eta SLJM>$ can be derived from the Racah–Wigner algebra. Here *S*, *L*, *J*, *M* stand for the quantum numbers of the spin, orbital momentum, total momentum and its projection, respectively; η denotes all other quantum numbers needed to distinguish between the states with identical *S*, *L*, *J*, *M* sets.

Table 6.2 Atomic numbers and electron configurations for the considered di-, tri-, and tetravalent 4*f* and 5*f* ions

Atomic number *Z*	Symbol	Electron configuration $4f^N$		
		Divalent	**Trivalent**	**Tetravalent**
57	La	$4f^1$	—	—
58	Ce	$4f^2$	$4f^1$	—
59	Pr	$4f^3$	$4f^2$	$4f^1$
60	Nd	$4f^4$	$4f^3$	$4f^2$
61	Pm	$4f^5$	$4f^4$	$4f^3$
62	Sm	$4f^6$	$4f^5$	$4f^4$
63	Eu	$4f^7$	$4f^6$	$4f^5$
64	Gd	$4f^8$ $(4f^7 5d)$	$4f^7$	$4f^6$
65	Tb	$4f^9$	$4f^8$	$4f^7$
66	Dy	$4f^{10}$	$4f^9$	$4f^8$
67	Ho	$4f^{11}$	$4f^{10}$	$4f^9$
68	Er	$4f^{12}$	$4f^{11}$	$4f^{10}$
69	Tm	$4f^{13}$	$4f^{12}$	$4f^{11}$
70	Yb	$4f^{14}$	$4f^{13}$	$4f^{12}$
71	Lu	—	$4f^{14}$	$4f^{13}$
		Electron configuration $5f^N$		
		Divalent	**Trivalent**	**Tetravalent**
89	Ac	$5f^1$ $(6d^1)$	—	—
90	Th	$5f^2$ $(6d^2)$	$5f^1$	—
91	Pa	$5f^3$ $(5f^2 6d)$	$5f^2$	$5f^1$
92	U	$5f^4$ $(5f^3 6d)$	$5f^3$	$5f^2$
93	Np	$5f^5$ $(5f^4 6d)$	$5f^4$	$5f^3$
94	Pu	$5f^6$	$5f^5$	$5f^4$
95	Am	$5f^7$	$5f^6$	$5f^5$
96	Cm	$5f^8$ $(5f^7 6d)$	$5f^7$	$5f^6$

97	Bk	$5f^9$	$5f^8$	$5f^7$
98	Cf	$5f^{10}$	$5f^9$	$5f^8$
99	Es	$5f^{11}$	$5f^{10}$	$5f^9$
100	Fm	$5f^{12}$	$5f^{11}$	$5f^{10}$
101	Md	$5f^{13}$	$5f^{12}$	$5f^{11}$
102	No	$5f^{14}$	$5f^{13}$	$5f^{12}$
103	Lr	—	$5f^{14}$	$5f^{13}$

Note: The ground state configurations (if different) are given in the parentheses.

The values of the radial fitting parameters can be obtained by minimizing the root-mean-square (rms) deviation between the theoretically calculated and experimentally observed energy levels. Such theoretical method has been systematically implemented for successful interpretation of spectra of trivalent lanthanide and actinide ions in $LaCl_3$ and LaF_3 by Carnall and his coworkers, and the trend of the obtained fitting parameter values across the whole *f*-element series was analyzed and discussed (see, for example, the review article.[13])

Later, those discovered trends were applied to other spectroscopic studies of trivalent lanthanide ions in solids to reduce the number of the fitting energy parameters.[14]

A similar trend of the energy parameters of trivalent ions in Cs_2NaYCl_6 across the whole lanthanide series[15] was also demonstrated. It is noticeable that in those above-mentioned publications the main studied objects are the trivalent lanthanide and actinide ions in crystals. However, some other oxidation states of these ions can be encountered as well. The divalent lanthanide ions (first of all, Eu^{2+}) can be stabilized in some materials and they are of paramount importance for persistent phosphors. A comparative study of the spectroscopic properties of four isoelectronic ions with the $4f^2$ configuration (La^+, Ce^{2+}, Pr^{3+}, Nd^{4+})[16] was published recently.

In the previous chapter, we presented a detailed study of the spectroscopic properties of the ions with the unfilled 3*d* and 4*d*/5*d* shells across the whole groups depending on the atomic number and/or oxidation state.[17,18] It was shown that the Racah parameters *A*, *B* and *C* and SO constant ζ_d for those 3*d*, 4*d*, and 5*d* ions are all related to the atomic number *Z*; these linear trends reduce the number of independent parameters needed for a proper description of the free ions spectra.

This chapter is in a certain sense a continuation and a further extension of the previous one. It concludes the studies of the spectroscopic properties of the free ions with the unfilled electron shells by the systematic analysis of the main energetic parameters of the 4*f* and 5*f* ions.[19,20] The doubly, triply, and quadruply charged lanthanide and actinide ions in a free state are considered. The averaged values of the second, fourth and sixth power of the 4*f* or 5*f* electrons radial coordinate $<r^2>$, $<r^4>$, $<r^6>$ are reported, because these quantities are necessary to calculate the crystal field parameters. The barycenter energies E_{avg} of all considered f^N electron configurations and the ionization energies E_I were calculated as well.

6.3 Free di-, Tri-, and Tetravalent 4*f*/5*f* Ions: The Hamiltonian Structure

The parameterized free-ion Hamiltonian H_{FI} of a f^N (N = 1...13) electron configuration can be written as 1[11,12]:

$$\mathbf{H}_{FI}(f^N) = E_{avg} + \sum_{k=2,4,6} F^k \mathbf{f_k} + \varsigma_f \sum_{i=1}^{N} \mathbf{l}_i \cdot \mathbf{s}_i + \alpha \mathbf{L}(\mathbf{L}+\mathbf{1}) + \beta \mathbf{G}(\mathbf{G_2})$$
$$+ \gamma \mathbf{G}(\mathbf{R_7}) + \sum_{i=2,3,4,6,7,8} T^i \mathbf{t_i} + \sum_{h=0,2,4} M^h \mathbf{m_h} + \sum_{k=2,4,6} P^k \mathbf{p_k}, \qquad (6.1)$$

where the italic and bold letters denote the radial fitting parameters and the angular operators, respectively. The first term E_{avg} is the f^N configuration barycenter energy, which does not cause any energy splitting. The second term is the Coulomb interaction between the *f* electrons, where F^k (k = 2,4 and 6) are the Slater parameters. The $\mathbf{f_k}$ operators can be calculated for a given electron configuration with N electrons[21]:

$$\mathbf{f_k} = \frac{1}{2} \sum_{i,j=1}^{N} \frac{4\pi}{2k+1} \sum_{q=-k}^{k} Y_{kq}^{*}(\theta_i, \varphi_i) Y_{kq}(\theta_j, \varphi_j), \qquad (6.2)$$

with Y_{kq} being the spherical functions of the *f*-electrons' coordinates θ, φ with $i \neq j$ for obvious reasons, since this is the Coulomb interaction between a pair of *f* electrons. The third term stands for the *f* electron's SO interaction, where ζ_f is the SO coupling constant and $\mathbf{l}_i \cdot \mathbf{s}_i$ is the SO interaction operator with the first and second terms in the scalar product being the operators of the angular and spin momenta of an individual electron with index *i*. The second and third terms dominate in the Hamiltonian (6.1) and to a large extent determine the order and sequence of the energy levels. The next four terms are the two- and three-electron Coulomb correlation contributions from the higher configurations with the same parity. The last two terms represent the magnetically and electrostatically correlated interactions. All these six last terms only produce very small corrections to the energy levels determined by the second and third terms of Hamiltonian (6.1).

We restrict our analysis to the second and third terms of the free ion Hamiltonian (6.1), since they are most important in the formation of the overall energy level scheme of a particular configuration. Besides, the calculations of those additional correction terms are very complicated. As a rule, the remaining parameters of Hamiltonian (6.1) are treated as the fitting parameters, which can be determined numerically from the procedure of non-linear fitting of the calculated energy level schemes to the experimental ones.

The studied ions were grouped as the di-, tri-, and tetravalent ions, whose atomic numbers and electron configurations are listed in Table 6.2. As seen from the table, increase of the atomic number by unity with keeping the same electric charge of an ion leads to appearance of one additional *f*-electron in the unfilled shell.

For some divalent ions, due to a particular order in the electron shells filling, the ground state configuration is not of pure *f* character but contains one 5*d* (Gd^{2+}) or 6*d* electron (Ac^{2+}, Th^{2+}, Pa^{2+}, U^{2+}, Np^{2+}, Cm^{2+}). However, the difference between the ground states of these "mixed *f*–*d*" and "pure *f*" configurations is not large. For example, for the Gd^{2+} ions such energy difference is about 2400 cm^{-1}.[22] So, being based on this fact, on one hand, and for the sake of consistency, on the other hand, we consider only pure *f* configurations for all ions studied in the present chapter.

6.4 Variation of the Slater Integrals and Spin–Orbit Coupling Constant with Atomic Number for the f^N Electronic Configurations of Free 4f/5f Ions

The numerical values of the Slater parameters F^2, F^4, F^6 and SO interaction constant ζ_f were calculated by us using the suite of the above-mentioned Cowan programs with relativistic effects taken into account. Similar to the previous chapter, we did not use any scaling coefficient in our analysis. All calculated parameters for the considered ions are collected in Table 6.3.

Figures 6.1 and 6.2 show the variation of the F^2, F^4, F^6 parameters against the atomic number Z in the considered groups of ions. All three parameters excellently follow the linear trend of the form $F^k(k = 2,4,6) = a_1+a_2Z$. The values of the a_1, a_2 coefficients obtained after a linear fit of the calculated data points are all collected in Table 6.4. It can be noticed that the absolute values of the a_1, a_2 coefficients all decrease in the "$F^2 \rightarrow F^4 \rightarrow F^6$" and "divalent-trivalent-tetravalent ions" directions for both 4f and 5f groups. The strongest dependence of the Slater parameters on Z in each of the di-, tri-, and tetravalent ions' groups is realized for the F^2 parameter; the weakest—for the F^6 parameter. If the 4f and 5f ions are compared, then the Slater parameters also behave in a similar way: The a_2 coefficient is the greatest for the F^2 parameter, and the difference between the a_2 coefficients for the 4f and 5f ions decreases when moving in the "$F^2 \rightarrow F^4 \rightarrow F^6$" direction.

Moreover, it is possible to reduce further the number of the independent electrostatic parameters by considering the so-called hydrogenic ratios F^4/F^2 and F^6/F^2 [23] (assuming that the 4f or the 5f radial wave functions are the hydrogen-like ones).

Figure 6.3 shows these calculated ratios for all considered series of isovalent ions. The F^4/F^2 and F^6/F^2 ratios are nearly constant across the considered groups, very slightly depending on the atomic number. In Fig. 6.3, we give the averaged values of those hydrogenic ratios for each group. Thus, for the di-, tri-, and tetravalent 4f ions $\langle F^4/F^2 \rangle = 0.622, 0.626, 0.629$; $\langle F^6/F^2 \rangle = 0.446, 0.450, 0.453$, correspondingly. For the di-, tri-, and tetravalent 5f elements, we obtained the following values: $\langle F^4/F^2 \rangle$ = 0.646, 0.652, 0.656; $\langle F^6/F^2 \rangle$ = 0.471, 0.478, and 0.482, correspondingly.

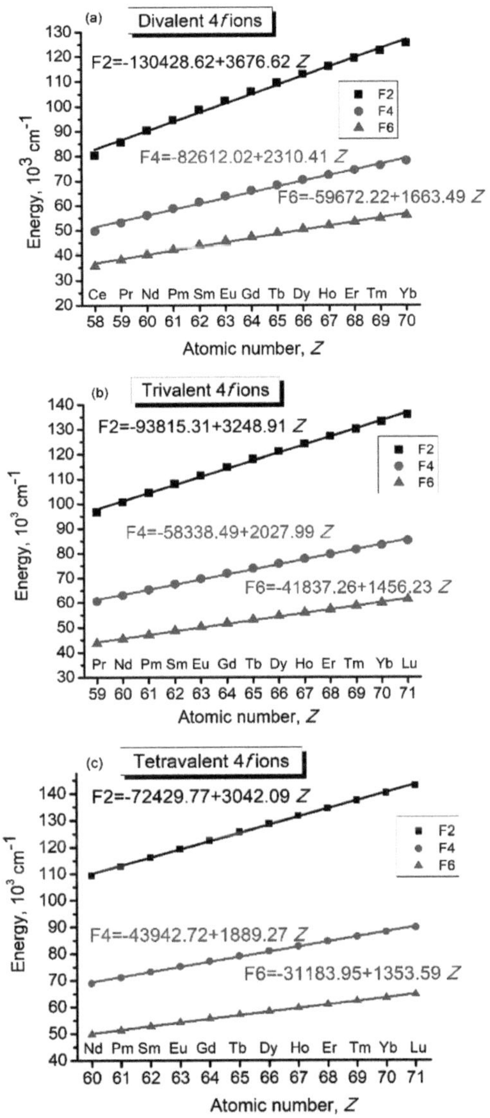

Figure 6.1 Dependence of the F^2, F^4, F^6 parameters on the atomic number Z for the di- (a), tri- (b), and tetravalent (c) $4f$ ions.

Table 6.3 Calculated energy parameters (in cm^{-1}), radial integrals (in $Å^k$) and first-ionization energies E_I (in eV) of the f^N configurations for the considered di-, tri-, and tetravalent $4f$ and $5f$ ions

Ion	La^{2+}	Ce^{2+}	Pr^{2+}	Nd^{2+}	Pm^{2+}	Sm^{2+}	Eu^{2+}	Gd^{2+}	Tb^{2+}	Dy^{2+}	Ho^{2+}	Er^{2+}	Tm^{2+}	Yb^{2+}	
$4f^N$	1	2	3	4	5	6	7	8	9	10	11	12	13	14	
F^2	—	80374	85636	90328	94615	98622	102408	106014	109486	112834	116084	119256	122343	125372	
F^4	—	49798	53142	56116	58825	61352	63734	65999	68176	70271	72303	74283	76208	78095	
F^6	—	35647	38065	40212	42167	43989	45705	47335	48901	50408	51867	53290	54672	56026	
ζ_f	470	580	698	825	960	1106	1264	1433	1615	1810	2019	2244	2484	2740	
r^2	0.5704	0.5041	0.4469	0.4038	0.3699	0.3419	0.3184	0.2984	0.2809	0.2653	0.2516	0.2393	0.2282	0.2181	
r^4	0.8618	0.6821	0.5366	0.4398	0.3711	0.3193	0.2792	0.2473	0.2212	0.1997	0.1815	0.1661	0.1529	0.1415	
r^6	2.6921	1.9440	1.3713	1.0301	0.8085	0.6548	0.5433	0.4598	0.3953	0.3445	0.3037	0.2703	0.2431	0.2200	
E_I	19.558	20.468	21.733	22.794	23.694	24.473	25.148	25.731	26.241	26.678	27.055	27.380	27.646	27.869	
Ion	La^{3+}	Ce^{3+}	Pr^{3+}	Nd^{3+}	Pm^{3+}	Sm^{3+}	Eu^{3+}	Gd^{3+}	Tb^{3+}	Dy^{3+}	Ho^{3+}	Er^{3+}	Tm^{3+}	Yb^{3+}	Lu^{3+}
$4f^N$	0	1	2	3	4	5	6	7	8	9	10	11	12	13	14
F^2	—	—	96681	100645	104389	107971	111416	114742	117981	121132	124214	127240	130201	133119	135996
F^4	—	—	60533	63030	65383	67630	69786	71865	73886	75850	77768	79650	81489	83300	85085

F^6	—	—	43509	45309	47003	48619	50169	51662	53113	54523	55899	57248	58566	59864	61142
ζ_f	—	689	808	937	1075	1225	1387	1561	1749	1950	2165	2396	2643	2906	3187
r^2	—	0.3724	0.3472	0.3215	0.2999	0.2812	0.2648	0.2504	0.2374	0.2258	0.2153	0.2057	0.1969	0.1888	0.1814
r^4	—	0.3156	0.2785	0.2409	0.2113	0.1874	0.1678	0.1514	0.1375	0.1256	0.1154	0.1064	0.0986	0.0916	0.0855
r^6	—	0.5070	0.4300	0.3505	0.2918	0.2473	0.2123	0.1846	0.1620	0.1435	0.1281	0.1151	0.1042	0.0948	0.0866
E_I	—	38.510	39.164	40.423	41.534	42.530	43.421	44.219	44.941	45.584	46.163	46.685	47.140	47.548	47.910

Ion	Ce^{4+}	Pr^{4+}	Nd^{4+}	Pm^{4+}	Sm^{4+}	Eu^{4+}	Gd^{4+}	Tb^{4+}	Dy^{4+}	Ho^{4+}	Er^{4+}	Tm^{4+}	Yb^{4+}	Lu^{4+}
$4f^N$	0	1	2	3	4	5	6	7	8	9	10	11	12	13
F^2	—	—	109481	112901	116217	119442	122583	125662	128677	131639	134560	137430	140266	143071
F^4	—	—	69009	71147	73217	75226	77180	79093	80965	82801	84611	86387	88142	89875
F^6	—	—	49736	51271	52757	54198	55599	56970	58311	59626	60921	62192	63447	64687
ζ_f	—	913	1047	1190	1345	1512	1692	1885	2093	2315	2553	2806	3077	3365
r^2	—	0.2876	0.2726	0.2568	0.2427	0.2301	0.2188	0.2085	0.1992	0.1906	0.1827	0.1754	0.1686	0.1623
r^4	—	0.1746	0.1590	0.1424	0.1284	0.1165	0.1063	0.0974	0.0897	0.0829	0.0769	0.0715	0.0668	0.0625
r^6	—	0.1916	0.1697	0.1456	0.1262	0.1105	0.0975	0.0867	0.0776	0.0699	0.0632	0.0575	0.0525	0.0481
E_I	—	59.890	60.476	61.788	62.993	64.098	65.110	66.046	66.902	67.690	68.419	69.078	69.685	70.243

(Continued)

Table 6.3 (*Continued*)

Ion	Ac^{2+}	Th^{2+}	Pa^{2+}	U^{2+}	Np^{2+}	Pu^{2+}	Am^{2+}	Cm^{2+}	Bk^{2+}	Cf^{2+}	Es^{2+}	Fm^{2+}	Md^{2+}	No^{2+}	
$5f^N$	1	2	3	4	5	6	7	8	9	10	11	12	13	14	
F^2	—	53181	58224	62604	66534	70161	73533	76733	79767	82689	85493	88219	90855	93433	
F^4	—	33835	37220	40164	42805	45240	47503	49648	51679	53633	55505	57324	59080	60796	
F^6	—	24509	27025	29217	31184	32998	34685	36283	37796	39251	40646	41999	43306	44583	
ζ_f	818	1087	1356	1631	1914	2209	2516	2838	3175	3530	3906	4290	4704	5133	
r^2	1.4234	1.1504	0.9669	0.8404	0.7466	0.6732	0.6138	0.5645	0.5228	0.4870	0.4559	0.4284	0.4041	0.3822	
r^4	4.5026	2.9579	2.0678	1.5511	1.2170	0.9849	0.8171	0.6893	0.5906	0.5119	0.4484	0.3960	0.3526	0.3157	
r^6	26.2187	14.3456	8.4146	5.4809	3.8230	2.7931	2.1168	1.6469	1.3118	1.0628	0.8753	0.7299	0.6162	0.5248	
E_I	14.279	15.621	17.174	18.558	19.816	20.988	22.079	23.116	24.091	25.027	25.913	26.767	27.576	28.359	
Ion	Ac^{3+}	Th^{3+}	Pa^{3+}	U^{3+}	Np^{3+}	Pu^{3+}	Am^{3+}	Cm^{3+}	Bk^{3+}	Cf^{3+}	Es^{3+}	Fm^{3+}	Md^{3+}	No^{3+}	Lr^{3+}
$5f^N$	0	1	2	3	4	5	6	7	8	9	10	11	12	13	14
F^2	—	—	65387	69091	72526	75769	78840	81792	84622	87370	90028	92626	95153	97635	100058
F^4	—	—	42303	44790	47094	49268	51324	53298	55187	57021	58792	60521	62200	63848	65456

F^6	—	—	30883	32737	34455	36076	37607	39078	40486	41851	43169	44456	45705	46931	48126
ζ_f	—	1329	1588	1860	2143	2440	2751	3078	3420	3780	4162	4553	4975	5411	5868
r^2	—	0.8614	0.7690	0.6903	0.6275	0.5755	0.5321	0.4945	0.4623	0.4338	0.4088	0.3862	0.3660	0.3478	0.3310
r^4	—	1.4303	1.1496	0.9261	0.7657	0.6447	0.5515	0.4773	0.4176	0.3684	0.3276	0.2930	0.2638	0.2386	0.2169
r^6	—	4.0689	2.9820	2.1695	1.6401	1.2745	1.0141	0.8210	0.6759	0.5630	0.4748	0.4040	0.3472	0.3004	0.2618
E_I	—	28.326	29.661	31.419	33.044	34.575	36.013	37.389	38.691	39.947	41.142	42.300	43.404	44.476	45.501

Ion	**Th^{4+}**	**Pa^{4+}**	**U^{4+}**	**Np^{4+}**	**Pu^{4+}**	**Am^{4+}**	**Cm^{4+}**	**Bk^{4+}**	**Cf^{4+}**	**Es^{4+}**	**Fm^{4+}**	**Md^{4+}**	**No^{4+}**	**Lr^{4+}**
$5f^N$	0	1	2	3	4	5	6	7	8	9	10	11	12	13
F^2	—	—	74359	77494	80497	83371	86158	88849	91478	94033	96541	98990	101402	103764
F^4	—	—	48597	50695	52702	54621	56480	58272	60022	61720	63386	65010	66609	68173
F^6	—	—	35658	37222	38718	40148	41533	42867	44170	45433	46672	47879	49068	50229
ζ_f	—	1812	2083	2369	2669	2985	3317	3665	4031	4421	4819	5249	5693	6158
r^2	—	0.6514	0.5995	0.5519	0.5116	0.4769	0.4464	0.4198	0.3960	0.3747	0.3554	0.3380	0.3220	0.3075
r^4	—	0.7625	0.6508	0.5532	0.4764	0.4150	0.3647	0.3232	0.2883	0.2589	0.2335	0.2117	0.1928	0.1763
r^6	—	1.4563	1.1640	0.9183	0.7385	0.6041	0.5007	0.4203	0.3562	0.3048	0.2626	0.2282	0.1994	0.1754
E_I	—	44.371	45.817	47.775	49.639	51.402	53.101	54.719	56.286	57.786	59.244	60.642	62.004	63.313

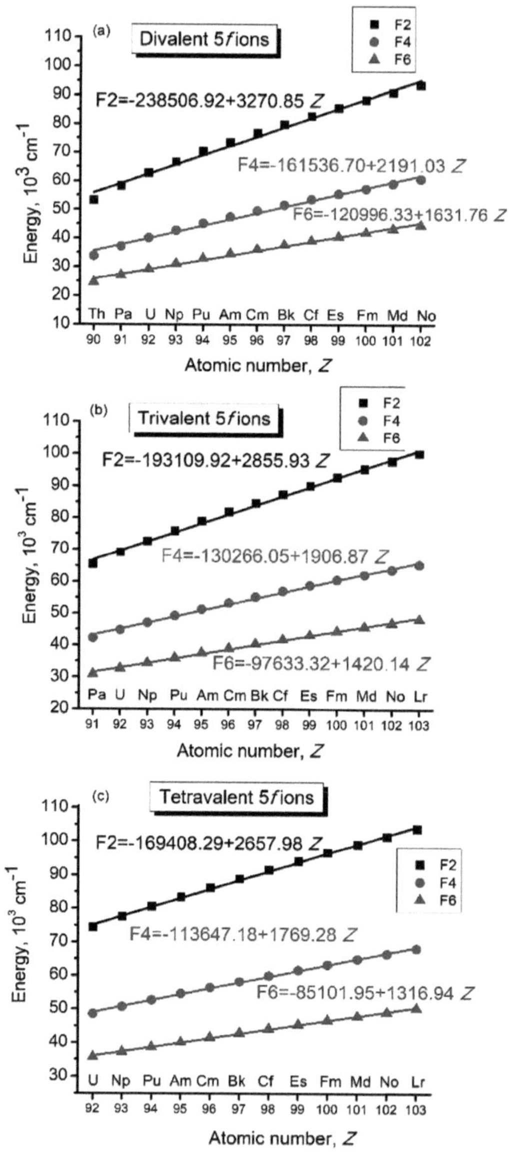

Figure 6.2 Dependence of the F^2, F^4, F^6 parameters on the atomic number Z for the di- (a), tri- (b), and tetravalent (c) 5*f* ions.

Table 6.4 Linear approximations $a_1 + a_2Z$ of the Slater parameters F^2, F^4, and F^6 as functions of the atomic number Z for the di-, tri-, and tetravalent $4f$ and $5f$ ions

		$4f$ ions		$5f$ ions	
		a_1, cm^{-1}	a_2, cm^{-1}/at.number	a_1, cm^{-1}	a_2, cm^{-1}/ at.number
Divalent ions	F^2	−130428.62 ± 3676.62	3676.62 ± 89.23	−238506.92 ± 9350.29	3270.85 ± 97.32
	F^4	−82612.02 ± 3737.27	2310.41 ± 58.30	−161536.70 ± 6360.04	2191.03 ± 66.20
	F^6	−59672.22 ± 2728.48	1663.49 ± 42.56	−120996.33 ± 4734.37	1631.76 ± 49.28
Trivalent ions	F^2	−93815.31 ± 3248.91	3248.91 ± 46.12	−193109.92 ± 5193.31	2855.93 ± 53.50
	F^4	−58338.49 ± 1969.81	2027.99 ± 30.25	−130266.05 ± 3575.50	1906.87 ± 36.83
	F^6	−41837.26 ± 1440.70	1456.23 ± 22.13	−97633.32 ± 2681.56	1420.14 ± 27.62
Tetravalent ions	F^2	−72429.77 ± 1900.76	3042.09 ± 28.98	−169408.29 ± 3488.87	2667.98 ± 35.76
	F^4	−43942.72 ± 1248.61	1889.27 ± 19.04	−113647.18 ± 2412.12	1769.28 ± 24.72
	F^6	−31183.951 ± 912.90	1353.59 ± 13.92	−85101.95 ± 1819.53	1316.94 ± 18.65

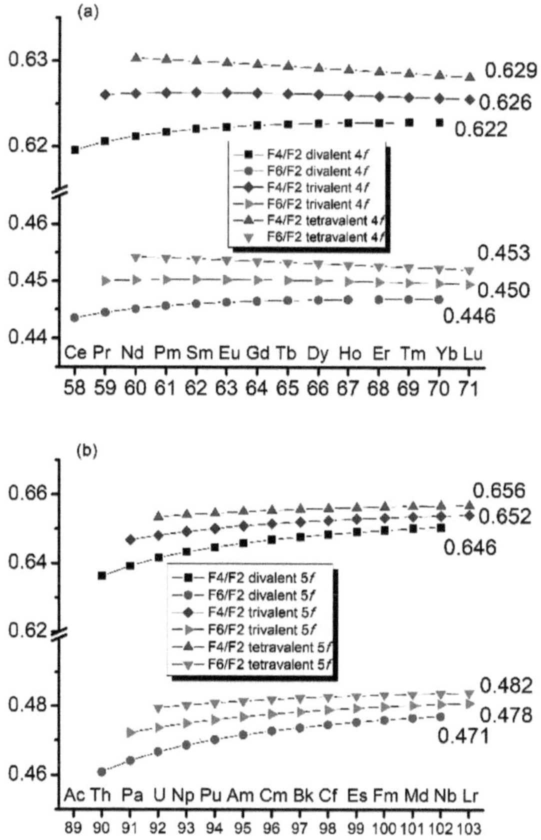

Figure 6.3 Dependence of the F^4/F^2 and F^6/F^2 ratios on the atomic number for the di-, tri-, and tetravalent 4*f* (a) and 5*f* (b) ions. The lines are the guides to the eye only and do not represent any fit. The averaged values of the ratios are given in each figure.

As far as the SO constant ζ variation is concerned, it is a well-established fact that it grows linearly with the fourth power of the atomic number $\zeta \sim Z^4$, or a plot of $\zeta^{1/4}$ against Z should be a straight line, as confirmed by Fig. 6.4. The derived linear equations for $\zeta^{1/4}$ (measured in $\text{cm}^{-1/4}$) are as follows (Z^* denotes an effective nuclear charge):

Divalent 4*f*: $\zeta^{1/4} = (-6.39991 \pm 0.17977) + (0.19556 \pm 0.00283)Z = 0.19556(Z-32.73) = 0.19556Z^*$

Trivalent 4*f*: $\zeta^{1/4} = (-5.43241 \pm 0.09300) + (0.18275 \pm 0.00144)Z = 0.18275(Z-29.73) = 0.18275Z^*$

Tetravalent 4*f*: $\zeta^{1/4} = (-4.86638 \pm 0.05956) + (0.17606 \pm 0.00009)Z = 0.17606(Z-27.64) = 0.17606Z^*$

Divalent 5*f*: $\zeta^{1/4} = (-14.97853 \pm 0.69153) + (0.23113 \pm 0.00723)Z = 0.23113(Z-64.81) = 0.23113Z^*$

Trivalent 5*f*: $\zeta^{1/4} = (-12.35019 \pm 0.36986) + (0.20563 \pm 0.00383)Z = 0.20563(Z-60.06) = 0.20563Z^*$

Tetravalent 5*f*: $\zeta^{1/4} = (-10.97840 \pm 0.24443) + (0.19305 \pm 0.00252) Z = 0.19305(Z-56.87) = 0.19305Z^*$

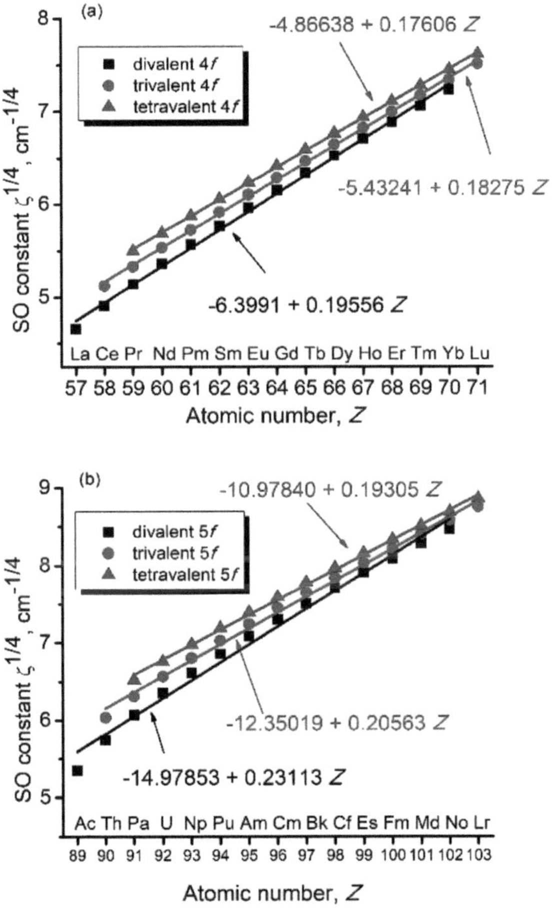

Figure 6.4 Linear dependencies of the $\zeta^{1/4}$ value on the atomic number for the di-, tri-, and tetravalent 4*f* (a) and 5*f* (b) ions.

Again, the slopes of these linear dependences all decrease when increasing the ions' oxidation state. It can be observed that the $\zeta^{1/4}$ dependence for the 5*f* ions is steeper than for the 4*f* ones.

Since the F^2, F^4 and F^6 parameters and the SO constant $\zeta^{1/4}$ are proportional to the atomic number *Z*, they should be also proportional to each other, as is evidenced by Fig. 6.5. The linear relations between them are given below. All calculated results are expressed in cm^{-1}; the coefficient before $\zeta^{1/4}$ is measured in $cm^{-3/4}$.

Divalent 4*f* ions:

$$\begin{aligned} F^2 &= (-12522.97 \pm 1443.08) + (19173.54 \pm 234.08)\zeta^{1/4} \\ F^4 &= (-8528.79 \pm 978.11) + (12050.29 \pm 158.66)\zeta^{1/4} \\ F^6 &= (-6334.88 \pm 723.06) + (8676.59 \pm 117.28)\zeta^{1/4} \end{aligned} \tag{6.3}$$

Trivalent 4*f* ions:

$$\begin{aligned} F^2 &= (1563.49 \pm 858.78) + (17953.47 \pm 132.41)\zeta^{1/4} \\ F^4 &= (1193.84 \pm 588.48) + (11207.31 \pm 90.74)\zeta^{1/4} \\ F^6 &= (909.56 \pm 437.00) + (8047.71 \pm 67.38)\zeta^{1/4} \end{aligned} \tag{6.4}$$

Tetravalent 4*f* ions:

$$\begin{aligned} F^2 &= (10878.51 \pm 586.41) + (17390.82 \pm 87.60)\zeta^{1/4} \\ F^4 &= (7793.47 \pm 403.93) + (10800.70 \pm 60.34)\zeta^{1/4} \\ F^6 &= (5882.61 \pm 300.08) + (7738.37 \pm 44.82)\zeta^{1/4} \end{aligned} \tag{6.5}$$

Divalent 5*f* ions:

$$\begin{aligned} F^2 &= (-31271.26 \pm 487.73) + (14769.30 \pm 367.03)\zeta^{1/4} \\ F^4 &= (-22725.27 \pm 360.42) + (9894.71 \pm 49.53)\zeta^{1/4} \\ F^6 &= (-17617.14 \pm 36.85) + (7368.99 \pm 36.85)\zeta^{1/4} \end{aligned} \tag{6.6}$$

Trivalent 5*f* ions:

$$\begin{aligned} F^2 &= (-24154.33 \pm 284.80) + (14210.35 \pm 37.27)\zeta^{1/4} \\ F^4 &= (-17463.93 \pm 232.40) + (9489.05 \pm 30.41)\zeta^{1/4} \\ F^6 &= (-13625.38 \pm 180.48) + (7067.13 \pm 23.62)\zeta^{1/4} \end{aligned} \tag{6.7}$$

Tetravalent 5*f* ions:

$$\begin{aligned} F^2 &= (-20023.65 \pm 213.43) + (13984.44 \pm 27.10)\zeta^{1/4} \\ F^4 &= (-14214.10 \pm 179.65) + (9309.26 \pm 22.81)\zeta^{1/4} \\ F^6 &= (-11091.25 \pm 143.93) + (6929.39 \pm 18.27)\zeta^{1/4} \end{aligned} \tag{6.8}$$

Therefore, if one out of four parameters (F^2, F^4, F^6, or ζ) is known for any of the di-, tri-, or tetravalent 4*f* or 5*f* ions, the remaining three parameters for that ion can all be reliably estimated using the above given linear equations and hydrogenic ratios.

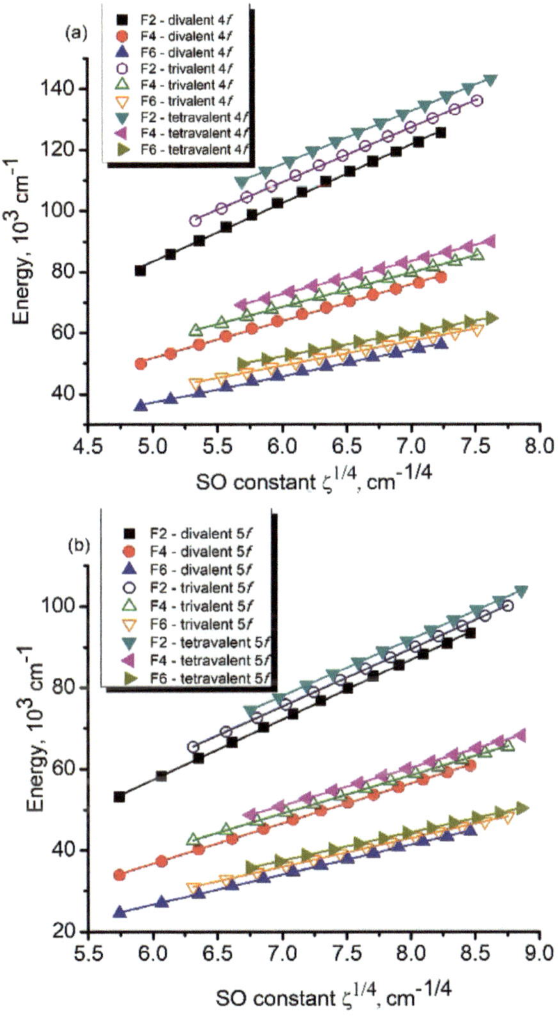

Figure 6.5 Dependence of the calculated F^2, F^4 and F^6 parameters on the $\zeta^{1/4}$ value for the di-, tri-, and tetravalent 4*f* (a) and 5*f* (b) ions. The straight lines are the linear fits, whose equations are given in the text.

The non-energetical quantities calculated here are the moments of the electron density of the *f*-electrons for all considered ions, they were found in a standard and a straightforward way by a numerical integration of the following expression:

$$\left\langle r^k \right\rangle = \int_0^\infty r^{k+2} \left| R_{nf}(r) \right|^2 dr , \qquad (6.9)$$

where k = 2, 4, 6; $R_{nf}(r)$ is the corresponding radial wave function with n being a principal quantum number (4 or 5), and f corresponds to the orbital quantum number l = 3 for the *f*-electrons. The data given in Table 6.3 show that the $\langle r^k \rangle$ values systematically decrease with increase of atomic number and oxidation state; the obtained results can be used for calculations of the parameters of crystal field acting upon the 4*f* and 5*f* ions in crystals. Considering the decreasing dependence of $\langle r^k \rangle$ related to the crystal field parameters, one can easily conclude that the crystal-field effects on the *f* electrons should exhibit a decreasing trend across the considered *nf* series, provided the interatomic distances are kept constant and only the central *f* ion is changed.

Additionally, the barycenter energies E_{avg} of all considered f^N electron configurations were calculated by employing the *f*-shell programs of Prof. M. F. Reid[24] with the obtained Slater and SO parameter values. The obtained results are listed in Table 6.3 and plotted in Fig. 6.6. There is a characteristic "Λ-shape," which is preserved for all considered electron configurations, with the maximal energy E_{avg} achieved for the ions in the middle of the considered series.

For the ions in the first half of the lanthanide and actinide series (before Gd and Cm) the barycenters of the divalent electronic configurations are higher than those of the tri- and tetravalent ones. For the ions following Gd and Cm, the order of the energies is changed and becomes (in the decreasing order): tetravalent-trivalent-divalent. A special case of the Gd and Cm ions is that the order of the barycenters is different from the described above: trivalent (the highest)—tetravalent (medium)—divalent (the lowest). It is also worthwhile noting that inside each group (di-, tri-, and tetravalent ions) the barycenter energy has a maximum value for the half-filled configurations, having 7 *f*-electrons.

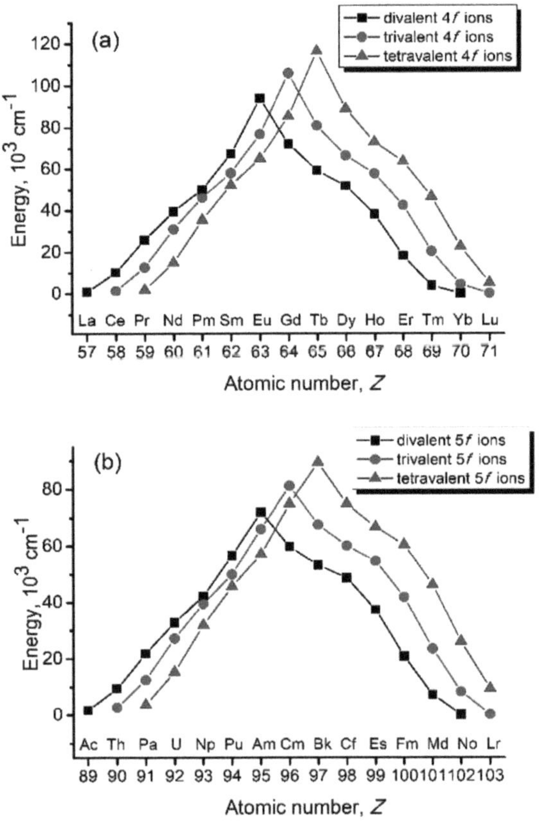

Figure 6.6 Calculated barycenter energies E_{avg} of the $4f^N$ (a) and $5f^N$ (b) electron configurations.

A particular feature of the more-than-half filled electron shells with N electrons is that they can be treated as the "filled" shells with 14-N positive holes. As such, the number of states for the f^N and f^{14-N} conjugate configurations is equal and is decreasing when moving to the left and to the right from $N = 7$. This decrease of number of states would compress the overall spitting of the f states after going over half-filled electron configurations with $N = 7$. At the same time, it can be noted that the Coulomb interaction parameters and SO constant increase with increasing atomic number. Then, due to increased Hamiltonian parameters, the barycenters of the more-than-half filled configurations are higher than those ones of the conjugate less-than-half filled configurations, thus producing an asymmetric Λ-shape, as shown in Fig. 6.6.

An additional energetic characteristic of the considered electronic configurations is the ionization energy, which is equal to the energy needed to remove one *f* electron to convert, for example, a neutral atom into a positively charged ion, etc. This energy has been also calculated (Table 6.3). Figure 6.7 shows the third, fourth, and fifth ionization energies against the atomic number *Z*. As seen from Fig. 6.7, for both lanthanide and actinide ions this ionization energy E_I is a practically perfect linear function of the atomic number; the equations of the linear fits are all given in the graph.

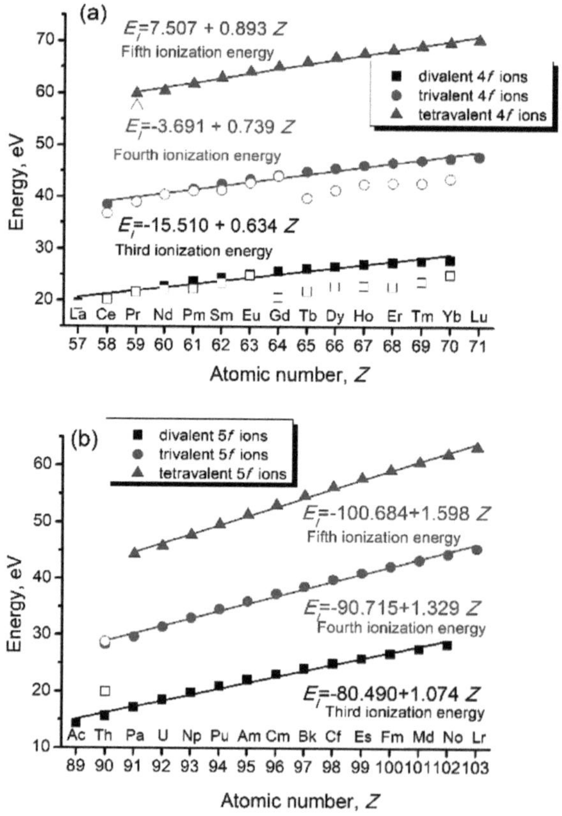

Figure 6.7 Calculated first ionization energies E_I for the $4f^N$ (a) and $5f^N$ (b) electron configurations. Experimental data (if available) are shown by empty symbols.

The slope of these linear fits of the ionization energies increase with the oxidation state (with moving to the ionization energies of higher order), and when moving from the 4*f* to the 5*f* ions. The corresponding experimental energies for some ions from the considered group were found; these data were taken from the well-known website http://www.webelements.com and are shown in Fig. 6.7 by the open symbols. As can be seen from that Figure, only for the lanthanides there is a complete set of the experimental data on the third and fourth ionization energies. No experimental data are available on the fifth ionization energy, except for Pr. Finally, the experimental ionization energies of the actinides were practically not found at all, except for the third and fourth ionization energies of Th. Overall agreement between the calculated and experimental ionization energies for lanthanides is good (Fig. 6.7), especially for the first half of the lanthanide series; it can be noticed that for the second half of this group there exists a small systematic overestimation of the calculated energy by about 4–5 eV, which may be related to the well-known overestimation of the Hartree–Fock calculated energies.

6.5 Energy Levels of the f^N Electronic Configurations of Free 4*f*/5*f* Ions

In this paragraph we demonstrate a practical application of the calculated parameters from Table 6.3 to the calculations of the energy level schemes of all considered 4*f* and 5*f* ions.[25] Figures 6.8–6.13 show the calculated energy level schemes obtained in the present work. Although the free ion Hamiltonians were diagonalized in the complete basis sets of all f^N configurations, only the energy levels up to 40000 cm^{-1} (42500 cm^{-1} in the case of tetravalent lanthanides) are shown in the figures. The complete energy level schemes with corresponding wave functions are available from the authors upon request.

The assignment of the calculated energy levels, as shown in all diagrams, was done using the main component of the corresponding wave function; the common notation $^{2S+1}L_J$ of the *J*-manifolds is used in all diagrams (with *S*, *L*, *J* being the total spin, orbital and angular momenta, respectively). If only value of *J* is given, then the corresponding ^{2S+1}L term can be found below those *J*-labeled levels.

For example, the second and third energy levels for Ce^{2+} are labeled as "5" and "6," and they correspond to the $^3H_5$ and $^3H_6$ states located above the $^3H_4$ level. Since for many electron configurations the density of the calculated energy levels is high, only the energy levels well separated from the others are assigned in all figures.

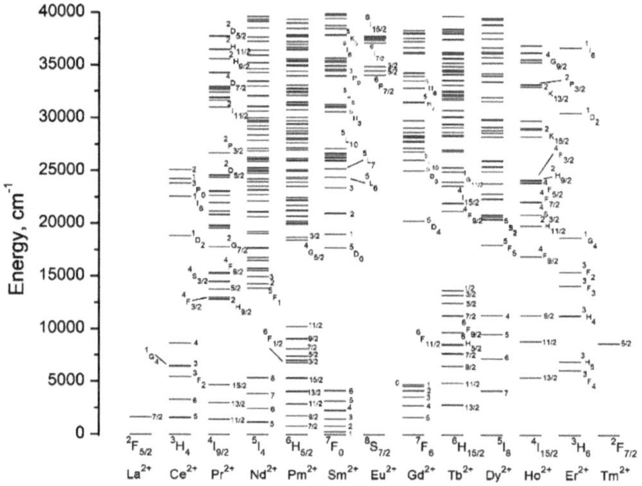

Figure 6.8 Calculated energy levels of divalent 4*f* ions in the energy range up to 40000 cm^{-1}.

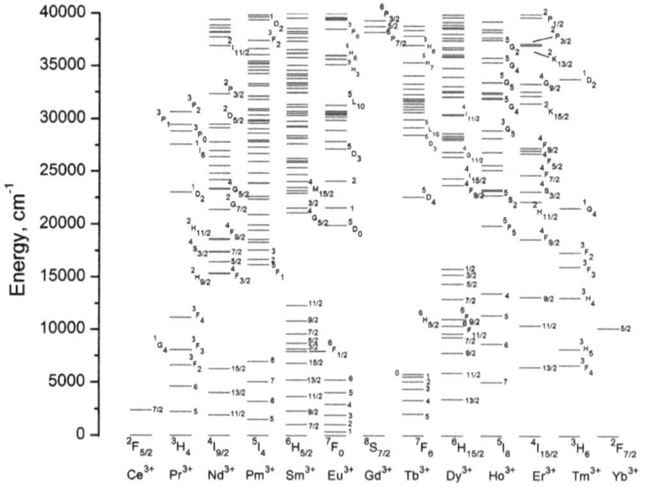

Figure 6.9 Calculated energy levels of trivalent 4*f* ions in the energy range up to 40000 cm^{-1}.

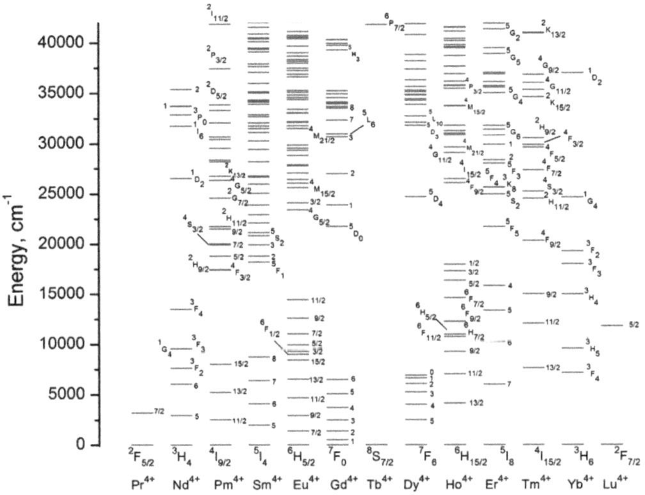

Figure 6.10 Calculated energy levels of tetravalent 4*f* ions in the energy range up to 42500 cm^{-1}.

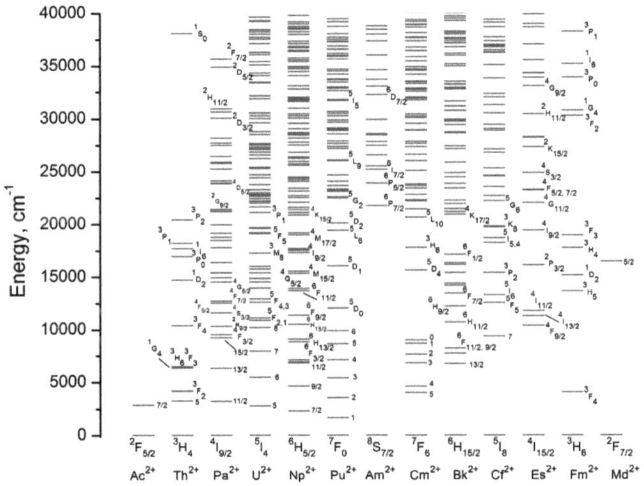

Figure 6.11 Calculated energy levels of divalent 5*f* ions in the energy range up to 40000 cm^{-1}.

Although at the first glance these energy level diagrams may seem to be similar, they do have distinct features that make them be clearly different from each other. Comparative description of the most noticeable characteristics of the calculated energy level schemes for each electron configuration is given below.

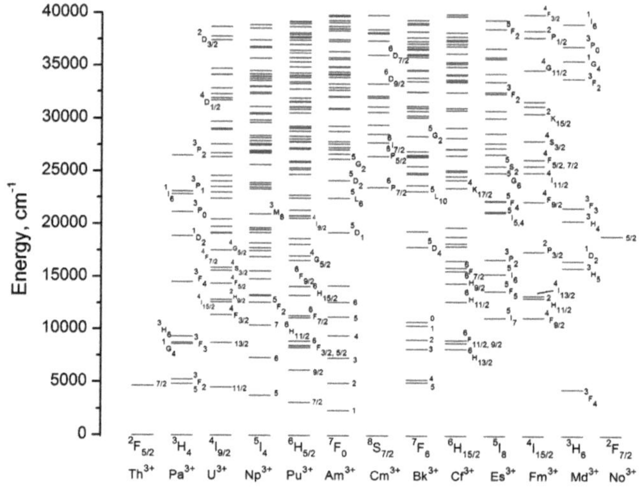

Figure 6.12 Calculated energy levels of trivalent 5f ions in the energy range up to 40000 cm^{-1}.

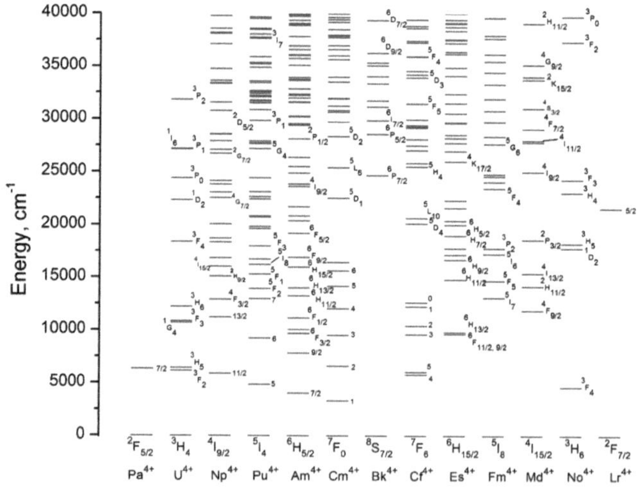

Figure 6.13 Calculated energy levels of tetravalent 5f ions in the energy range up to 40000 cm^{-1}.

Even the calculated energy levels are slightly overestimated due to the above-mentioned overestimation of the electrostatic interaction parameters, which is characteristic for the Hartree–Fock calculations, the observed trends in the energy level location remain valid and can serve as reliable guides for the choice of a particular 4*f* (5*f*) ion for the desired purposes. Besides, the discussion of energy levels of the conjugate electron configurations is also very useful for a deeper understanding of similar and different spectroscopic properties of these configurations.

6.5.1 f^1 and f^{13} Configurations

These two electron configurations are the simplest ones among all considered; they are characteristic for the following ions: La^{2+}, Ce^{3+}, Pr^{4+} ($4f^1$), Ac^{2+}, Th^{3+}, Pa^{4+} ($5f^1$), Tm^{2+}, Yb^{3+}, Lu^{4+} ($4f^{13}$), Md^{2+}, No^{3+}, Lr^{4+} ($5f^{13}$). In these configurations, there is only one term ^2F, which is split by the SO coupling into the $^2F_{5/2}$ (ground state) and $^2F_{7/2}$ (excited state) for the f^1 configuration. For the f^{13} configuration the order of these two energy levels is opposite. The magnitude of this splitting is increased in the following directions: $La^{2+} < Ce^{3+} < Pr^{4+}$ and $Ac^{2+} < Th^{3+} < Pa^{4+}$, reaching about 6300 cm^{-1} for Pa^{4+}, if the f^1 ions are considered. For the f^{13} ions the value of the SO splitting is considerably greater due to the increased atomic numbers of those ions. The position of the excited $^2F_{5/2}$ state in this case changes as follows: $Tm^{2+} < Yb^{3+} < Lu^{4+}$ and $Md^{2+} < No^{3+} < Lr^{4+}$, varying from about 8000 cm^{-1} for Tm^{2+} to about 21500 cm^{-1} for Lr^{4+}. The corresponding energy level schemes are shown in Fig. 6.14.

6.5.2 f^2 and f^{12} Configurations

These are the electron configurations of the following ions: Ce^{2+}, Pr^{3+}, Nd^{4+} ($4f^2$), Th^{2+}, Pa^{3+}, U^{4+} ($5f^2$), Er^{2+}, Tm^{3+}, Yb^{4+} ($4f^{12}$), Fm^{2+}, Md^{3+}, No^{4+} ($5f^{12}$) (Fig. 6.15). The ground states are $^3H_4$ (f^2) and $^3H_6$ (f^{12}). For the $4f^2$ ions the three states $^3H_{4,5,6}$ and three states $^3F_{2,3,4}$ are well separated from each other. However, the situation is different for the $4f^{12}$ ions: Enhanced SO coupling pushes the $^3F_4$ state further down so it becomes the first excited state located between the $^3H_6$ and $^3H_{5,4}$ states. This shows importance of the *J–J* coupling for the heavy lanthanides. For the $5f^2$ actinides, the $^3F_2$ state—the lowest out of three $^3F_{2,3,4}$ states—is also located between the *J*-states arising from the ^3H term. It can be also noticed that the overall splitting of the ^3H and ^3F terms produced by the SO coupling increases with increased charge of the 4*f* or 5*f* ions.

The calculated splitting of the ground ^3H term for the $4f^2$ ions varies from about 3300 cm^{-1} (Ce^{2+}) to 6000 cm^{-1} (Nd^{4+}); the same range for the $5f^2$ ions is considerably wider and extends from about 6300 cm^{-1} (Th^{2+}) to 12200 cm^{-1} (U^{4+}). The splitting of the ^3H term for the f^{12} configuration varies from about 12200 cm^{-1} (Er^{2+}, $4f^{12}$) to almost 22000 cm^{-1} (No^{4+}, $5f^{12}$).

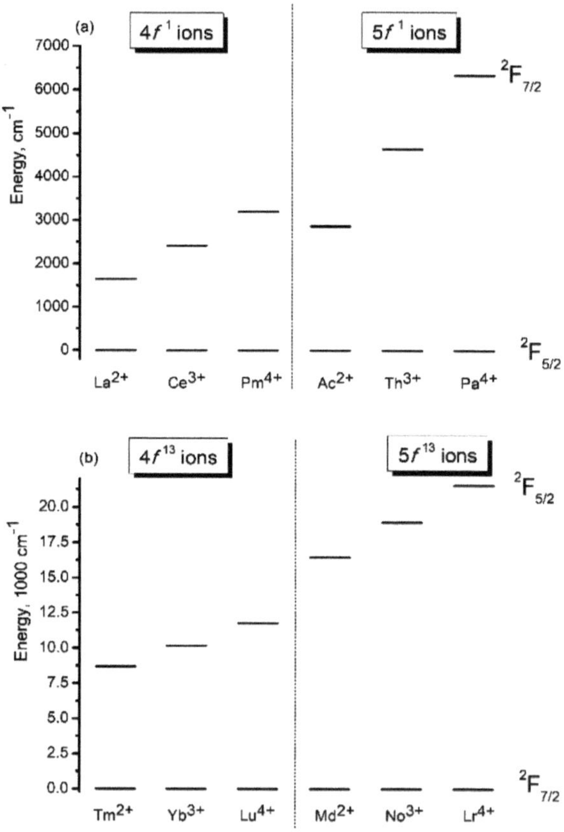

Figure 6.14 Energy level schemes of the f^1 (a) and f^{13} (b) configurations.

The $^3P_0$ state of the Pr^{3+} ions is of importance for getting visible laser emission,[26] its calculated value (although overestimated) is at about 28000 cm^{-1}. However, such an overestimation does not hide the following trend: For the Ce^{2+} ions, this level is moved down to about 24000 cm^{-1}, whereas for the Nd^{4+} ions, it goes up to approximately 33000 cm^{-1}. For the 5f^2 ions this $^3P_0$ state is located in a wide interval between 17000 cm^{-1} (Th^{2+}) and 25000 cm^{-1} (U^{4+}).

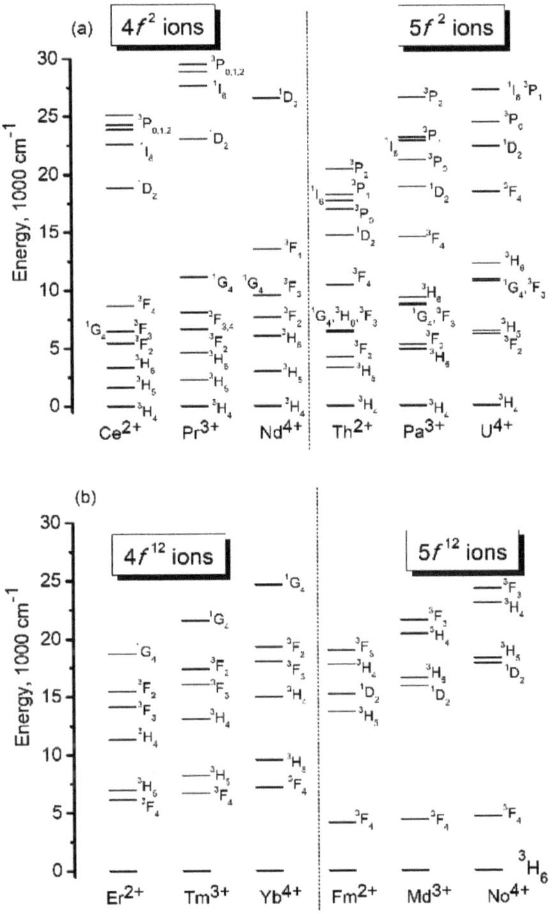

Figure 6.15 Energy level schemes of the f^2 (a) and f^{12} (b) configurations.

The $^1D_2$ and $^1G_4$ levels of the Tm^{3+} ions are used for getting laser emission in the visible range on the transitions to the ^3H_J states, since these $^1D_2$ and $^1G_4$ states are well separated from the lower lying levels, which considerably reduces the non-radiative relaxation. In comparison with the Tm^{3+} ions, for the Er^{2+} (Yb^{4+}) ions these spin-singlet states are located lower (higher), respectively.

The states with the same J values are strongly mixed up by the SO coupling, especially for the $5f^{12}$ ions. For some states (e.g. $^1D_2$ and $^3F_2$, $^1D_2$ and $^3P_2$) the contributions of these states into the total wave function are comparable, which hinders energy levels assignment.

6.5.3 f^3 and f^{11} Configurations

This group consists of the following ions: Pr^{2+}, Nd^{3+}, Pm^{4+} ($4f^3$), Pa^{2+}, U^{3+}, Np^{4+} ($5f^3$), Ho^{2+}, Er^{3+}, Tm^{4+} ($4f^{11}$), Es^{2+}, Fm^{3+}, Md^{4+} ($5f^{11}$). The energy levels of these ions are shown in Fig. 6.16.

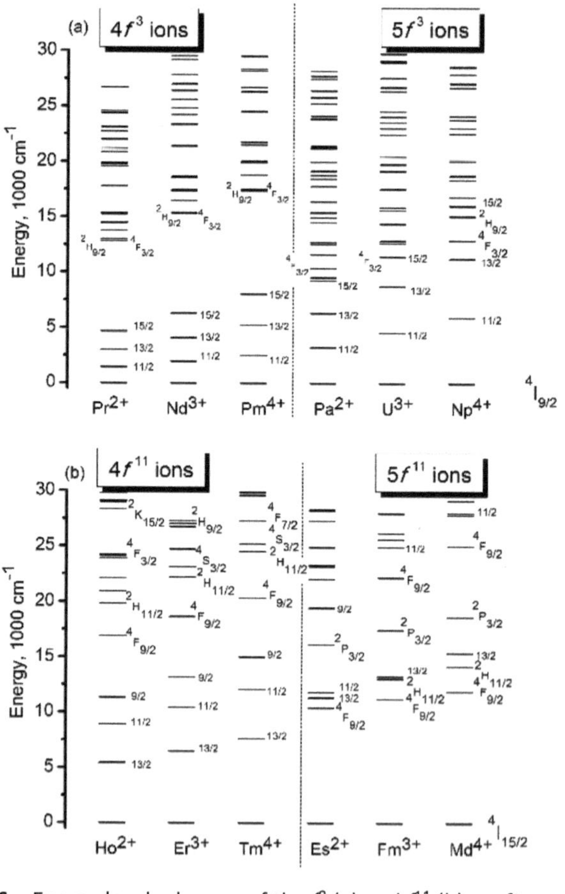

Figure 6.16 Energy level schemes of the f^3 (a) and f^{11} (b) configurations.

The ground states are $^4I_{9/2}$($^4I_{15/2}$) for the f^3(f^{11}) configurations. The $^4F_{3/2}$ state, which is a lasing level of the Nd^{3+} ion in the infrared region, is blue-shifted for the Pm^{4+} ion and red-shifted for the Pr^{2+} ion. For the $5f^2$ ions the $^4F_{3/2}$ state is located between the J manifolds of the ^4I term due to strong SO coupling and effects of J-mixing. If the isoelectronic $4f^3$ and $5f^3$ ions are compared, this level for the former ions is located higher than for the latter.

The overall splitting of the ^4I term (energy separation between the $^4I_{9/2}$ and $^4I_{15/2}$) is greatly increased for the f^{11} ions, in comparison with the ions with the f^3 electron configuration; it is also greater for the actinide ions. Thus, for the Md^{4+} ions it reaches the value of about 25000 cm^{-1} (compare with about 6000 cm^{-1} for Nd^{3+}). Effects of J-mixing are also much stronger for the f^{11} configuration, than for the f^3 configuration.

6.5.4 f^4 and f^{10} Configurations

The following ions: Nd^{2+}, Pm^{3+}, Sm^{4+} ($4f^4$), U^{2+}, Np^{3+}, Pu^{4+} ($5f^4$), Dy^{2+}, Ho^{3+}, Er^{4+} ($4f^{10}$), Cf^{2+}, Es^{3+}, Fm^{4+} ($5f^{10}$) are in this group (Fig. 6.17).

The ground states for the f^4(f^{10}) configurations are the $^5I_4$($^5I_8$) levels. The splitting of the ^5I term reaches about 8000 cm^{-1} for the Sm^{4+} ion and goes even much higher to about 17000 for the Pu^{4+} ion. Due to radioactivity of the Pm^{3+} ions, their optical applications are restricted. However, the Nd^{2+} and Sm^{4+} ions have the $^5F_1$ state well separated from the states of the ^5I term; such situation is similar to the case of the Nd^{3+} ion with its $^4F_{3/2}$ state used for lasing in the infrared region. As a result, the $^5F_1$ states of the Nd^{2+} and Sm^{4+} ions may be interesting for getting emission in the infrared/visible ranges.

A strict and well-ordered sequence of the ^5I_J (J = 8...4) manifolds in the case of $5f^{10}$ ions is interrupted by the states of the ^5F_J (J = 5...1) manifolds, due to enhanced SO coupling and effects of the J-mixing. For the Fm^{4+} ion the $^5F_4$ state can be a suitable emitting level in the visible/infrared regions, depending on the terminating lower level.

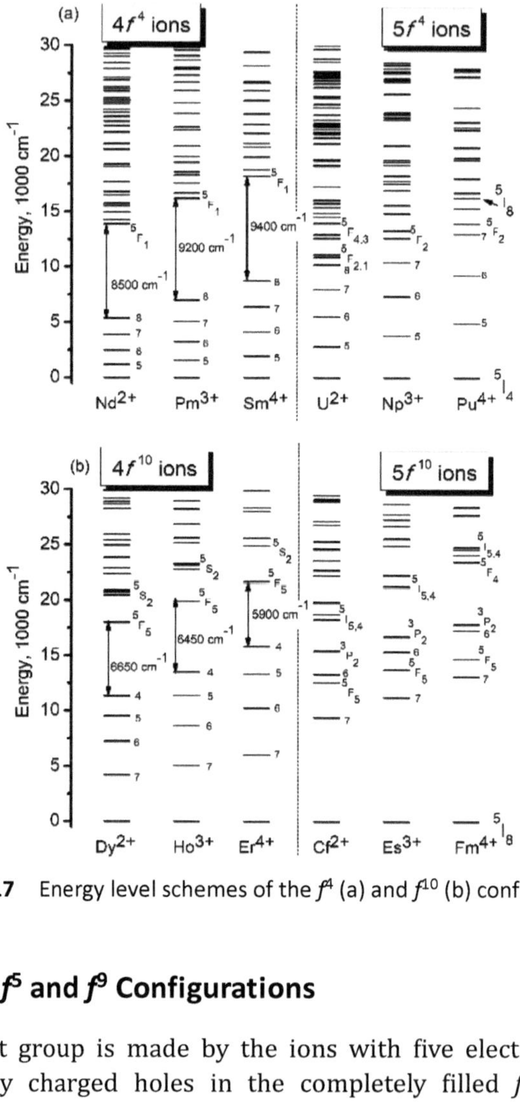

Figure 6.17 Energy level schemes of the f^4 (a) and f^{10} (b) configurations.

6.5.5 f^5 and f^9 Configurations

The next group is made by the ions with five electrons and five positively charged holes in the completely filled f shell: Pm^{2+}, Sm^{3+}, Eu^{4+} ($4f^5$), Np^{2+}, Pu^{3+}, Am^{4+} ($5f^5$), Tb^{2+}, Dy^{3+}, Ho^{4+} ($4f^9$), Bk^{2+}, Cf^{3+}, Es^{4+} ($5f^9$). The energy level diagram for these ions is given in Fig. 6.18. The ground state is $^6H_{5/2}$ ($^6H_{15/2}$) for the f^5 (f^9) configurations; six J states (with J varying from 5/2 to 15/2 with a step of unity) arising from the ^{6}H term are grouped together for all $4f^5$ ions; they are followed by another group of six levels coming from the ^{6}F term with J varying from 1/2 to 11/2. The next excited state separated from those levels is $^4G_{5/2}$, which is responsible for visible emission of the Sm^{3+} ions. For this ion, all energy levels located higher are very close to each other, so the non-radiative relaxation from those levels would easily populate the $^4G_{5/2}$ state, leading to efficient emission. In the case of the Eu^{4+} ions, an additional channel of emission can become possible: The $^4M_{15/2}$ state at about 25000 cm^{-1} is separated by a ~1500 cm^{-1} gap from the lower states; in the hosts with low phonon energy the non-radiative relaxation from this state would be inefficient and emission lines corresponding to the transitions from the $^4M_{15/2}$ state to lower located levels of the ^6H_J and ^6F_J manifolds can be expected.

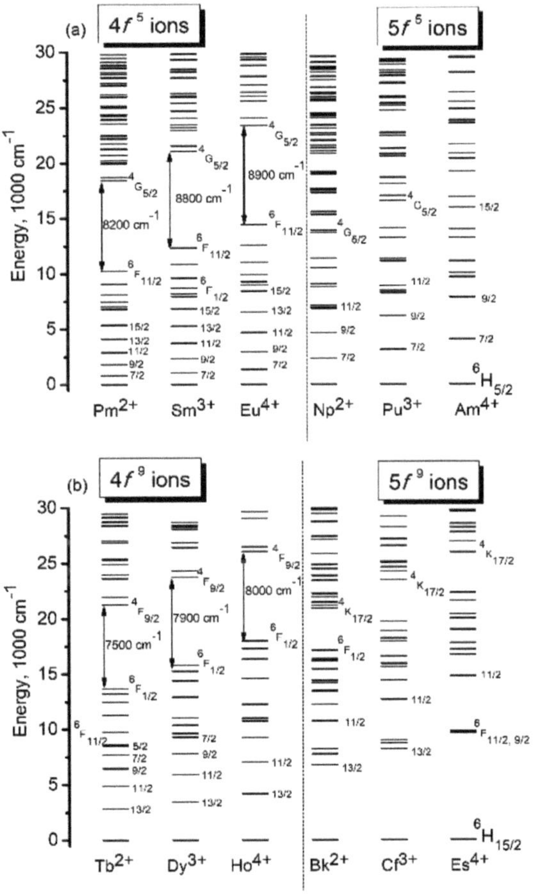

Figure 6.18 Energy level schemes of the f^5 (a) and f^9 (b) configurations.

For the $5f^5$ ions the $^4G_{5/2}$ state is very close to the lower levels, and no emission transitions from it can be expected. However, another excited state $^4I_{9/2}$ is well separated from the lower states. Therefore, it can produce an infrared and visible emission on account of the transitions to the ^6H_J, ^6F_J manifolds.

In the case of the $4f^9$ ions the $^4F_{9/2}$ level plays an analogous role to the $^4G_{5/2}$ state: The gap from it to the lower states is very wide, and efficient emission can be observed. Again, for the $5f^9$ ions, this state comes very close to lower levels, and emission from it is expected to be quenched. However, the $^4K_{17/2}$ level at about 22500 cm^{-1} (Bk^{2+}) or 24000 cm^{-1} (Cf^{3+}) or 25500 cm^{-1} (Es^{4+}) becomes a potentially emitting state, with the band gap to the next below-lying levels to be about 3000–5000 cm^{-1}.

6.5.6 f^6 and f^8 Configurations

These electronic configurations are represented by the following ions: Sm^{2+}, Eu^{3+}, Gd^{4+} ($4f^6$), Pu^{2+}, Am^{3+}, Cm^{4+} ($5f^6$), Gd^{2+}, Tb^{3+}, Dy^{4+} ($4f^8$), Cm^{2+}, Bk^{3+}, Cf^{4+} ($5f^8$). The ground state is $^7F_0$ ($^7F_6$) for the f^6 (f^8) configurations (Fig. 6.19). The group of seven states arising from the ^7F term (with J varying from 0 to 6) is well separated from the emitting $^5D_0$ level for all $4f^6$ ions and from the emitting $^5D_4$ state for all $4f^8$ and $5f^8$ ions. This holds true for all $4f^6$ ions (the only difference is the gradual upward shift of the $^5D_0$ term with an increased charge of the corresponding ions). The case of the $5f^6$ ions is different: For the Pu^{2+} ions, the $^5D_0$ state is too close to the $^7F_6$ level; for the Am^{3+} and Cm^{4+} ions the $^5D_1$ state emerges as a potentially emitting level.

For the $4f^8$ ions emission can be expected from the $^5D_4$ state. The $5f^8$ ions have more complicated energy level schemes, with many closely located energy levels. However, the gap between the $^5I_5$, $^5I_4$ terms and lower lying levels is about 4000–5000 cm^{-1} for the Cm^{2+}, Bk^{3+} ions.

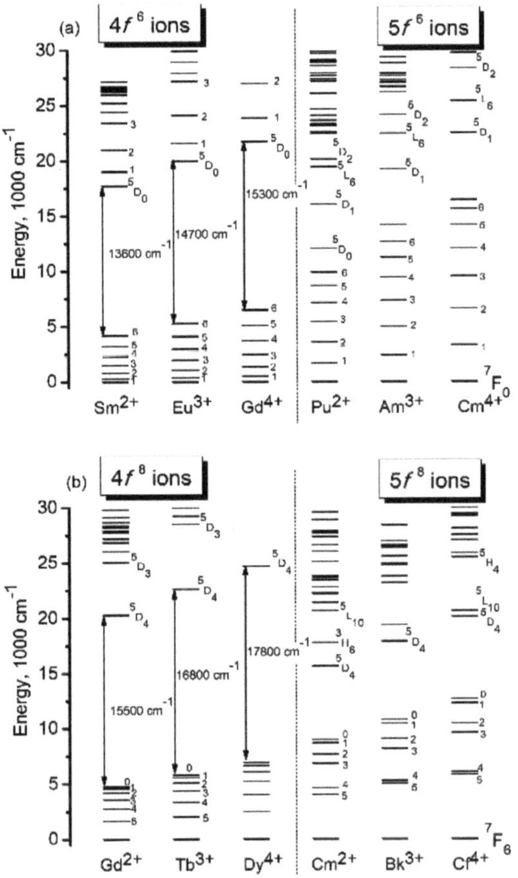

Figure 6.19 Energy level schemes of the f^6 (a) and f^8 (b) configurations.

6.5.7 f^7 Configurations

This configuration corresponds to the half-filled *f*-shell of the following ions: Eu^{2+}, Gd^{3+}, Tb^{4+} ($4f^7$), Am^{2+}, Cm^{3+}, Bk^{4+} ($5f^7$). The ground state of this configuration is $^8S_{7/2}$ (Fig. 6.20), and all excited states are separated by a wide gap, which is more than 30000 cm^{-1} for the 4*f* ions and more than 20000 cm^{-1} for the 5*f* ions. The width of this gap is increasing with increased ionic charge in both groups of the 4*f* and 5*f* ions. This gap for the actinides ions is smaller if compared with the lanthanide ions with the same charge.

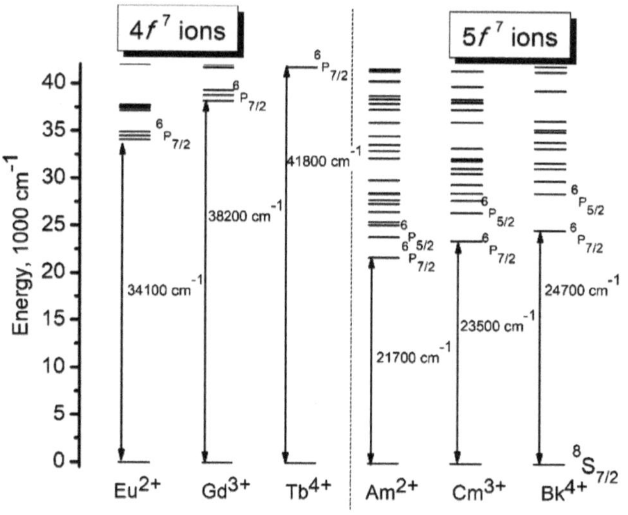

Figure 6.20 Energy level schemes of the f^7 configuration.

6.6 Variation of the Slater Integrals and Spin–Orbit Coupling Constant with Atomic Number for the $f^{N-1}d$ Electronic Configurations of Free 4*f*/5*f* Ions

In this paragraph we consider the excited $4f^{N-1}5d$ (lanthanides) or $5f^{N-1}6d$ (actinides) electronic configurations that can be formed by promoting one of the 4*f* or 5*f* electrons to one of the five 5*d* or 6*d* orbitals, respectively. A completely different feature of such excited $f^{N-1}d$ electron configurations from the ground f^N electron configurations is that the excited 5*d* or 6*d* orbitals are much more extended than the 4*f* or 5*f* orbitals and strongly interact with the crystalline environment, whereas the 4*f* or 5*f* orbitals of the inner f^{N-1} core slightly contract relative to those in the ground f^N electron configurations.

The ground and excited electron configurations f^N and $f^{N-1}d$ are characterized by a very large number of highly degenerated electronic states, which span an extremely wide spectral range from infrared to vacuum ultraviolet. To illustrate such a statement, Table

6.5 collects the numbers of possible states of the $f^{N-1}d$ electron configurations. As seen from this Table, a possible number of states can well exceed 20000–30000, which would lead to a formation of extremely rich and dense energy level schemes that can be used for optical applications.

Table 6.5 Electronic configurations and the number of possible states for the $f^N + f^{N-1}d$ configurations

Electron configuration	Number of states for $f^N + f^{N-1}d$ configurations	Number of J manifolds for $f^N + f^{N-1}d$ configurations
$f^1 + f^0d$	24	4
$f^2 + f^1d$	231	33
$f^3 + f^2d$	1274	148
$f^4 + f^3d$	4641	493
$f^5 + f^4d$	12012	1175
$f^6 + f^5d$	23023	2173
$f^7 + f^6d$	33462	3052
$f^8 + f^7d$	37323	3401
$f^9 + f^8d$	32032	2923
$f^{10} + f^9d$	21021	1985
$f^{11} + f^{10}d$	10374	1018
$f^{12} + f^{11}d$	3731	399
$f^{13} + f^{12}d$	924	109
$f^{14} + f^{13}d$	141	21

Among all possible transitions between any two of these energy levels, the inter-configurational $f^N \leftrightarrow f^{N-1}d$ (i.e., f–d) transitions have been paid considerable attention in the last years due to constantly growing demand for new luminescent phosphors for mercury-free fluorescent lamps and plasma display panels, wide applications in ultraviolet lasers, fast scintillators, etc.[1,2,4]. These examples emphasize importance of a thorough research focused on these excited configurations of the lanthanide and actinide ions. These transitions are parity allowed, and their intensities are much higher than those of the parity forbidden f–f transitions.

There has been found much systematics in the $4f$–$5d$ transitions of the di- and trivalent lanthanides in crystals.[27] However, consideration of the spectroscopic properties of the $f^{N-1}d$ configurations of these ions in a free state is still scarce in the literature. In addition, it is very interesting to extend the fundamental knowledge of the spectroscopic parameters determining the properties of the $5d$ and $6d$ orbitals across these considered two f-elements series.

The semi-empirical calculation technique of energy levels for the $4f^N$ configuration of lanthanide ions in crystals[11,12] has been successfully extended to the analysis of the $4f^{N-1}5d$ configurations spectral properties[28] (and can be applied in the same form to the $5f^{N-1}6d$ configurations of actinides[29]) by accounting for some additional effective operators acting within the $f^{N-1}d$ excited configuration in the framework of the "effective Hamiltonian" theory.[30] The values of the radial fitting parameters can be initially estimated by employing the empirical rules proposed by Reid et al.[31] (for example, the atomic interaction parameters involving the $5d$ or $6d$ electron are directly taken from ab initio calculations of free lanthanide or actinide ions), and then appropriate scaling coefficients may be used to fit the measured f–d transition spectra.

Such a theoretical method has been systematically implemented for successful interpretation of the trivalent lanthanide ions spectra in $LiYF_4$, CaF_2, and YPO_4 crystals[28,32,33] by van Pieterson et al., and the trend of the obtained fitting parameter values across the whole $4f$-element series has been also analyzed and discussed.[34] Recently, similar works have been also carried out for the trivalent lanthanide ions in K_3YF_6, Cs_2NaYCl_6, and Cs_2NaYF_6 crystals.[14,35,36]

It is noticeable that in those above-mentioned publications the main studied objects are the trivalent lanthanide ions in crystals. However, as it was mentioned above, some other oxidation states of these ions can be encountered. The divalent lanthanide ions (first of all, Eu^{2+}) can be found in some materials; these ions are of paramount importance for the applications of white-light LEDs due to the match of their $4f$–$5d$ transition energies with the blue light emitted by the LED chips. It is also instructive to make systematic studies of the main spectroscopic parameters for the $f^{N-1}d$ electronic configurations of the ions in a free state and solids across the actinide series. Even more important, such trend studies of all the $4f$ and $5f$ ions with different oxidation states in a free state are the necessary starting points for the crystal field calculations of the f-d transition spectra of these ions in solids.

In what follows, we discuss the Slater parameters of the Coulomb interaction inside of the f^{N-1} core and the direct and exchange Coulomb interactions between the f and d electrons, the SO interaction constants $\zeta(ff)$ and $\zeta(dd)$ for the f and d electrons, and the averaged values of the second, fourth and sixth power of thc corresponding electrons radial coordinates. The last entries are needed to calculate the crystal field parameters. The energy difference between the barycenters of the ground and excited configurations f^N and $f^{N-1}d$ (i.e., $\Delta_E(fd)$) for all considered $4f$ and $5f$ ions is also reported.

6.6.1 Description of the Free Ion Hamiltonian for the f^{N-1}d Electron Configurations

A brief summary of parameterized calculations for the $4f^{N-1}5d$ and $5f^{N-1}6d$ (for the sake of brevity denoted in what follows as $f^{N-1}d$) energy levels is now given as follows. The effective Hamiltonian for the $f^{N-1}d$ configuration generally consists of three parts: (i) interactions within the f^{N-1} core, (ii) SO interaction of a single d electron, and (iii) Coulomb and exchange interaction between the f and d electrons:

$$\mathrm{H}(f^{N-1}d) = \mathrm{H}(ff) + \mathrm{H}(dd) + \mathrm{H}(fd), \tag{6.10}$$

where the first and last terms do not exist for $N = 1$. The first term $\mathrm{H}(ff)$ comprises the quasi-free ion and crystal field (CF) interactions of the f^{N-1} core and can be further expressed as

$$\mathrm{H}(ff) = \sum_{k=2,4,6} F^k(ff)f_k + \varsigma_f(f)A_{so} + \alpha(ff)L(L+1) + \beta(ff)G(G_2) + \gamma(ff)G(R_7)$$
$$+ \sum_{i=2,3,4,6,7,8} T^i(ff)t_i + \sum_{h=0,2,4} M^h(ff)m_h + \sum_{k=2,4,6} P^k(ff)p_k + \sum_{\substack{k=2,4,6\\ q=-k\cdots k}} B_q^k(f)C_q^{(k)}, \tag{6.11}$$

where the first two terms are two strongest interactions within the f^{N-1} core, i.e., the Coulomb and SO interactions, which produce the so-called fine structure of energy levels and split the f^{N-1} core states into the $^{2S+1}L_J$ multiplets; the last term stands for the anisotropic components of the CF interaction, which splits each multiplet into CF states; and the other terms between the first two and the last one are some effective corrections appearing due to the configuration interactions, mainly including two-body Coulomb correlation, three-body Coulomb correlation (effectively contributing only when $3 < N < 13$), and magnetically and electrostatically correlated interactions (two-body interactions). It is noted that for the case of $N = 2$ or $N = 14$ there are only the SO and CF interactions in Eq. (6.11).

The interaction felt by the d electron H(dd) is described by the single electron Hamiltonian, which can be written as

$$\mathrm{H}(dd) = \sum_{\substack{k=2,4 \\ q=-k,\cdots k}} B_q^k(d) C_q^{(k)} + \varsigma_{\mathrm{d}}(d) A_{so(\mathrm{d})}, \tag{6.12}$$

where the first term is the anisotropic CF interaction and the second term is the SO interaction for a single d electron. The former leads to a very large—about 20000 cm^{-1}—splitting of the outer d orbitals due to their extended character. The latter produces an important contribution to the spin-forbidden transitions due to "stealing" or "borrowing" intensities from the spin-allowed transitions by effects of mixing of states with the same total momenta J but different spins S. The last term in Eq. (6.10) H(fd) can be defined and shown as follows:

$$\mathrm{H}(fd) = \sum_{k=2,4} F^k(fd) f_k + \sum_{j=1,3,5} G^j(fd) g_j + \Delta_E(fd), \tag{6.13}$$

where the first two terms are the direct and exchange Coulomb interactions between the f and d electrons, which induces the well-known energy splitting between the high- and low-spin states [28, 37]. The last term represents the energy separation between the barycenters of the two configurations f^N and $f^{N-1}d$, which obviously incorporates the contributions from both atomic and lattice effects, such as spherically symmetric part of the Coulomb and CF interactions within the $f^{N-1}d$ configuration ($F^0(ff)$, $F^0(fd)$, $B^0(d)$ and $B^0(f)$). Any CF effects in this paragraph will not be considered; the main emphasis will be placed on the free ion parameters entering the Hamiltonian (6.10).

6.6.2 Results of Calculations: Direct and Exchange Coulomb Interaction Parameters and SO Interaction Constants for the f^{N-1} Core and $f^{N-1}d$ Electron Configurations

The numerical values of the Slater parameters $F^2(ff)$, $F^4(ff)$, $F^6(ff)$, $F^2(fd)$, $F^4(fd)$, $G^1(dd)$, $G^3(dd)$, $G^5(dd)$, and SO interaction constants $\zeta(ff)$, $\zeta(dd)$ were calculated using the Cowan programs taking into account the relativistic effects. All calculated parameters are collected in Table 6.6.

Analysis of the data presented in Table 6.6 allows one to find several trends that link together the main parameters of the Hamiltonian of free lanthanides and actinides.

Thus, Figs. 6.21 and 6.22 show the variation of the $F^2(ff)$, $F^4(ff)$, and $F^6(ff)$ parameters for the f^{N-1}cores against the atomic number Z in the considered groups of ions. It can be easily seen from both figures that all these parameters are increasing with the atomic number Z and are excellently described by the linear equations $a_1 + a_2 Z$, whose equations are given in the figures. The values of the fitting coefficients obtained after a linear fit of the calculated data points are all collected in Table 6.7. It is seen that the absolute values of the a_1, a_2 coefficients all decrease in the "$F^2(ff) \rightarrow F^4(ff) \rightarrow F^6(ff)$" direction and with increase of the charge of the ions for both lanthanides and actinides. The strongest dependence of the Slater parameters on Z in each of the di-, tri-, and tetravalent ions' groups is realized for the $F^2(ff)$ parameter; the weakest—for the $F^6(ff)$ parameter. If the 4*f* and 5*f* ions are compared, then the Slater parameters also behave in a similar way: The a_2 coefficient is the greatest for the $F^2(ff)$ parameter, and the difference between the a_2 coefficients for the 4*f* and 5*f* ions decreases when moving in the "$F^2(ff) \rightarrow F^4(ff) \rightarrow F^6(ff)$" direction.

Table 6.6 Calculated energy parameters (in cm^{-1}) and radial integrals (in $Å^k$) of the $f^{N-1}d$ configurations for the considered di-, tri-, and tetravalent 4*f* and 5*f* ions

Ion	La^{2+}	Ce^{2+}	Pr^{2+}	Nd^{2+}	Pm^{2+}	Sm^{2+}	Eu^{2+}	Gd^{2+}	Tb^{2+}	Dy^{2+}	Ho^{2+}	Er^{2+}	Tm^{2+}	Yb^{2+}
$4f^{N-1}5d$	1	2	3	4	5	6	7	8	9	10	11	12	13	14
$\Delta_E(fd)$	1184	9535	16134	21940	26945	31339	35170	38502	41436	43951	46112	47987	49500	50756
$F^2(ff)$	—	—	95375	99500	103365	107042	110565	113955	117248	120445	1235666	126627	129618	132562
$F^4(ff)$	—	—	59562	62258	64692	67004	69213	71335	73393	75388	77333	79239	81099	82928
$F^6(ff)$	—	—	42859	44738	46493	48157	49746	51272	52750	54183	55579	56946	58280	59591
$\zeta(ff)$	—	679	798	927	1067	1217	1379	1553	1741	1942	2158	2389	2636	2899
$F^2(fd)$	—	26027	25903	25644	25395	25151	24910	24670	24430	24189	23948	23704	23458	23210
$F^4(fd)$	—	13064	12896	12634	12394	12169	11956	11753	11556	11366	11181	10999	10821	10646
$G^1(fd)$	—	12927	12691	12261	11905	11601	11337	11104	10893	10702	10525	10360	10206	10059
$G^3(fd)$	—	10362	10182	9876	9613	9380	9171	8980	8802	8636	8479	8328	8185	8046
$G^5(fd)$	—	7869	7731	7506	7309	7133	6973	6825	6686	6555	6431	6311	6196	6085
$\zeta(dd)$	730	778	825	870	915	959	1003	1045	1088	1130	1171	1212	1251	1291
$r^2(ff)$	—	0.3867	0.3593	0.3313	0.3080	0.2882	0.2707	0.2555	0.2420	0.2299	0.2190	0.2091	0.2000	0.1917
$r^4(ff)$	—	0.3552	0.3114	0.2663	0.2317	0.2043	0.1819	0.1635	0.1481	0.1349	0.1236	0.1138	0.1053	0.0978
$r^6(ff)$	—	0.6493	0.5452	0.4361	0.3584	0.3002	0.2556	0.2209	0.1929	0.1702	0.1514	0.1358	0.1226	0.1114
$r^2(dd)$	2.2475	2.1767	2.1148	2.0633	2.0187	1.9795	1.9448	1.9143	1.8868	1.8625	1.8409	1.8216	1.8045	1.7891
$r^4(dd)$	8.1945	7.7342	7.3484	7.0386	6.7783	6.5564	6.3666	6.2035	6.0616	5.9392	5.8334	5.7416	5.6640	5.5973

Ion	Ce^{3+}	Pr^{3+}	Nd^{3+}	Pm^{3+}	Sm^{3+}	Eu^{3+}	Gd^{3+}	Tb^{3+}	Dy^{3+}	Ho^{3+}	Er^{3+}	Tm^{3+}	Yb^{3+}	Lu^{3+}
$4f^{N-1}5d$	1	2	3	4	5	6	7	8	9	10	11	12	13	14
$\Delta_E(fd)$	57811	65390	71828	77596	82785	87426	91536	95254	98529	101436	104036	106249	108198	109862
$F^2(ff)$	—	—	108418	111920	115305	118587	121777	124899	127951	130946	133896	136792	139652	142478
$F^4(ff)$	—	—	68285	70478	72595	74643	76631	78574	80471	82331	84161	85955	87726	89475
$F^6(ff)$	—	—	49198	50775	52296	53766	55192	56586	57945	59277	60588	61873	63140	64391
$\zeta(ff)$	—	903	1036	1180	1335	1502	1682	1875	2083	2305	2543	2796	3067	3355
$F^2(fd)$	—	30271	30301	30208	30111	30011	29906	29796	29682	29563	29440	29313	29180	29045
$F^4(fd)$	—	15095	15039	14891	14748	14611	14478	14348	14219	14093	13968	13844	13721	13598
$G^1(fd)$	—	12904	12915	12741	12591	12462	12349	12247	12157	12075	11998	11928	11862	11799
$G^3(fd)$	—	11160	11135	10986	10851	10728	10615	10509	10409	10314	10222	10133	10048	9964
$G^5(fd)$	—	8692	8660	8541	8432	8330	8236	8146	8061	7979	7899	7822	7746	7673
$\zeta(dd)$	1095	1161	1229	1297	1366	1434	1504	1573	1644	1714	1786	1857	1929	2001
$r^2(ff)$	—	0.2946	0.2788	0.2621	0.2474	0.2342	0.2225	0.2118	0.2022	0.1933	0.1852	0.1777	0.1708	0.1644
$r^4(ff)$	—	0.1876	0.1706	0.1520	0.1365	0.1235	0.1125	0.1029	0.0946	0.0873	0.0809	0.0752	0.0701	0.0656
$r^6(ff)$	—	0.2222	0.1965	0.1674	0.1444	0.1258	0.1107	0.0981	0.0876	0.0787	0.0711	0.0646	0.0590	0.0541
$r^2(dd)$	1.6595	1.6116	1.5679	1.5290	1.4934	1.4612	1.4315	1.4043	1.3791	1.3556	1.3341	1.3139	1.2954	1.2778
$r^4(dd)$	4.2611	4.0376	3.8392	3.6675	3.5154	3.3805	3.2590	3.1500	3.0512	2.9618	2.8802	2.8057	2.7383	2.6756

(Continued)

Table 6.6 (*Continued*)

Ion	Pr^{4+}	Nd^{4+}	Pm^{4+}	Sm^{4+}	Eu^{4+}	Gd^{4+}	Tb^{4+}	Dy^{4+}	Ho^{4+}	Er^{4+}	Tm^{4+}	Yb^{4+}	Lu^{4+}
$4f^{N-1}5d$	1	2	3	4	5	6	7	8	9	10	11	12	13
$\Delta_E(fd)$	123628	130830	137075	142859	148118	152875	157239	161159	164711	167965	170807	173377	175670
$F^2(ff)$	—	—	119622	122813	125933	128983	131985	134932	137835	140704	143529	146325	149093
$F^4(ff)$	—	—	75723	77710	79649	81542	83403	85229	87025	88799	90544	92270	93979
$F^6(ff)$	—	—	54671	56095	57484	58840	60172	61478	62763	64032	65279	66512	67733
$\zeta(ff)$	—	1143	1293	1454	1627	1812	2012	2226	2455	2700	2961	3239	3535
$F^2(fd)$	—	34087	34202	34206	34202	34192	34173	34149	34118	34081	34038	33989	33935
$F^4(fd)$	—	16969	16972	16889	16806	16725	16643	16561	16480	16398	16315	16232	16149
$G^1(fd)$	—	13150	13265	13216	13176	13145	13119	13099	13083	13069	13059	13050	13043
$G^3(fd)$	—	11982	12031	11965	11904	11849	11795	11745	11697	11650	11605	11560	11516
$G^5(fd)$	—	9495	9516	9455	9398	9344	9291	9241	9192	9143	9096	9050	9004
$\zeta(dd)$	1496	1585	1675	1767	1860	1955	2051	2149	2248	2349	2451	2554	2658
$r^2(ff)$	—	0.2413	0.2304	0.2188	0.2082	0.1986	0.1898	0.1817	0.1743	0.1674	0.1610	0.1551	0.1495
$r^4(ff)$	—	0.1190	0.1100	0.1001	0.0914	0.0839	0.0773	0.0715	0.0663	0.0617	0.0576	0.0539	0.0505
$r^6(ff)$	—	0.1049	0.0949	0.0832	0.0735	0.0654	0.0585	0.0527	0.0477	0.0433	0.0395	0.0362	0.0333
$r^2(dd)$	1.3237	1.2876	1.2543	1.2237	1.1954	1.1691	1.1445	1.1215	1.1002	1.0801	1.0610	1.0431	1.0263
$r^4(dd)$	2.6363	2.5038	2.3838	2.2772	2.1808	2.0929	2.0137	1.9408	1.8734	1.8114	1.7550	1.7024	1.6538

Ion	Ac^{2+}	Th^{2+}	Pa^{2+}	U^{2+}	Np^{2+}	Pu^{2+}	Am^{2+}	Cm^{2+}	Bk^{2+}	Cf^{2+}	Es^{2+}	Fm^{2+}	Md^{2+}	No^{2+}
$5f^{N-1}6d$	1	2	3	4	5	6	7	8	9	10	11	12	13	14
$\Delta_E(fd)$	-16431	-7740	230.4	8062	15702	23066	30305	37172	43928	50430	56694	63106	69507	75060
$F^2(ff)$	—	—	63094	67198	70919	74378	77616	80704	83643	86484	89219	91884	94468	97001
$F^4(ff)$	—	—	40659	43428	45934	48262	50437	52509	54477	56377	58204	59982	61703	63388
$F^6(ff)$	—	—	29632	31698	33569	35307	36930	38475	39943	41359	42720	44044	45326	46579
$\zeta(ff)$	—	1260	1528	1808	2097	2398	2712	3042	3386	3748	4133	4525	4948	5386
$F^2(fd)$	—	30285	30223	30039	29822	29582	29326	29056	28776	28485	28188	27882	27572	27255
$F^4(fd)$	—	17194	16885	16516	16163	15821	15493	15172	14862	14558	14263	13972	13688	13408
$G^1(fd)$	—	22121	20862	19597	18530	17603	16793	16066	15416	14820	14277	13773	13308	12871
$G^3(fd)$	—	15749	15030	14285	13638	13063	12549	12078	11649	11249	10878	10530	10203	9892
$G^5(fd)$	—	11643	11157	10649	10202	9799	9436	9100	8791	8502	8231	7976	7735	7505
$\zeta(dd)$	1689	1835	1966	2083	2188	2287	2380	2468	2551	2631	2707	2779	2848	2914
$r^2(ff)$	—	0.9445	0.8283	0.7326	0.6589	0.5998	0.5514	0.5102	0.4752	0.4447	0.4178	0.3940	0.3727	0.3537
$r^4(ff)$	—	1.8436	1.4233	1.1049	0.8900	0.7348	0.6193	0.5295	0.4587	0.4013	0.3544	0.3152	0.2824	0.2542
$r^6(ff)$	—	6.5459	4.5059	3.0852	2.2310	1.6761	1.2991	1.0294	0.8325	0.6834	0.5690	0.4789	0.4076	0.3496
$r^2(dd)$	2.9151	2.7510	2.6272	2.5315	2.4533	2.3875	2.3318	2.2834	2.2416	2.2045	2.1725	2.1439	2.1187	2.0966
$r^4(dd)$	13.2445	11.8565	10.8684	10.1392	9.5746	9.1198	8.7512	8.4297	8.1709	7.9435	7.7530	7.5883	7.4487	7.3280

(Continued)

Table 6.6 (*Continued*)

Ion	Th^{3+}	Pa^{3+}	U^{3+}	Np^{3+}	Pu^{3+}	Am^{3+}	Cm^{3+}	Bk^{3+}	Cf^{3+}	Es^{3+}	Fm^{3+}	Md^{3+}	No^{3+}	Lr^{3+}
$5f^{N-1}6d$	1	2	3	4	5	6	7	8	9	10	11	12	13	14
$\Delta_E(fd)$	15353	25121	34465	43514	52358	60589	69383	77398	85398	92720	100908	108135	115677	122355
$F^2(ff)$	—	—	72778	76100	79252	82246	85133	87908	90609	93226	95788	98283	100738	103137
$F^4(ff)$	—	—	47442	49675	51790	53795	55727	57581	59383	61127	62832	64491	66121	67713
$F^6(ff)$	—	—	34770	36437	38016	39512	40953	42334	43677	44975	46245	47479	48691	49875
$\zeta(ff)$	—	1755	2029	2319	2623	2941	3275	3625	3993	4385	4783	5214	5660	6126
$F^2(fd)$	—	34160	34231	34203	34148	34070	33973	33861	33734	33595	33446	33286	33118	32940
$F^4(fd)$	—	19117	18953	18723	18496	18272	18050	17831	17613	17398	17184	16973	16763	16554
$G^1(fd)$	—	20944	20251	19451	18744	18119	17551	17040	16568	16136	15733	15360	15009	14680
$G^3(fd)$	—	15964	15546	15052	14605	14201	13827	13484	13161	12861	12576	12308	12053	11810
$G^5(fd)$	—	12124	11833	11488	11172	10884	10614	10365	10129	9907	9696	9496	9304	9120
$\zeta(dd)$	2498	2664	2825	2976	3123	3265	3404	3539	3672	3804	3931	4059	4183	4305
$r^2(ff)$	—	0.6844	0.6259	0.5729	0.5284	0.4906	0.4581	0.4296	0.4044	0.3820	0.3618	0.3436	0.3268	0.3117
$r^4(ff)$	—	0.8728	0.7352	0.6158	0.5242	0.4525	0.3946	0.3475	0.3083	0.2756	0.2476	0.2237	0.2031	0.1852
$r^6(ff)$	—	1.8841	1.4761	1.1366	0.8966	0.7222	0.5909	0.4908	0.4119	0.3496	0.2993	0.2585	0.2246	0.1965
$r^2(dd)$	2.1837	2.0901	2.0103	1.9428	1.8840	1.8325	1.7860	1.7449	1.7071	1.6732	1.6418	1.6135	1.5872	1.5631
$r^4(dd)$	7.1178	6.5454	6.0780	5.7001	5.3809	5.1096	4.8736	4.6681	4.4870	4.3262	4.1819	4.0533	3.9365	3.8322

Ion	Pa^{4+}	U^{4+}	Np^{4+}	Pu^{4+}	Am^{4+}	Cm^{4+}	Bk^{4+}	Cf^{4+}	Es^{4+}	Fm^{4+}	Md^{4+}	No^{4+}	Lr^{4+}
$5f^{N-1}6d$	1	2	3	4	5	6	7	8	9	10	11	12	13
$\Delta_E(fd)$	54011	64683	74950	85170	94846	104505	114105	123465	132131	141022	149653	158193	165835
$F^2(ff)$	—	—	80550	83498	86327	89075	91733	94333	96863	99350	101779	104175	106524
$F^4(ff)$	—	—	52917	54890	56780	58614	60386	62117	63799	65451	67063	68651	70207
$F^6(ff)$	—	—	38940	40412	41821	43188	44508	45797	47049	48278	49476	50657	51812
$\zeta(ff)$	—	2251	2540	2848	3171	3511	3867	4241	4640	5046	5485	5939	6413
$F^2(fd)$	—	37617	37769	37836	37879	37901	37904	37890	37863	37822	37770	37707	37633
$F^4(fd)$	—	20894	20812	20665	20516	20365	20213	20060	19906	19751	19596	19440	19283
$G^1(fd)$	—	20294	19883	19337	18850	18404	18000	17625	17281	16959	16660	16378	16114
$G^3(fd)$	—	16289	16032	15683	15363	15065	14790	14530	14287	14056	13837	13628	13429
$G^5(fd)$	—	12625	12442	12193	11963	11746	11544	11352	11170	10996	10831	10671	10518
$\zeta(dd)$	3362	3559	3752	3942	4128	4314	4497	4679	4862	5041	5222	5402	5579
$r^2(ff)$	—	0.5511	0.5133	0.4775	0.4467	0.4192	0.3951	0.3733	0.3540	0.3363	0.3204	0.3055	0.2921
$r^4(ff)$	—	0.5347	0.4673	0.4052	0.3551	0.3137	0.2792	0.2501	0.2253	0.2039	0.1854	0.1692	0.1551
$r^6(ff)$	—	0.8397	0.6939	0.5632	0.4647	0.3876	0.3274	0.2789	0.2396	0.2074	0.1808	0.1585	0.1398
$r^2(dd)$	1.7636	1.6981	1.6401	1.5889	1.5432	1.5018	1.4640	1.4295	1.3976	1.3682	1.3411	1.3159	1.2921
$r^4(dd)$	4.5215	4.2047	3.9349	3.7036	3.5044	3.3280	3.1727	3.0339	2.9092	2.7963	2.6944	2.6011	2.5164

Table 6.7 Linear approximations $a_1 + a_2Z$ of the Slater parameters $F^2(ff)$, $F^4(ff)$ and $F^6(ff)$ as functions of the atomic number Z for the f^{N-1} core configurations of the di-, tri-, and tetravalent lanthanides and actinides

		Lanthanides		Actinides	
		a_1, cm^{-1}	a_2, cm^{-1}/at.number	a_1, cm^{-1}	a_2, cm^{-1}/ at.number
Divalent ions	$F^2(ff)$	−101177.56±3437.61	3351.42±53.22	−211797.17±6717.78	3038.15±69.57
	$F^4(ff)$	−63508.06±2353.15	2100.18±36.43	−143538.13±4640.26	2036.11±48.05
	$F^6(ff)$	−45484.88±1655.02	1506.99±25.62	−107710.77±3483.82	1518.23±36.08
Trivalent ions	$F^2(ff)$	−75890.02±2079.57	3083.08±31.71	−178369.18±4126.18	2739.85±42.29
	$F^4(ff)$	−46324.66±1368.23	1917.58±20.86	−120186.05±2867.86	1828.98±29.40
	$F^6(ff)$	−32958.82±1002.29	1374.72±15.28	−90124.36±2166.20	1362.78±22.20
Tetravalent ions	$F^2(ff)$	−59384.89±1421.93	2940.75±21.52	−159546.38±2913.98	2587.40±29.72
	$F^4(ff)$	−35125.24±935.02	1821.28±14.15	−106867.18±2030.97	1722.09±20.71
	$F^6(ff)$	−24633.80±683.87	1303.07±10.35	−80010.62±1540.10	1282.05±15.71

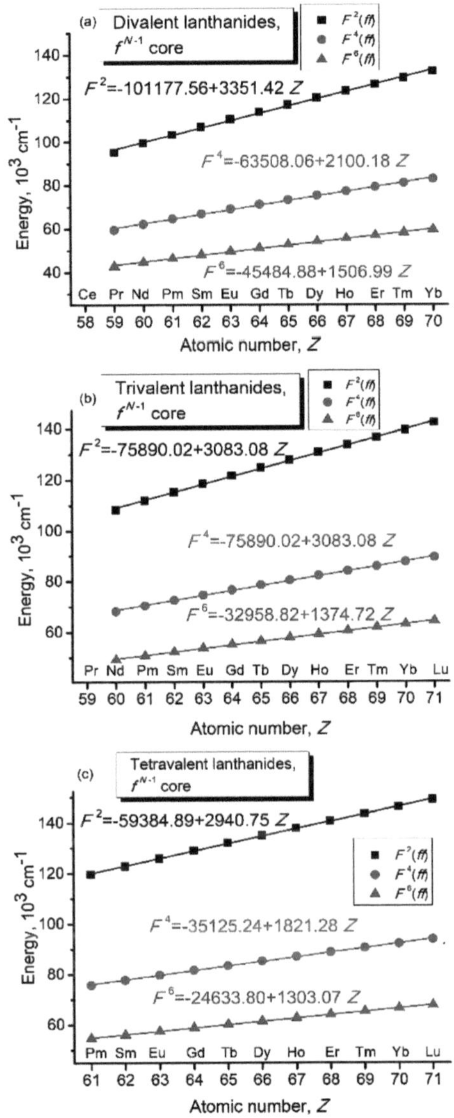

Figure 6.21 Dependence of the calculated direct Coulomb interaction parameters $F^2(ff)$, $F^4(ff)$, $F^6(ff)$ (symbols) on the atomic number Z for the f^{N-1} core electron configurations of di- (a), tri- (b), and tetravalent (c) lanthanides. The linear fits and their equations are shown.

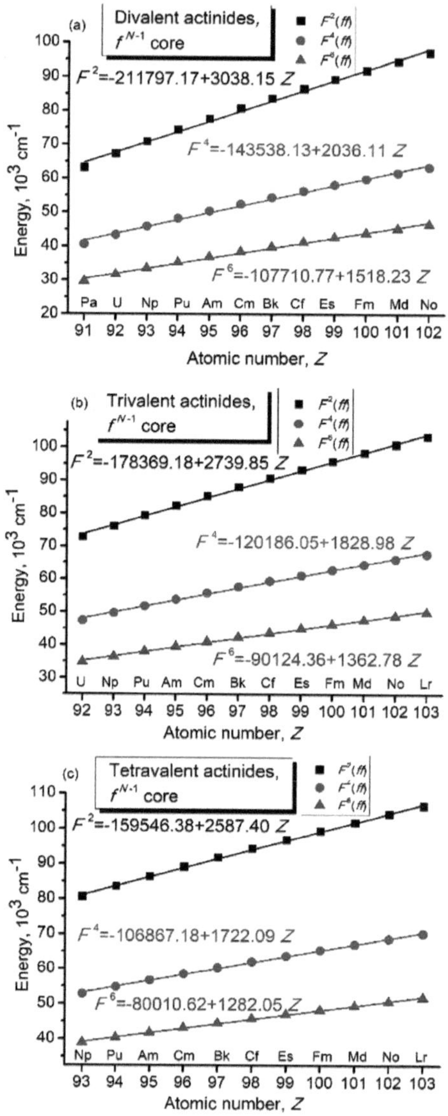

Figure 6.22 Dependence of the calculated direct Coulomb interaction parameters $F^2(ff)$, $F^4(ff)$, $F^6(ff)$ (symbols) on the on atomic number Z for the f^{N-1} core electron configurations of di- (a), tri- (b), and tetravalent (c) actinides. The linear fits and their equations are shown.

The behavior of the $F^2(fd)$, $F^4(fd)$, $G^1(dd)$, $G^3(dd)$, $G^5(dd)$ parameters as the functions of the atomic number is exhibited in Figs. 6.23 and

6.24. In the contrast to the behavior of the core configurations electrostatic parameters, these ones are the decreasing functions

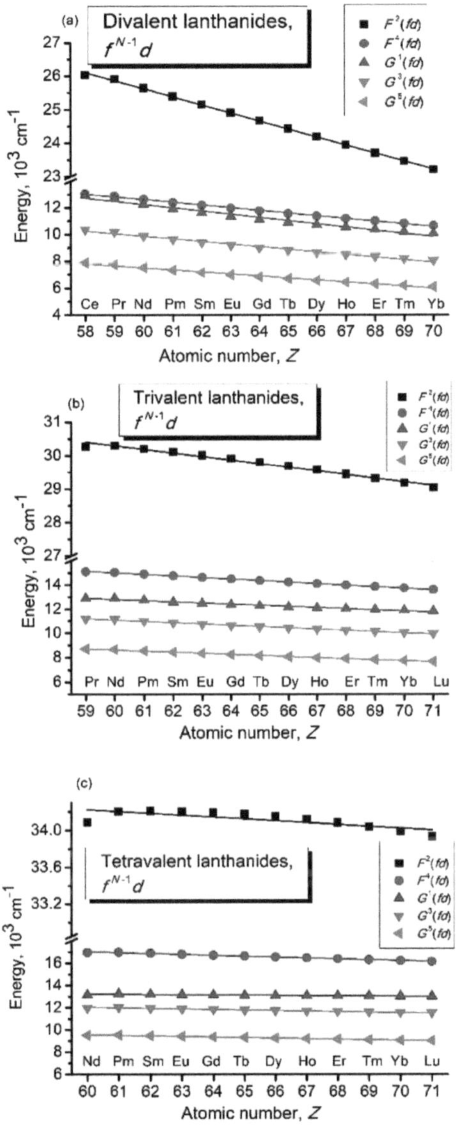

Figure 6.23 Dependence of the calculated direct $F^2(fd)$, $F^4(fd)$ and exchange $G^1(dd)$, $G^3(dd)$, $G^5(dd)$ Coulomb interaction parameters (symbols) on the atomic number Z for the $f^{N-1}d$ electron configurations of di- (a), tri- (b), and tetravalent (c) lanthanides.

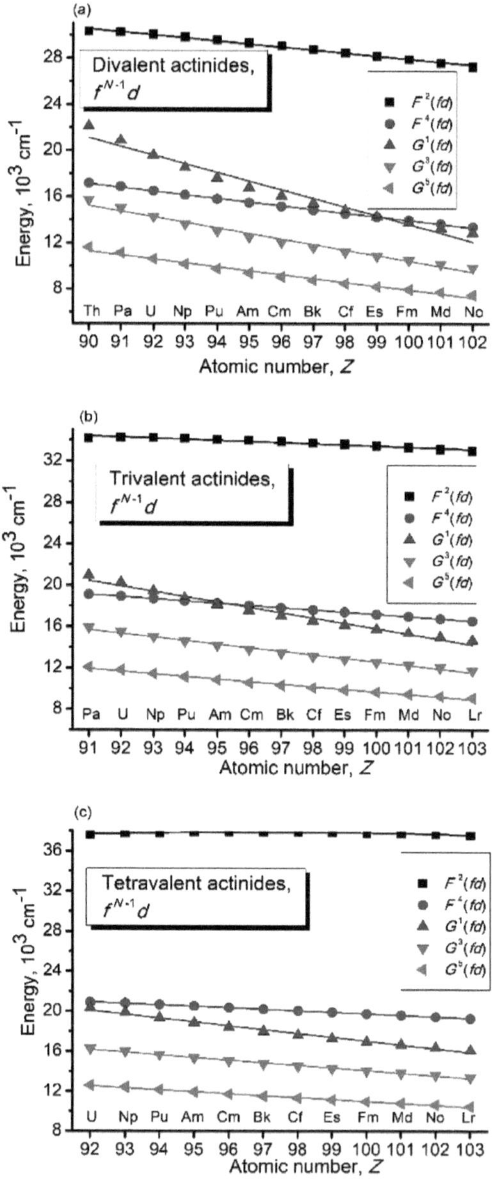

Figure 6.24 Dependence of the calculated direct $F^2(fd)$, $F^4(fd)$ and exchange $G^1(dd)$, $G^3(dd)$, $G^5(dd)$ Coulomb interaction parameters (symbols) on the atomic number Z for the $f^{N-1}d$ electron configurations of di- (a), tri- (b), and tetravalent (c) actinides.

of the atomic number *Z*. The decreasing trend of the *f–d* Coulomb interaction parameters indicates the fact that the contraction of the inner 4*f* (5*f*) orbitals is much stronger than that of the outmost 5*d* (6*d*) orbitals with the nuclear charge increasing. The coefficients of the $b_1 + b_2Z$ linear approximations are all collected in Table 6.8. The absolute values of these fitting coefficients decrease in the $F^2(fd) \rightarrow F^4(fd)$ and $G^1(dd) \rightarrow G^3(dd) \rightarrow G^5(dd)$ sequences. We also note that some of the $F^2(fd)$, $F^4(fd)$, $G^1(dd)$, $G^3(dd)$ and $G^5(dd)$ values for the tetravalent 4*f* and 5*f* ions slightly increase in the very beginning of the lanthanide and actinide series, and then monotonously decrease. Such behavior can be attributed to the interplay between the inner *f* orbitals shrinking and simultaneous increase of the effective charge and size of nuclei with increasing atomic number. In other words, the relative *f*-orbitals shrinking is stronger for the early members of the lanthanide/actinide groups, which effectively increases the separation between the 4*f*/5*f* and 5*d*/6*d* electrons, leading to a weakened Coulomb interaction.

It is also noted that the $F^4(ff)/F^2(ff)$ and $F^6(ff)/F^2(ff)$ ratios for the di, tri- and tetravalent lanthanides are pretty much constant across the whole series of the lanthanides and actinides. Besides, the $F^4(fd)/F^2(fd)$, $G^3(fd)/G^1(fd)$ and $G^5(fd)/G^1(fd)$ ratios vary smoothly in a rather narrow intervals, as displayed by Fig. 6.25. With a good accuracy, they can be taken equal to the mid-range values, which are shown in this figure.

As long as the SO interaction constants $\zeta(ff)$ and $\zeta(dd)$ are concerned, they also perfectly follow the Z^4 law, yielding $\zeta(ff)^{1/4}$ and $\zeta(dd)^{1/4}$ be the linear functions of the atomic number *Z* (Figs. 6.26 and 6.27), whose equations are all presented in these two figures. It is interesting to note that the $\zeta(ff) < \zeta(\mathrm{dd})$ relation between both SO constants is valid for the early members of the lanthanides/actinides series and becomes inverted with increasing atomic number. The point of intersection of the $\zeta(ff)^{1/4}$ and $\zeta(dd)^{1/4}$ linear fits moves to the elements with greater *Z* when increasing the charge of the ions. For the di-, tri-, and tetravalent lanthanides such a shift is as follows: Pr → Sm → Tb; for the di-, tri-, and tetravalent actinides—Np→ Cm → Fm. An overall observation is that the $\zeta(ff)^{1/4}$ dependence on *Z* is considerably stronger than that one of the $\zeta(dd)^{1/4}$ constant, with the coefficient of linear approximation being about 2–3 times greater.

Table 6.8 Linear approximations $b_1 + b_2Z$ of the Slater parameters $F^2(fd)$, $F^4(fd)$, $G^1(fd)$, $G^3(fd)$, $G^5(fd)$ as functions of the atomic number Z for the $f^{N-1}d$ configurations of the di-, tri-, and tetravalent lanthanides and actinides

		Lanthanides		Actinides	
		b_1, cm^{-1}	b_2, cm^{-1}/at.number	b_1, cm^{-1}	b_2, cm^{-1}/ at.number
Divalent ions	$F^2(fd)$	40007.66±131.20	−239.74±2.05	54126.03±787.86	−262.14±8.15
	$F^4(fd)$	24837.59±217.65	−203.67±3.39	45683.48±372.89	−317.22±3.88
	$G^1(fd)$	26613.24±763.66	−239.66±11.91	88495.86±3930.58	−748.73±40.91
	$G^3(fd)$	21504.44±444.29	−194.13±6.93	58314.19±2039.24	−478.60±21.23
	$G^5(fd)$	16470.57±301.85	−149.65±4.71	41874.92±1300.65	−339.46±13.54
Trivalent ions	$F^2(fd)$	36782.35±256.59	−108.10±3.94	44329.48±841.65	−109.05±8.67
	$F^4(fd)$	22713.28±86.74	−128.54±1.33	38865.55±103.78	−216.75±1.07
	$G^1(fd)$	18699.85±279.83	−98.31±4.30	67526.29±2161.86	−517.25±22.27
	$G^3(fd)$	17309.87±164.48	−104.09±2.53	47119.11±1134.69	−345.04±11.69
	$G^5(fd)$	13897.68±114.20	−88.11±1.75	34723.82±725.38	−250.02±7.47
Tetravalent ions	$F^2(fd)$	35400.52±323.93	−19.64±4.94	38227.77±856.55	−4.40±8.78
	$F^4(fd)$	21760.48±123.34	−78.86±1.88	34752.47±176.81	−150.02±1.81
	$G^1(fd)$	14256.95±181.06	−17.31±2.76	55299.01±1301.51	−382.74±13.34
	$G^3(fd)$	14903.57±129.25	−47.77±1.97	40367.38±675.30	−262.75±6.92
	$G^5(fd)$	12449.62±93.23	−48.56±1.42	30392.32±425.14	−193.72±4.36

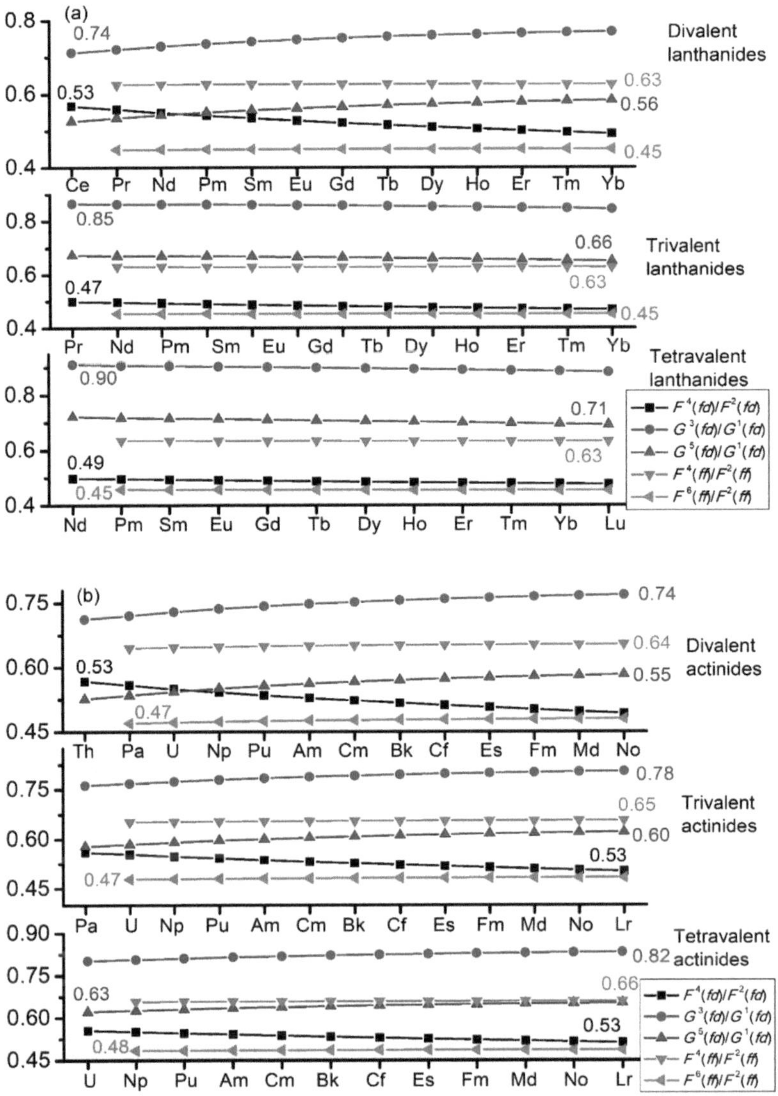

Figure 6.25 Calculated $F^4(ff)/F^2(ff)$, $F^6(ff)/F^2(ff)$, $F^4(fd)/F^2(fd)$, $G^3(fd)/G^1(fd)$ and $G^5(fd)/G^1(fd)$ ratios for the considered groups of the lanthanide (a) and actinide (b) ions.

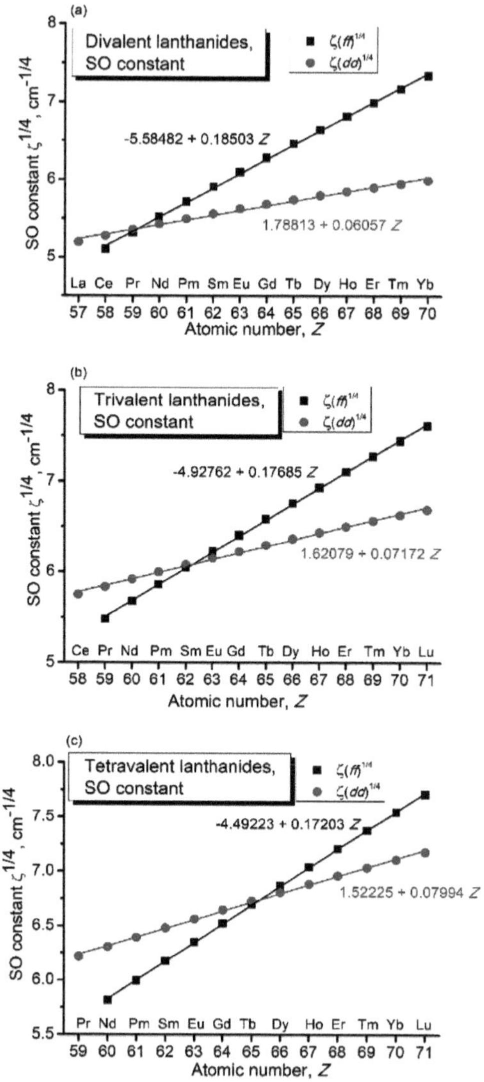

Figure 6.26 Dependence of the SO constants $\zeta(ff)^{1/4}$ and $\zeta(dd)^{1/4}$ on atomic number Z for the di- (a), tri- (b), and tetravalent (c) lanthanides.

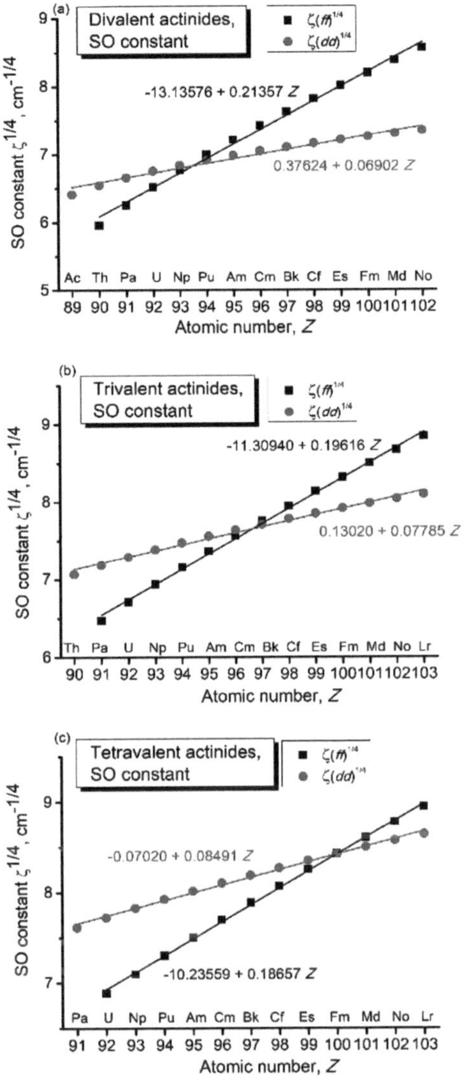

Figure 6.27 Dependence of the SO constants $\zeta(ff)^{1/4}$ and $\zeta(dd)^{1/4}$ on atomic number Z for the di- (a), tri- (b), and tetravalent (c) actinides.

The derived linear equations for the $\zeta(ff)^{1/4}$ value (expressed in $cm^{-1/4}$) are as follows (Z^* denotes an effective nuclear charge):

Divalent lanthanides:

$$\zeta(ff)^{1/4} = (-5.58482 \pm 0.10287) + (0.18503 \pm 0.00160)Z = 0.18503\,(Z-30.18) = 0.18503\,Z^*$$

Trivalent lanthanides:

$$\zeta(ff)^{1/4} = (-4.92762\pm0.06299) + (0.17685\pm0.00097)Z = 0.17685\,(Z-27.86) = 0.17685\,Z^*$$

Tetravalent lanthanides:

$$\zeta(ff)^{1/4} = (-4.49223\pm0.04361) + (0.17203\pm0.00066)Z = 0.17203\,(Z-26.11) = 0.17203\,Z^*$$

Divalent actinides:

$$\zeta(ff)^{1/4} = (-13.13576\pm0.44907) + (0.21357\pm0.00467)Z = 0.21357\,(Z-61.51) = 0.21357\,Z^*$$

Trivalent actinides:

$$\zeta(ff)^{1/4} = (-11.30490\pm0.27410) + (0.19616\pm0.00282)Z = 0.19616\,(Z-57.63) = 0.19616\,Z^*$$

Tetravalent actinides:

$$\zeta(ff)^{1/4} = (-10.23559\pm0.19000) + (0.18657\pm0.00195)Z = 0.18657\,(Z-54.86) = 0.18657\,Z^*$$

The derived linear equations for the $\zeta(dd)^{1/4}$ value (expressed in $cm^{-1/4}$) are as follows (Z^* denotes an effective nuclear charge):

Divalent lanthanides:

$$\zeta(dd)^{1/4} = (1.78813\pm0.09530) + (0.06057\pm0.00150)Z = 0.06057\,(Z+29.52) = 0.06057\,Z^*$$

Trivalent lanthanides:

$$\zeta(dd)^{1/4} = (1.62079\pm0.06771) + (0.07172\pm0.00105)Z = 0.07172\,(Z+22.60) = 0.07172\,Z^*$$

Tetravalent lanthanides:

$$\zeta(dd)^{1/4} = (1.52225\pm0.05427) + (0.07994\pm0.00083)Z = 0.07994\,(Z+19.04) = 0.07994\,Z^*$$

Divalent actinides:

$$\zeta(dd)^{1/4} = (0.37624\pm0.32947) + (0.06902\pm0.00345)Z = 0.06902\,(Z+5.45) = 0.06902\,Z^*$$

Trivalent actinides:

$$\zeta(dd)^{1/4} = (0.13020\pm0.21784) + (0.07785\pm0.00226)Z = 0.07785\,(Z+1.67) = 0.07785\,Z^*$$

Tetravalent actinides:

$$\zeta(dd)^{1/4} = (-0.07020\pm0.16396) + (0.08491\pm0.00169)Z = 0.08491\,(Z-0.83) = 0.08491\,Z^*$$

The effective nuclear charge Z^* is less than Z for the $4f/5f$ SO interaction constant and is greater than Z for the $5d/6d$ SO interaction constant, which can be explained by different character of the orbitals involved: The $4f/5f$ states are shielded by the outer completely filled orbitals, whereas the $5d/6d$ states are very much extended and are not screened by the completely filled shells.

Since the direct and exchange Coulomb interaction parameters and the SO interaction constants $\zeta(ff)^{1/4}$, $\zeta(dd)^{1/4}$ are proportional to the atomic number Z, there should be a linear dependence linking all these parameters as well. Figure 6.28 and Table 6.9 confirm and illustrate such a relation. Indeed, the $F^2(ff)$, $F^4(ff)$, and $F^6(ff)$ parameters all grow linearly with $\zeta(ff)^{1/4}$ (equations of linear fits are collected in Table 6.9). The steepest dependence on $\zeta(ff)^{1/4}$ is realized for the $F^2(ff)$ parameter for both lanthanides and actinides; the coefficients c_2 of the linear approximations $c_1 + c_2\zeta(ff)^{1/4}$ decrease with increase of the oxidation state.

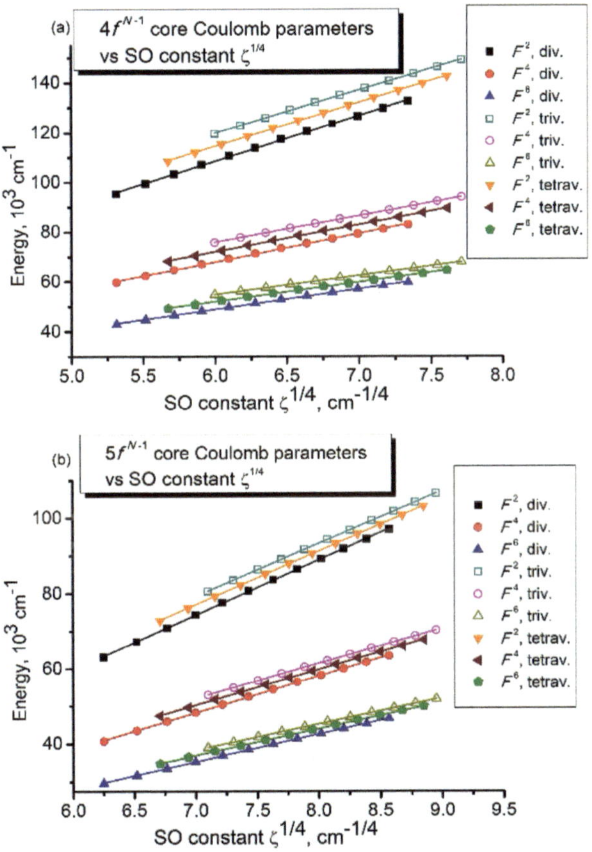

Figure 6.28 Dependence of the direct Coulomb interaction parameters $F^2(ff)$, $F^4(ff)$, $F^6(ff)$ on the SO constant $\zeta(ff)^{1/4}$ for the di-, tri-, and tetravalent lanthanides (a) and actinides (b). The linear fits are shown.

Table 6.9 Linear approximations $c_1 + c_2\zeta(ff)^{1/4}$ of the Slater parameters $F^2(ff)$, $F^4(ff)$ and $F^6(ff)$ as functions of the SO interaction constant $\zeta(ff)^{1/4}$ for the di-, tri-, and tetravalent lanthanides and actinides

		Lanthanides		Actinides	
		c_1, cm^{-1}	c_2, cm$^{-3/4}$	c_1, cm^{-1}	c_2, cm$^{-3/4}$
Divalent ions	$F^2(ff)$	−1354.04 ± 974.62	18313.59 ± 152.66	−28130.92 ± 471.18	14633.38 ± 62.67
	$F^4(ff)$	−961.20 ± 718.72	11477.55 ± 112.58	−20457.83 ± 367.43	9808.31 ± 48.87
	$F^6(ff)$	−602.85 ± 497.18	8235.55 ± 77.88	−15937.34 ± 282.98	7313.80 ± 37.64
Trivalent ions	$F^2(ff)$	16813.79 ± 475.42	17177.97 ± 69.06	−19033.43 ± 237.07	14039.52 ± 29.36
	$F^4(ff)$	12065.51 ± 327.48	10638.89 ± 47.57	−13349.71 ± 197.77	9344.69 ± 24.50
	$F^6(ff)$	9129.50 ± 243.37	7611.84 ± 35.35	−10390.21 ± 159.63	6957.03 ± 19.77
Tetravalent ions	$F^2(ff)$	9200.79 ± 641.98	17549.81 ± 96.02	−22466.17 ± 304.99	14212.46 ± 38.83
	$F^4(ff)$	6597.20 ± 442.66	10915.77 ± 66.21	−16119.36 ± 250.42	9488.26 ± 31.88
	$F^6(ff)$	4980.48 ± 329.75	7825.65 ± 49.32	−12585.35 ± 198.83	7069.94 ± 25.31

Correlation between the Coulomb interaction parameters $F^2(fd)$, $F^4(fd)$, $G^1(dd)$, $G^3(dd)$, $G^5(dd)$ for the $f^{N-1}d$ configuration and the SO interaction constant is not that straightforward, since they link together electronic states from different orbitals. To overcome this difficulty, we introduce a new combination of the $\zeta(ff)$ and $\zeta(dd)$ constants, namely, $[\zeta(ff)\zeta(dd)]^{1/8}$, which seems to be a natural choice, also consistent with the dimension and parameterization of the Coulomb parameters of the f^N and f^{N-1} core configurations. The obtained correlations between the calculated values of $F^2(fd)$, $F^4(fd)$, $G^1(dd)$, $G^3(dd)$, $G^5(dd)$ and $[\zeta(ff)\zeta(dd)]^{1/8}$ are shown in Figs. 6.29 and 6.30. As seen, the Coulomb interaction parameters behave linearly with $[\zeta(ff)\zeta(dd)]^{1/8}$; the linear fits coefficients are all collected in Table 6.10.

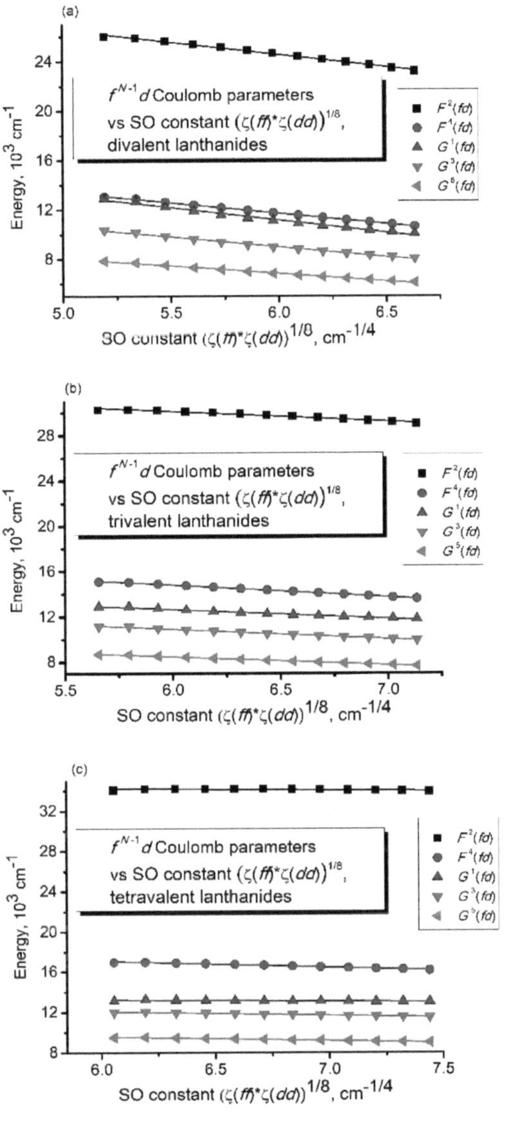

Figure 6.29 Dependence of the calculated values of the direct $F^2(fd)$, $F^4(fd)$ and exchange $G^1(dd)$, $G^3(dd)$, $G^5(dd)$ Coulomb parameters on the $[\zeta(ff)\zeta(dd)]^{1/8}$ parameter for di- (a), tri- (b), and tetravalent (c) lanthanides. The linear fits are shown.

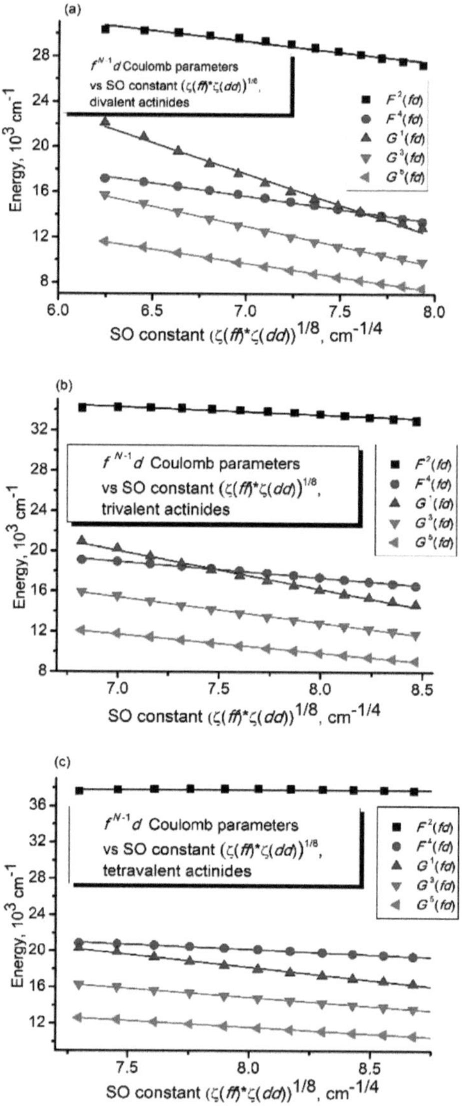

Figure 6.30 Dependence of the calculated values of the direct $F^2(fd)$, $F^4(fd)$ and exchange $G^1(dd)$, $G^3(dd)$, $G^5(dd)$ Coulomb parameters on the $[\zeta(ff)\zeta(dd)]^{1/8}$ parameter for di- (a), tri- (b), and tetravalent (c) actinides. The linear fits are shown.

Table 6.10 Linear approximations $d_1 + d_2[\zeta(ff)\zeta(dd)]^{1/8}$ of the Slater parameters $F^2(fd)$, $F^4(fd)$, $G^1(dd)$, $G^3(dd)$, and $G^5(dd)$ as functions of the SO interaction constant $[\zeta(ff)\zeta(dd)]^{1/8}$ for the di-, tri-, and tetravalent lanthanides and actinides

		Lanthanides		Actinides	
		d_1, cm^{-1}	d_2, cm$^{-3/4}$	d_1, cm^{-1}	d_2, cm$^{-3/4}$
Divalent ions	F^2	36600.49 ± 248.81	−2005.64 ± 41.69	42364.03 ± 826.66	−1866.34 ± 114.81
	F^4	21975.92 ± 69.40	−1709.45 ± 11.63	31677.76 ± 313.31	−2290.24 ± 43.51
	G^1	23309.59 ± 426.48	−2022.27 ± 71.46	55987.39 ± 965.09	−5482.01 ± 134.04
	G^3	18806.76 ± 212.46	−1634.42 ± 35.60	37449.78 ± 368.74	−3492.45 ± 51.21
	G^5	14386.04 ± 133.16	−1259.07 ± 22.31	27050.18 ± 188.12	−2473.51 ± 26.13
Trivalent ions	F^2	35387.83 ± 257.55	−876.99 ± 40.00	39809.10 ± 610.86	−786.63 ± 79.15
	F^4	21079.74 ± 89.10	−1046.71 ± 13.84	30091.27 ± 234.81	−1590.77 ± 30.42
	G^1	17468.26 ± 189.56	−803.26 ± 29.44	46849.01 ± 698.84	−3830.12 ± 90.55
	G^3	15996.10 ± 103.01	−848.97 ± 16.00	33287.23 ± 292.33	−2549.94 ± 37.88
	G^5	12783.99 ± 69.52	−718.40 ± 10.80	24688.96 ± 158.86	−1846.13 ± 20.58
Tetravalent ions	F^2	35154.45 ± 273.59	−153.73 ± 40.35	37987.43 ± 526.29	−23.29 ± 65.02
	F^4	20838.16 ± 123.33	−627.16 ± 18.19	29028.34 ± 222.21	−1101.81 ± 27.35
	G^1	14055.08 ± 149.42	−137.79 ± 22.04	40836.02 ± 486.03	−2828.24 ± 60.05
	G^3	14346.00 ± 112.12	−380.01 ± 16.54	30419.27 ± 209.72	−1939.23 ± 25.91
	G^5	11883.10 ± 83.65	−386.41 ± 12.34	23051.33 ± 116.68	−1428.99 ± 14.42

Finally, an important parameter of the free ion Hamiltonian (6.10) is the so-called the configurational barycenter difference between the f^N and $f^{N-1}d$ electron configurations, i.e., $\Delta_E(fd)$. Visualization of its behavior for all $f^{N-1}d$ electron configurations is offered in Fig. 6.31 for both lanthanides and actinides.

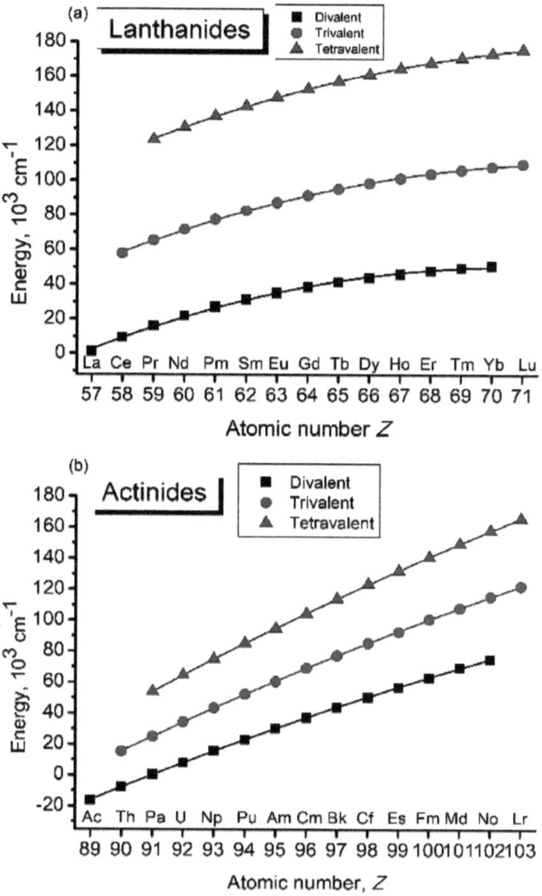

Figure 6.31 Calculated barycenters $\Delta_E(fd)$ of all $f^{N-1}d$ electron configurations for the di-, tri-, and tetravalent lanthanides (a) and actinides (b).

The overall observation is that these energies grow up monotonically with increased atomic number; moreover, the parameter $\Delta_E(fd)$ also shift up with increasing the ionic charge.

However, their variation is not linear for the lanthanides and is almost linear for the actinides. It is described best of all by the second order polynomials, which are given below.

Divalent lanthanides:

$$\Delta_E(fd) = (-1.22652 \pm 0.051680) \times 10^6 + (36152.18 \pm 1632.33)Z + (-255.96 \pm 12.85)\,Z^2$$

Trivalent lanthanides:

$$\Delta_E(fd) = (-1.09888 \pm 0.034139) \times 10^6 + (33076.54 \pm 1061.50)Z + (-226.21 \pm 8.23)\,Z^2$$

Tetravalent lanthanides:

$$\Delta_E(fd) = (-1.00565 \pm 0.025591) \times 10^6 + (31514.42 \pm 789.24)Z + (-209.59 \pm 6.07)\,Z^2$$

Divalent actinides:

$$\Delta_E(fd) = (-1.56086 \pm 0.033327) \times 10^6 + (26386.74 \pm 698.82)Z + (-101.46 \pm 3.66)\,Z^2$$

Trivalent actinides:

$$\Delta_E(fd) = (-1.67460 \pm 0.039889) \times 10^6 + (28015.23 \pm 827.73)Z + (-102.61 \pm 4.29)\,Z^2$$

Tetravalent actinides:

$$\Delta_E(fd) = (-1.84554 \pm 0.030516) \times 10^6 + (32072.62 \pm 629.86)Z + (-112.07 \pm 3.25)\,Z^2$$

6.7 Calculated Energy Levels and Wave Functions of the Trivalent Lanthanides in a Free State

Due to particular importance of the trivalent rare earth ions, in this paragraph we pay more attention to their energy levels in a free state, in relation with the experimental data. Carnall et al [11] listed the parameters of the free ion Hamiltonian (6.1) for all trivalent lanthanides in LaF_3. These values were obtained by fitting the calculated energy levels to the experimental ones with minimizing the deviation between these data sets. The free ion parameters from Ref. 11 are still considered as one of the best empirically identified sets for the $4f$ ions. All they are listed in Table 6.11; they were used

Table 6.11 Parameters of the free ion Hamiltonian (Eq. (6.1)) (in cm^{-1})[11]

	Ce^{3+}	Pr^{3+}	Nd^{3+}	Pm^{3+}	Sm^{3+}	Eu^{3+}	Gd^{3+}	Tb^{3+}	Dy^{3+}	Ho^{3+}	Er^{3+}	Tm^{3+}	Yb^{3+}
F^2		68878	73018	76400	79805	83125	85669	88995	91903	94564	97483	100134	
F^4		50347	52789	54900	57175	59268	60825	62919	64372	66397	67904	69613	
F^6		32901	35757	37700	40250	42560	44776	47252	49386	52022	54010	55975	
ζ	647	751.7	885.3	1025	1176	1338	1508	1707	1913	2145	2376	2636	2928
α		16.23	21.34	20.50	20.16	20.16	18.92	18.40	18.02	17.15	17.79	17.26	
β		−567	−593	−560	−567	−567	−600	−591	−633	−608	−582	−625	
γ		1371	1445	1475	1500	1500	1575	1650	1790	1800	1800	1820	
T^2			298	300	300	300	300	320	329	400	400	400	
T^3			35	35	36	40	42	40	36	37	43		
T^4			59	58	56	60	62	50	127	107	73		
T^6			−285	−310	−347	−300	−295	−395	−314	−264	−271		
T^7			332	350	373	370	350	303	404	316	308		
T^8			305	320	348	320	310	317	315	336	299		
M^0		2.08	2.11	2.4	2.6	2.1	3.22	2.39	3.39	2.54	3.86	3.81	
P^2		−88.6	192	275	357	360	676	373	719	605	594	695	

Note: $M^2 = 0.56M^0$, $M^4 = 0.31M^0$, $P^4 = 0.5P^2$, $P^6 = 0.1P^2$.

to diagonalize the free ion Hamiltonians in the whole basis set for each $4f^N$ electron configuration. In this way, complete energy level scheme of all trivalent lanthanides in the free state was obtained. It should be emphasized that many attempts in the past were made to get such energy level schemes; the reader is kindly advised to the following publications[25,38,39,40] for more details. Below in a series of tables we list the calculated energy levels of all trivalent lanthanides with their corresponding wave functions that were obtained by diagonalizing the free ion with all free ion's parameters taken from well-known Carnall paper [11]. If the number of cnergy levels of free lanthanides is more than fifty, only the first fifty levels are shown. The wave functions list two or three leading contributions. It can be seen that in some cases the energy levels are located very close to each other and the states are mixed up by the SO coupling so strongly, that the first two-three coefficients of the corresponding wave functions are comparable.

Careful inspection of Tables 6.12–6.24 allows us to see the importance of the *J*-states mixing by the SO coupling. In addition, the difference between the direct and inverted muptiplets of the conjugate electron configurations, which was highlighted when discussing the fine structure of the multielectron configurations, can be easily seen as well.

Table 6.12 Energy levels of free Ce^{3+} ion ($4f^1$-configuration) calculated with the Hamiltonian parameters from Table 6.11

	Energy, cm^{-1}	Value of *J*	Leading components of the corresponding wave functions (coefficients and LS terms are given).	
1	0	5/2	1.0000	^{2}F
2	2266	7/2	1.0000	^{2}F

Table 6.13 Energy levels of free Pr^{3+} ion ($4f^2$-configuration) calculated with the Hamiltonian parameters from Table 6.11

	Energy, cm^{-1}	Value of J	Two leading components of the corresponding wave functions (coefficients and LS terms are given).			
1	0	4	0.98619	^3H	0.16295	^1G
2	2081	5	1.0000	^3H		
3	4252	6	0.99846	^3H	−0.05547	^1I
4	5185	2	0.98824	^3F	0.15227	^1D
5	6572	3	1.0000	^3F		
6	6957	4	−0.79603	^3F	0.59289	^1G
7	9934	4	−0.78862	^1G	−0.60454	^3F
8	17201	2	0.94412	^1D	−0.29338	^3P
9	21100	6	0.99846	^1I	0.05547	^3H
10	21104	0	0.99473	^3P	0.10254	^1S
11	21724	1	1.0000	^3P		
12	22911	2	0.95589	^3P	0.29207	^1D
13	47181	0	0.99472	^1S	−0.10254	^3P

Table 6.14 Energy levels of free Nd^{3+} ion ($4f^3$-configuration) calculated with the Hamiltonian parameters from Table 6.11

	Energy, cm^{-1}	Value of J	Three leading components of the corresponding wave functions (coefficients and LS terms are given). The subscript distinguishes between the terms with the same L and S quantum numbers.					
1	0	9/2	-0.9850	^4I	0.16171	$^2H_{(2)}$	-0.05555	$^2H_{(1)}$
2	1857	11/2	-0.99487	^4I	0.09324	$^2H_{(2)}$	-0.03540	$^2H_{(1)}$
3	3836	13/2	0.99780	^4I	0.06211	^2K	-0.02315	^2I
4	5896	15/2	0.99355	^4I	0.11308	^2K	-0.00872	^2L
5	12021	3/2	0.97091	^4F	0.22056	$^2D_{(1)}$	-0.06077	$^2D_{(2)}$
6	12851	9/2	0.76735	$^2H_{(2)}$	0.32670	$^2G_{(1)}$	-0.31961	^4F
7	13055	5/2	0.98816	^4F	0.14518	$^2D_{(1)}$	-0.03217	$^2F_{(2)}$
8	14004	7/2	0.96198	^4F	-0.20798	$^2G_{(1)}$	0.16448	$^2G_{(2)}$
9	14314	3/2	-0.97191	^4S	-0.22113	^2P	-0.06581	^4F
10	15200	9/2	0.88256	^4F	0.40243	$^2H_{(2)}$	-0.15496	$^2G_{(1)}$
11	16114	11/2	0.89725	$^2H_{(2)}$	-0.35868	$^2H_{(1)}$	-0.23238	^4G
12	17645	5/2	-0.99298	^4G	-0.08139	$^2F_{(1)}$	-0.08119	$^2F_{(2)}$
13	17694	7/2	0.61970	^4G	-0.56981	$^2G_{(1)}$	0.48661	$^2G_{(2)}$
14	19236	13/2	-0.99321	^2K	0.09705	^2I	0.06408	^4I

(*Continued*)

Table 6.14 (*Continued*)

	Energy, cm^{-1}	Value of *J*	Three leading components of the corresponding wave functions (coefficients and LS terms are given). The subscript distinguishes between the terms with the same *L* and *S* quantum numbers.					
15	19509	7/2	−0.78032	^4G	−0.47689	$^2G_{(1)}$	0.37356	$^2G_{(2)}$
16	19907	9/2	−0.85788	^4G	0.28361	$^2G_{(1)}$	−0.26669	$^2G_{(2)}$
17	21205	15/2	−0.96948	^2K	0.21794	^2L	0.11225	^4I
18	21498	9/2	−0.62111	$^2G_{(1)}$	0.49401	$^2G_{(2)}$	−0.48791	^4G
19	21914	11/2	−0.96404	^4G	0.20707	$^2H_{(1)}$	−0.16558	$^2H_{(2)}$
20	21971	3/2	0.70062	$^2D_{(1)}$	−0.62722	^2P	−0.19732	^4F
21	24062	1/2	0.96373	^2P	0.26686	^4D	0.0000	^2L
22	24529	5/2	−0.98841	$^2D_{(1)}$	0.14540	^4F	−0.03219	$^2D_{(2)}$
23	26985	3/2	0.72559	^2P	0.63123	$^2D_{(1)}$	−0.18463	$^2D_{(2)}$
24	28911	3/2	−0.90719	^4D	−0.37670	$^2D_{(2)}$	0.13667	^2P
25	29060	5/2	0.89270	^4D	0.42429	$^2D_{(2)}$	−0.10638	$^2F_{(2)}$
26	29422	11/2	0.91868	^2I	−0.37157	$^2H_{(1)}$	−0.11428	$^2H_{(2)}$
27	29465	1/2	0.96373	^4D	−0.26686	^2P	0.0000	^2L
28	29597	15/2	−0.97592	^2L	−0.21752	^2K	0.01619	^4I
29	30752	13/2	−0.99501	^2I	−0.09832	^2K	−0.01696	^4I

	Energy, cm^{-1}	Value of *J*	Three leading components of the corresponding wave functions (coefficients and LS terms are given). The subscript distinguishes between the terms with the same *L* and *S* quantum numbers.					
30	31053	17/2	1.0000	^2L	0.0000	^2L	0.0000	^2L
31	31179	7/2	0.99281	^4D	-0.09700	$^2F_{(1)}$	-0.06717	$^2F_{(2)}$
32	33131	9/2	-0.93431	$^2H_{(1)}$	-0.33466	$^2H_{(2)}$	0.10330	$^2G_{(2)}$
33	34135	3/2	-0.89318	$^2D_{(2)}$	0.38520	^4D	-0.21010	$^2D_{(1)}$
34	34463	11/2	0.83015	$^2H_{(1)}$	0.39113	^2I	0.38179	$^2H_{(2)}$
35	35012	5/2	-0.77928	$^2D_{(2)}$	0.44445	^4D	0.34539	$^2F_{(2)}$
36	39159	5/2	0.67668	$^2F_{(2)}$	0.56135	$^2F_{(1)}$	0.45939	$^2D_{(2)}$
37	40524	7/2	-0.80929	$^2F_{(2)}$	-0.56613	$^2F_{(1)}$	-0.11220	^4D
38	48167	9/2	0.76732	$^2G_{(2)}$	0.63455	$^2G_{(1)}$	0.08142	$^2H_{(1)}$
39	49150	7/2	-0.76884	$^2G_{(2)}$	-0.62969	$^2G_{(1)}$	-0.1022	$^2F_{(2)}$
40	67128	7/2	0.81556	$^2F_{(1)}$	-0.57033	$^2F_{(2)}$	0.06999	$^2G_{(2)}$
41	68326	5/2	0.77175	$^2F_{(1)}$	-0.63545	$^2F_{(2)}$	0.01753	^4D

Table 6.15 Fifty lowest energy levels of free Pm^{3+} ion ($4f^4$-configuration) calculated with the Hamiltonian parameters from Table 6.11

	Energy, cm^{-1}	Value of J	Three leading components of the corresponding wave functions (coefficients and LS terms are given). The subscript distinguishes between the terms with the same L and S quantum numbers.					
1	0	4	0.98586	^5I	−0.12330	$^3H_{(4)}$	0.0820	$^3H_{(3)}$
2	1476	5	0.99213	^5I	−0.09011	$^3H_{(4)}$	0.06149	$^3H_{(3)}$
3	3084	6	0.99384	^5I	0.064730	$^3K_{(2)}$	−0.05293	$^3H_{(4)}$
4	4775	7	−0.98964	^5I	−0.11964	$^3K_{(2)}$	0.06372	$^3K_{(1)}$
5	6513	8	−0.97932	^5I	−0.17848	$^3K_{(2)}$	0.08936	$^3K_{(1)}$
6	12980	1	−0.97667	^5F	0.18705	$^3D_{(1)}$	0.10094	$^3D_{(2)}$
7	13415	2	0.96763	^5F	−0.17510	$^3D_{(1)}$	−0.09864	$^3F_{(3)}$
8	14251	3	0.97608	^5F	−0.11429	$^3F_{(3)}$	−0.10567	$^3D_{(1)}$
9	15069	2	0.96825	^5S	−0.22454	$^3P_{(2)}$	0.075600	^5F
10	15184	4	0.94922	^5F	0.17853	$^3G_{(2)}$	−0.10617	$^3F_{(3)}$
11	16504	5	0.95769	^5F	0.19944	$^3G_{(2)}$	0.12695	$^3G_{(1)}$
12	16548	6	0.86411	$^3K_{(2)}$	−0.48387	$^3K_{(1)}$	−0.07662	^5I
13	17793	7	0.78797	$^3K_{(2)}$	−0.44227	$^3K_{(1)}$	−0.39519	^3L
14	18165	2	0.96192	^5G	−0.20560	$^3F_{(4)}$	−0.11558	$^3F_{(3)}$
15	18219	4	−0.51593	$^3H_{(4)}$	−0.43004	^5G	0.36726	$^3H_{(3)}$
16	18696	3	−0.89784	^5G	0.30507	$^3G_{(2)}$	−0.22364	$^3G_{(3)}$

	Energy, cm^{-1}	Value of J	Three leading components of the corresponding wave functions (coefficients and LS terms are given). The subscript distinguishes between the terms with the same L and S quantum numbers.					
17	19337	8	0.74118	$^3K_{(2)}$	−0.39613	^3L	−0.38377	$^3K_{(1)}$
18	20400	5	−0.62805	^5G	−0.47675	$^3H_{(4)}$	0.32988	$^3H_{(3)}$
19	20637	4	0.84129	^5G	−0.36609	$^3H_{(4)}$	0.29313	$^3H_{(3)}$
20	22213	3	0.71779	$^3G_{(2)}$	−0.46779	$^3G_{(3)}$	0.39523	^5G
21	22549	5	−0.73872	^5G	0.45805	$^3H_{(4)}$	−0.40326	$^3H_{(3)}$
22	22834	6	0.86946	^5G	0.37428	$^3H_{(4)}$	−0.22920	$^3H_{(1)}$
23	23045	7	−0.91309	^3L	−0.36153	$^3K_{(2)}$	0.16489	$^3K_{(1)}$
24	23681	2	0.68132	$^3D_{(1)}$	−0.44819	$^3P_{(2)}$	0.32858	$^3D_{(2)}$
25	24082	8	−0.73537	^3L	0.51287	^3M	−0.39361	$^3K_{(2)}$
26	24774	6	−0.59481	$^3H_{(4)}$	0.53912	$^3H_{(3)}$	0.47991	^5G
27	24790	1	−0.66001	$^3P_{(2)}$	0.62525	$^3D_{(1)}$	0.29801	$^3D_{(2)}$
28	24982	9	−0.85423	^3L	0.51990	^3M	0.0000	^1N
29	25097	4	0.63815	$^3G_{(2)}$	−0.41808	$^3G_{(3)}$	0.30160	$^3H_{(4)}$
30	25192	8	−0.66252	^3M	−0.53348	$^1L_{(2)}$	−0.48480	^3L
31	26404	3	0.81867	$^3D_{(1)}$	0.39571	$^3D_{(2)}$	0.26000	$^3F_{(4)}$
32	26421	0	0.94551	$^3P_{(2)}$	0.31772	$^3P_{(3)}$	0.05738	$^3P_{(1)}$
33	27083	2	0.51766	$^3P_{(2)}$	0.44490	$^1D_{(3)}$	0.34521	$^3F_{(3)}$

(*Continued*)

Table 6.15 (*Continued*)

	Energy, cm^{-1}	Value of J	Three leading components of the corresponding wave functions (coefficients and LS terms are given). The subscript distinguishes between the terms with the same L and S quantum numbers.					
34	27564	5	0.67197	$^3G_{(2)}$	−0.42846	$^3G_{(3)}$	0.30975	$^3G_{(1)}$
35	27820	4	0.58878	$^3F_{(4)}$	0.55739	$^3F_{(2)}$	0.26272	$^1G_{(4)}$
36	28491	9	−0.85423	^3M	−0.51990	^3L	0.0000	^1N
37	28612	1	0.68292	$^3P_{(2)}$	0.60751	$^3D_{(1)}$	0.30761	$^3D_{(2)}$
38	28931	5	0.81486	$^3I_{(1)}$	−0.46833	$^3I_{(2)}$	−0.21399	$^1H_{(1)}$
39	29003	1	0.99285	^3M	0.11940	^1N	0.0000	^1N
40	29472	2	0.55260	$^3P_{(2)}$	0.42927	$^3F_{(2)}$	−0.37246	$^3F_{(3)}$
41	29705	2	0.59160	$^3F_{(3)}$	0.56271	$^3F_{(4)}$	0.34125	$^3F_{(1)}$
42	29990	3	−0.57238	$^3F_{(2)}$	−0.54565	$^3F_{(4)}$	0.38728	$^3F_{(3)}$
43	30027	6	−0.83808	$^3I_{(1)}$	0.47439	$^3I_{(2)}$	0.14819	$^3H_{(2)}$
44	30231	8	−0.73727	$^1L_{(2)}$	0.51307	^3M	0.25377	^3L
45	30493	7	−0.86292	$^3I_{(1)}$	0.44016	$^3I_{(2)}$	−0.21912	^1K
46	30766	0	−0.98329	^5D	−0.14640	$^3P_{(1)}$	0.10527	$^3P_{(3)}$
47	31239	1	−0.98404	^5D	−0.11893	$^3P_{(1)}$	−0.10430	$^3D_{(1)}$
48	32018	2	0.97255	^5D	0.14447	$^3F_{(2)}$	0.10667	$^3D_{(1)}$
49	32554	3	−0.60385	$^3F_{(3)}$	0.57906	^5D	−0.34089	$^3F_{(4)}$
50	33159	3	−0.75885	^5D	−0.38507	$^3F_{(3)}$	−0.37312	$^3F_{(4)}$

Table 6.16 Fifty lowest energy levels of free Sm^{3+} ion ($4f^5$-configuration) calculated with the Hamiltonian parameters from Table 6.11

	Energy, cm^{-1}	Value of J	Three leading components of the corresponding wave functions (coefficients and LS terms are given). The subscript distinguishes between the terms with the same L and S quantum numbers.					
1	0	5/2	–0.97876	^6H	0.15203	$^4G_{(4)}$	0.12538	$^4G_{(1)}$
2	1031	7/2	–0.98557	^6H	0.12350	$^4G_{(4)}$	0.10090	$^4G_{(1)}$
3	2235	9/2	–0.98998	^6H	0.089867	$^4G_{(4)}$	0.072805	$^4G_{(1)}$
4	3558	11/2	0.99054	^6H	0.084663	$^4I_{(3)}$	0.056643	$^4H_{(3)}$
5	4959	13/2	0.98655	^6H	0.13213	$^4I_{(3)}$	0.065661	$^4I_{(1)}$
6	6403	15/2	0.97756	^6H	0.18684	$^4I_{(3)}$	0.088041	$^4I_{(1)}$
7	6655	1/2	–0.98474	^6F	–0.10901	$^4D_{(2)}$	0.10668	$^4D_{(3)}$
8	6905	3/2	–0.97811	^6F	0.10900	$^4F_{(3)}$	–0.10179	$^4D_{(2)}$
9	7406	5/2	–0.97200	^6F	0.14907	$^4F_{(3)}$	–0.088704	$^4D_{(2)}$
10	8272	7/2	0.97522	^6F	–0.15264	$^4F_{(3)}$	0.090565	$^4F_{(1)}$
11	9449	9/2	–0.98228	^6F	0.12300	$^4F_{(3)}$	–0.076200	$^4G_{(3)}$
12	10840	11/2	–0.98701	^6F	–0.10693	$^4G_{(3)}$	0.098687	$^4G_{(2)}$
13	18186	5/2	–0.53828	$^4G_{(4)}$	–0.45128	$^4F_{(3)}$	–0.39970	$^4G_{(1)}$
14	19255	3/2	0.81962	$^4F_{(3)}$	–0.39769	$^4F_{(1)}$	0.35086	$^4F_{(4)}$
15	19665	15/2	–0.87081	^4M	0.27257	^4L	0.26086	$^2L_{(3)}$
16	20306	7/2	0.57904	$^4G_{(4)}$	0.43924	$^4F_{(3)}$	0.40827	$^4G_{(1)}$

(Continued)

Table 6.16 (*Continued*)

	Energy, cm^{-1}	Value of J	Three leading components of the corresponding wave functions (coefficients and LS terms are given). The subscript distinguishes between the terms with the same L and S quantum numbers.					
17	20317	9/2	0.88888	$^4I_{(3)}$	0.39992	$^4I_{(1)}$	0.13060	$^2H_{(6)}$
18	20732	11/2	0.76318	$^4I_{(3)}$	−0.47249	$^4K_{(1)}$	0.34681	$^4I_{(1)}$
19	21114	13/2	0.62187	$^4I_{(3)}$	−0.50046	$^4K_{(1)}$	−0.42172	4L
20	21372	17/2	−0.92473	^4M	0.30818	^4L	0.15854	$^2L_{(3)}$
21	22455	15/2	−0.64509	$^4I_{(3)}$	0.42696	$^4K_{(1)}$	0.36169	^4M
22	22506	5/2	0.62790	$^4F_{(3)}$	−0.41536	$^4G_{(4)}$	−0.32566	$^4G_{(1)}$
23	22861	19/2	−0.95106	^4M	0.27400	^4L	−0.13242	^2N
24	22996	9/2	0.62798	$^4G_{(4)}$	0.41590	$^4G_{(1)}$	−0.40394	$^4G_{(2)}$
25	23925	13/2	0.80120	^4L	0.52803	$^4I_{(3)}$	0.21226	$^4I_{(1)}$
26	24117	21/2	0.96995	^4M	0.23907	^2N	0.045151	^2O
27	24568	5/2	0.63737	^6P	0.43501	$^4P_{(2)}$	0.29601	$^4D_{(2)}$
28	24655	11/2	0.77587	$^4K_{(1)}$	0.47565	$^4I_{(3)}$	−0.33080	$^4K_{(2)}$
29	24899	15/2	0.75156	^4L	0.41851	$^4I_{(3)}$	0.31477	$^4K_{(1)}$
30	25205	7/2	0.64869	$^4F_{(3)}$	−0.38490	$^4G_{(4)}$	−0.30595	$^4F_{(1)}$
31	25546	3/2	0.84492	^6P	0.37320	$^4P_{(2)}$	−0.20721	$^4D_{(3)}$
32	25893	17/2	−0.72944	^4L	−0.54512	$^4K_{(1)}$	−0.32565	^4M
33	25907	11/2	−0.63760	$^4G_{(4)}$	0.50188	$^4G_{(2)}$	−0.37838	$^4G_{(1)}$

	Energy, cm^{-1}	Value of *J*	**Three leading components of the corresponding wave functions (coefficients and LS terms are given). The subscript distinguishes between the terms with the same *L* and *S* quantum numbers.**					
34	26332	13/2	−0.74843	$^4K_{(1)}$	0.33990	^4L	−0.31876	$^4I_{(3)}$
35	26822	1/2	−0.54753	$^4D_{(3)}$	0.53664	$^4D_{(2)}$	0.35901	$^2P_{(4)}$
36	27201	7/2	0.74353	^6P	0.39216	$^4D_{(2)}$	−0.37026	$^4D_{(3)}$
37	27646	9/2	0.69814	$^4F_{(3)}$	−0.32431	$^4F_{(1)}$	−0.32199	$^4H_{(1)}$
38	28000	3/2	−0.57269	$^4D_{(2)}$	0.55387	$^4D_{(3)}$	0.40282	^6P
39	28056	15/2	0.73452	$^4K_{(1)}$	−0.52438	^4L	−0.27082	$^4K_{(2)}$
40	28142	5/2	0.66837	^6P	−0.45400	$^4D_{(2)}$	0.41453	$^4D_{(3)}$
41	28223	7/2	−0.75301	$^4H_{(1)}$	−0.58545	$^4H_{(3)}$	−0.14461	$^2G_{(5)}$
42	28346	17/2	−0.67204	$^4K_{(1)}$	0.52307	^4L	−0.38324	$^2L_{(3)}$
43	28400	19/2	0.93521	^4L	0.28019	^4M	−0.18562	$^2M_{(1)}$
44	28939	9/2	0.65580	$^4H_{(1)}$	0.47169	$^4H_{(3)}$	−0.31762	$^4G_{(4)}$
45	29245	11/2	−0.66782	$^4H_{(1)}$	−0.46050	$^4H_{(3)}$	0.36658	$^4G_{(4)}$
46	29500	13/2	−0.77371	$^4H_{(1)}$	−0.53252	$^4H_{(3)}$	0.24586	$^4I_{(2)}$
47	29534	7/2	0.65251	^6P	−0.51383	$^4D_{(2)}$	0.47337	$^4D_{(3)}$
48	30291	7/2	0.71696	$^4G_{(2)}$	−0.55186	$^4G_{(3)}$	0.28001	$^4G_{(1)}$
49	30307	9/2	0.60306	$^4G_{(2)}$	−0.48843	$^4G_{(3)}$	0.27884	$^4G_{(1)}$
50	30543	5/2	−0.77962	$^4G_{(2)}$	0.54659	$^4G_{(3)}$	−0.22841	$^4G_{(1)}$

(Continued)

Table 6.17 Fifty lowest energy levels of free Eu^{3+} ion ($4f^6$-configuration) calculated with the Hamiltonian parameters from Table 6.11

	Energy, cm^{-1}	Value of J	Three leading components of the corresponding wave functions (coefficients and LS terms are given). The subscript distinguishes between the terms with the same L and S quantum numbers.					
1	0	0	0.96676	^7F	−0.18698	$^5D_{(1)}$	0.16663	$^5D_{(3)}$
2	381	1	0.97321	^7F	−0.16955	$^5D_{(1)}$	0.14750	$^5D_{(3)}$
3	1048	2	0.98105	^7F	−0.14025	$^5D_{(1)}$	0.11629	$^5D_{(3)}$
4	1907	3	0.98679	^7F	−0.10350	$^5D_{(1)}$	0.07978	$^5D_{(3)}$
5	2888	4	−0.98895	^7F	−0.07809	$^5F_{(2)}$	−0.06184	$^5G_{(1)}$
6	3942	5	0.98708	^7F	0.092586	$^5G_{(1)}$	−0.08792	$^5G_{(3)}$
7	5035	6	0.98102	^7F	0.12745	$^5G_{(1)}$	−0.12468	$^5G_{(3)}$
8	17421	0	0.67812	$^5D_{(3)}$	−0.54514	$^5D_{(1)}$	0.25024	$^3P_{(6)}$
9	19174	1	0.71444	$^5D_{(3)}$	−0.57029	$^5D_{(1)}$	−0.21342	^7F
10	21661	2	−0.73843	$^5D_{(3)}$	0.58909	$^5D_{(1)}$	0.19430	$^5D_{(2)}$
11	24119	6	0.94709	^5L	0.17169	$^3K_{(5)}$	0.14499	$^3K_{(1)}$
12	24569	3	−0.74262	$^5D_{(3)}$	0.59737	$^5D_{(1)}$	0.17707	$^5D_{(2)}$
13	25160	7	−0.96698	^5L	0.13195	^5K	−0.12928	$^3K_{(5)}$
14	26077	8	−0.97674	^5L	0.13746	^5K	−0.07862	$^3K_{(5)}$
15	26256	2	0.62360	$^5G_{(3)}$	−0.59979	$^5G_{(1)}$	0.42579	$^5G_{(2)}$
16	26448	3	−0.58746	$^5G_{(3)}$	0.56493	$^5G_{(1)}$	−0.40860	$^5G_{(2)}$

	Energy, cm^{-1}	Value of J	Three leading components of the corresponding wave functions (coefficients and LS terms are given). The subscript distinguishes between the terms with the same L and S quantum numbers.					
17	26523	6	0.48676	$^5G_{(3)}$	0.48169	$^5G_{(2)}$	−0.43095	$^5G_{(1)}$
18	26541	4	0.54728	$^5G_{(3)}$	−0.51949	$^5G_{(1)}$	0.42536	$^5G_{(2)}$
19	26547	5	−0.51308	$^5G_{(3)}$	0.47611	$^5G_{(1)}$	−0.45045	$^5G_{(2)}$
20	26827	9	−0.97895	^5L	0.12021	^5K	−0.12001	$^3M_{(3)}$
21	27321	10	0.97053	^5L	0.18605	$^3M_{(3)}$	−0.13331	$^3M_{(1)}$
22	27836	4	0.72984	$^5D_{(3)}$	−0.59636	$^5D_{(1)}$	−0.18350	$^5F_{(2)}$
23	30658	3	0.68269	$^5H_{(1)}$	0.59529	$^5H_{(2)}$	0.23467	$^5G_{(2)}$
24	30774	7	0.70093	$^5H_{(1)}$	0.48047	$^5H_{(2)}$	−0.43764	$^5I_{(2)}$
25	31066	4	0.59224	$^5H_{(1)}$	0.51433	$^5H_{(2)}$	−0.37228	$^5I_{(2)}$
26	31250	6	−0.55489	$^5H_{(1)}$	0.47278	$^5I_{(2)}$	−0.43235	$^5H_{(2)}$
27	31260	5	0.52713	$^5H_{(1)}$	−0.46946	$^5I_{(2)}$	0.44605	$^5H_{(2)}$
28	33135	0	−0.55245	$^3P_{(6)}$	−0.48197	$^3P_{(3)}$	0.36552	$^3P_{(1)}$
29	33183	2	0.76342	$^5F_{(2)}$	0.49381	$^5F_{(1)}$	0.22201	$^5G_{(2)}$
30	33211	3	−0.73335	$^5F_{(2)}$	−0.47566	$^5F_{(1)}$	−0.29116	$^5G_{(2)}$
31	33498	1	0.81790	$^5F_{(2)}$	0.52187	$^5F_{(1)}$	−0.15030	$^3D_{(5)}$
32	33585	4	−0.55295	$^5I_{(2)}$	0.53739	$^5F_{(2)}$	0.32873	$^5F_{(1)}$
33	33690	4	−0.65135	$^5I_{(2)}$	−0.49972	$^5F_{(2)}$	−0.30833	$^5F_{(1)}$

(Continued)

Table 6.17 (*Continued*)

	Energy, cm^{-1}	Value of J	Three leading components of the corresponding wave functions (coefficients and LS terms are given). The subscript distinguishes between the terms with the same L and S quantum numbers.					
34	33941	5	0.56837	$^5I_{(2)}$	−0.55283	^5K	0.42256	$^5H_{(1)}$
35	34155	5	−0.79855	$^5F_{(2)}$	−0.42680	$^5F_{(1)}$	−0.26300	$^5G_{(2)}$
36	34487	8	−0.92035	$^5I_{(2)}$	0.25737	^5K	−0.15662	$^5I_{(1)}$
37	34544	6	−0.63120	$^5I_{(2)}$	−0.46222	$^5H_{(1)}$	0.45741	^5K
38	34998	7	−0.74719	$^5I_{(2)}$	−0.41482	$^5H_{(1)}$	0.40745	^5K
39	35620	5	−0.73433	^5K	−0.57688	$^5I_{(2)}$	−0.19196	$^3I_{(2)}$
40	36748	6	−0.81581	^5K	−0.49696	$^5I_{(2)}$	−0.15539	$^3I_{(2)}$
41	37654	7	0.86921	^5K	0.41401	$^5I_{(2)}$	0.14558	$^3K_{(5)}$
42	38045	8	0.90364	^5K	0.22420	$^3K_{(5)}$	0.21169	$^5I_{(2)}$
43	38347	6	0.38552	$^3K_{(5)}$	0.28293	$^3K_{(1)}$	−0.27687	$^3I_{(5)}$
44	38501	1	0.60719	$^3P_{(6)}$	0.55030	$^3P_{(3)}$	−0.39305	$^3P_{(1)}$
45	38683	2	−0.70795	$^5G_{(2)}$	−0.37010	$^5G_{(1)}$	0.30326	$^5G_{(3)}$
46	38910	9	0.97274	^5K	−0.13257	$^3L_{(3)}$	−0.13220	$^3L_{(1)}$
47	38996	10	0.79931	^3O	0.34481	^3N	0.26673	$^3M_{(3)}$
48	39185	3	0.71225	$^5G_{(2)}$	0.33897	$^5G_{(1)}$	−0.27898	$^5G_{(3)}$
49	39744	4	0.67370	$^5G_{(2)}$	0.33174	$^5G_{(1)}$	−0.32923	$^5F_{(2)}$
50	40173	2	0.58176	$^5D_{(2)}$	−0.54288	^5P	0.27799	$^5G_{(2)}$

Table 6.18 Fifty lowest energy levels of free Gd^{3+} ion ($4f^7$-configuration) calculated with the Hamiltonian parameters from Table 6.11

	Energy, cm^{-1}	Value of J	**Three leading components of the corresponding wave functions (coefficients and LS terms are given). The subscript distinguishes between the terms with the same L and S quantum numbers.**					
1	0	7/2	0.98583	^8S	0.16596	^6P	0.01447	$^4D_{(6)}$
2	32138	7/2	0.86473	^6P	−0.36687	^6D	0.19796	$^4D_{(6)}$
3	32743	5/2	0.90835	^6P	−0.37791	^6D	0.11175	$^4D_{(6)}$
4	33347	3/2	0.94609	^6P	−0.28633	^6D	−0.10874	$^4S_{(2)}$
5	35157	7/2	0.96145	^6I	0.19589	$^4H_{(2)}$	0.16440	$^4H_{(3)}$
6	35499	9/2	−0.97160	^6I	−0.16254	$^4H_{(2)}$	−0.13714	$^4H_{(3)}$
7	35528	17/2	0.97850	^6I	−0.14995	$^4K_{(1)}$	−0.13579	$^4K_{(2)}$
8	35771	11/2	−0.97965	^6I	−0.12467	$^4H_{(2)}$	−0.10385	$^4H_{(3)}$
9	35878	15/2	0.98530	^6I	−0.10679	$^4K_{(1)}$	−0.09852	$^4K_{(2)}$
10	35919	13/2	0.98448	^6I	0.08036	$^4H_{(2)}$	−0.07197	^6H
11	39563	9/2	−0.95692	^6D	0.20163	^6F	−0.18168	$^4F_{(4)}$
12	40494	1/2	−0.99253	^6D	0.07602	$^4P_{(1)}$	0.07138	^6F
13	40612	7/2	0.88217	^6D	0.35961	^6P	−0.20853	^6F
14	40759	3/2	−0.93945	^6D	−0.28714	^6P	0.13350	^6F
15	40907	5/2	0.89077	^6D	0.38151	^6P	−0.18117	^6F
16	48906	7/2	−0.68821	^6G	0.38500	^6F	−0.26931	$^4F_{(4)}$

(Continued)

Table 6.18 (*Continued*)

	Energy, cm^{-1}	Value of J	Three leading components of the corresponding wave functions (coefficients and LS terms are given). The subscript distinguishes between the terms with the same L and S quantum numbers.					
17	49304	9/2	0.78678	^6G	−0.47787	^6F	−0.21921	^6H
18	49360	11/2	0.77532	^6G	−0.54081	^6F	−0.20908	^6H
19	49391	5/2	0.75008	^6G	−0.40665	^6F	0.30527	$^4F_{(4)}$
20	50179	3/2	−0.80914	^6G	0.40973	^6F	−0.30388	$^4F_{(4)}$
21	50976	13/2	−0.93830	^6G	0.23382	^6H	−0.22615	$^4H_{(2)}$
22	52141	7/2	0.44583	^6G	0.33831	$^4D_{(6)}$	0.28242	$^4D_{(1)}$
23	52507	1/2	−0.86364	^6F	−0.34250	$^4D_{(1)}$	−0.33061	$^4D_{(6)}$
24	53511	3/2	−0.74498	^6F	−0.50107	^6G	−0.28236	$^4D_{(1)}$
25	53977	11/2	0.74204	^6F	0.51937	^6G	0.27199	$^4G_{(6)}$
26	54085	17/2	0.86191	^4N	−0.32725	$^2M_{(2)}$	0.18197	$^4L_{(2)}$
27	54091	5/2	−0.57080	^6G	−0.54071	^6F	−0.39449	$^4D_{(6)}$
28	54248	9/2	−0.69691	^6F	−0.30864	$^4F_{(4)}$	0.26499	^6H
29	54447	5/2	0.64626	^6F	0.36949	$^4F_{(4)}$	−0.34744	$^4D_{(6)}$
30	54460	7/2	−0.77451	^6F	−0.41254	^6G	0.26244	^6H
31	55080	23/2	0.97175	^4N	−0.22259	^2O	−0.07843	^2Q
32	55099	19/2	−0.94302	^4N	0.21261	$^2M_{(2)}$	−0.16092	$^4L_{(2)}$
33	55327	9/2	0.45296	^6G	0.38918	^6F	−0.36770	$^4H_{(3)}$

	Energy, cm^{-1}	Value of *J*	Three leading components of the corresponding wave functions (coefficients and LS terms are given). The subscript distinguishes between the terms with the same *L* and *S* quantum numbers.					
34	55494	21/2	−0.98289	^4N	−0.13079	^4M	0.12262	^2O
35	55862	7/2	−0.46251	$^4H_{(2)}$	−0.38535	$^4H_{(3)}$	−0.36873	$^4D_{(6)}$
36	56093	3/2	−0.59607	$^4D_{(6)}$	0.48017	^6F	−0.47654	$^4D_{(1)}$
37	56277	13/2	0.52527	$^4H_{(2)}$	0.49727	^6H	0.42443	$^4H_{(3)}$
38	56675	5/2	0.75666	^6H	−0.30518	$^4G_{(7)}$	−0.25973	$^4D_{(6)}$
39	56914	17/2	0.52997	$^4L_{(2)}$	−0.45112	$^4K_{(1)}$	−0.38899	^4N
40	56938	11/2	0.58004	$^4H_{(2)}$	0.52715	$^4H_{(3)}$	0.33766	$^4G_{(7)}$
41	57038	1/2	0.65116	$^4D_{(6)}$	0.51952	$^4D_{(1)}$	−0.48568	^6F
42	57049	15/2	−0.63595	$^4L_{(2)}$	0.45430	$^4L_{(3)}$	0.43348	$^4K_{(1)}$
43	57130	13/2	−0.64622	$^4L_{(2)}$	0.41880	$^4L_{(3)}$	0.38400	$^4K_{(1)}$
44	57583	7/2	0.76763	^6H	0.30023	^6G	0.27714	$^4H_{(2)}$
45	57644	15/2	−0.93591	^6H	−0.23804	$^4I_{(5)}$	0.16164	$^4I_{(4)}$
46	57997	11/2	0.84538	^6H	0.27647	^6G	0.24709	$^4H_{(5)}$
47	58044	9/2	−0.82947	^6H	−0.27801	$^4H_{(2)}$	−0.24735	^6G
48	58676	19/2	0.80490	$^4L_{(2)}$	−0.50058	$^4L_{(3)}$	−0.25287	^4N`
49	59137	13/2	0.72297	^6H	−0.39774	$^4H_{(2)}$	−0.37711	$^4H_{(3)}$
50	59317	11/2	0.72438	$^4K_{(1)}$	0.55877	$^4K_{(2)}$	−0.17117	$^2I_{(9)}$

(Continued)

Table 6.19 Fifty lowest energy levels of free Tb^{3+} ion ($4f^8$-configuration) calculated with the Hamiltonian parameters from Table 6.11

	Energy, cm^{-1}	Value of J	Three leading components of the corresponding wave functions (coefficients and LS terms are given). The subscript distinguishes between the terms with the same L and S quantum numbers.					
1	0	6	−0.97760	^7F	0.14564	$^5G_{(1)}$	−0.13573	$^5G_{(3)}$
2	2058	5	0.98436	^7F	−0.10065	$^5G_{(1)}$	0.096001	$^5G_{(3)}$
3	3332	4	0.97640	^7F	0.10896	$^5D_{(1)}$	−0.10520	$^5D_{(3)}$
4	4319	3	0.97456	^7F	0.14079	$^5D_{(1)}$	−0.12244	$^5D_{(3)}$
5	5012	2	−0.97240	^7F	−0.16427	$^5D_{(1)}$	0.13203	$^5D_{(3)}$
6	5470	1	−0.97171	^7F	−0.17824	$^5D_{(1)}$	0.13554	$^5D_{(3)}$
7	5695	0	0.97127	^7F	0.18533	$^5D_{(1)}$	−0.13689	$^5D_{(3)}$
8	20558	4	−0.73238	$^5D_{(3)}$	0.58075	$^5D_{(1)}$	0.19610	$^5D_{(2)}$
9	25898	10	−0.96645	^5L	0.19160	$^3M_{(3)}$	−0.14855	$^3M_{(1)}$
10	26217	6	−0.51973	$^5G_{(3)}$	0.49875	$^5G_{(1)}$	−0.33215	$^5G_{(2)}$
11	26362	3	0.73111	$^5D_{(3)}$	−0.57827	$^5D_{(1)}$	0.20016	^7F
12	27351	9	−0.97025	^5L	−0.16504	^5K	0.12988	$^3M_{(3)}$
13	27592	5	0.52570	$^5G_{(3)}$	−0.48947	$^5G_{(1)}$	0.41203	$^5G_{(2)}$
14	28126	4	0.51464	$^5G_{(3)}$	0.48455	$^5G_{(2)}$	−0.45899	$^5G_{(1)}$
15	28142	8	0.95506	^5L	0.18757	^5K	−0.10940	$^3K_{(5)}$
16	28236	2	0.64115	$^5D_{(3)}$	−0.53939	$^5D_{(1)}$	−0.27552	$^3P_{(6)}$

	Energy, cm^{-1}	Value of *J*	Three leading components of the corresponding wave functions (coefficients and LS terms are given). The subscript distinguishes between the terms with the same *L* and *S* quantum numbers.					
17	28457	7	0.91416	^5L	0.18368	^5K	−0.18362	$^3K_{(5)}$
18	28661	6	0.90669	^5L	−0.23195	$^3K_{(5)}$	−0.16495	$^3K_{(1)}$
19	28841	3	0.54983	$^5G_{(2)}$	0.54261	$^5G_{(3)}$	−0.46252	$^5G_{(1)}$
20	29412	2	−0.60719	$^5G_{(2)}$	−0.57756	$^5G_{(3)}$	0.46888	$^5G_{(1)}$
21	30753	1	0.67507	$^5D_{(3)}$	−0.57287	$^5D_{(1)}$	−0.24404	$^3P_{(6)}$
22	31018	7	0.64038	$^5H_{(1)}$	0.58725	$^5H_{(2)}$	0.35174	$^5I_{(2)}$
23	31381	0	0.65130	$^5D_{(3)}$	−0.56696	$^5D_{(1)}$	−0.28150	$^3P_{(6)}$
24	32584	6	0.58359	$^5H_{(1)}$	0.46782	$^5H_{(2)}$	0.46752	$^5I_{(2)}$
25	33483	5	0.56587	$^5H_{(1)}$	0.47799	$^5I_{(2)}$	0.40205	$^5H_{(2)}$
26	34069	4	−0.60398	$^5H_{(1)}$	−0.45947	$^5I_{(2)}$	−0.39614	$^5H_{(2)}$
27	34579	8	0.89643	$^5I_{(2)}$	0.22979	^5K	0.21062	$^5I_{(1)}$
28	34727	3	0.78093	$^5H_{(1)}$	0.47887	$^5H_{(2)}$	−0.22488	$^5G_{(2)}$
29	34933	5	−0.76500	$^5F_{(2)}$	−0.49192	$^5F_{(1)}$	0.26946	$^5G_{(2)}$
30	35381	4	−0.73950	$^5F_{(2)}$	−0.45208	$^5F_{(1)}$	0.19772	$^5G_{(2)}$
31	36106	7	−0.82660	$^5I_{(2)}$	0.31294	$^5H_{(1)}$	−0.25648	^5K
32	36576	3	−0.78939	$^5F_{(2)}$	−0.44021	$^5F_{(1)}$	−0.21624	$^5D_{(1)}$
33	37163	6	0.76242	$^5I_{(2)}$	−0.43507	$^5H_{(1)}$	−0.26786	$^5H_{(2)}$

(*Continued*)

Table 6.19 (*Continued*)

	Energy, cm^{-1}	Value of J	Three leading components of the corresponding wave functions (coefficients and LS terms are given). The subscript distinguishes between the terms with the same L and S quantum numbers.					
34	37170	2	0.83406	$^5F_{(2)}$	0.41360	$^5F_{(1)}$	0.15335	$^5D_{(1)}$
35	37201	4	−0.79410	$^5I_{(2)}$	0.41862	$^5H_{(1)}$	0.24216	$^5H_{(2)}$
36	37519	1	0.87611	$^5F_{(2)}$	0.38873	$^5F_{(1)}$	0.17503	$^3D_{(5)}$
37	37565	5	−0.77440	$^5I_{(2)}$	0.43996	$^5H_{(1)}$	0.25080	$^5H_{(2)}$
38	38441	9	−0.96118	^5K	0.16695	^5L	−0.14756	$^3L_{(1)}$
39	39569	2	0.42845	$^5D_{(2)}$	−0.41243	$^3P_{(6)}$	−0.41185	$^3P_{(3)}$
40	39684	6	0.42523	^5K	−0.33120	$^3I_{(5)}$	−0.31901	$^5G_{(2)}$
41	40104	8	−0.92195	^5K	0.23187	$^5I_{(2)}$	−0.12348	$^3K_{(1)}$
42	40791	5	0.73947	^5K	−0.31774	$^3I_{(5)}$	−0.29269	$^3I_{(2)}$
43	41008	7	0.86457	^5K	−0.27116	$^5I_{(2)}$	−0.19472	$^3I_{(5)}$
44	41167	6	−0.77166	$^5G_{(2)}$	−0.32165	$^5G_{(1)}$	−0.24392	^5K
45	41354	10	0.60039	^3O	0.41098	$^3M_{(3)}$	−0.38570	^3N
46	42237	5	0.66655	$^5G_{(2)}$	0.30501	^5K	0.29646	$^5G_{(1)}$
47	42392	3	−0.70045	$^5D_{(2)}$	−0.53290	^5P	0.31008	$^5D_{(1)}$
48	42642	6	−0.76347	^5K	0.21438	$^3H_{(9)}$	−0.21262	$^3K_{(5)}$
49	42963	4	0.47725	$^3F_{(8)}$	0.37921	$^5D_{(2)}$	0.27643	$^3F_{(3)}$
50	43063	12	0.95775	^3O	−0.28759	^1Q	0.0000	^1Q

Table 6.20 Fifty lowest energy levels of free Dy^{3+} ion ($4f^9$-configuration) calculated with the Hamiltonian parameters from Table 6.11

	Energy, cm^{-1}	Value of J	Three leading components of the corresponding wave functions (coefficients and LS terms are given). The subscript distinguishes between the terms with the same L and S quantum numbers.					
1	0	15/2	–0.96827	^6H	0.21627	$^4I_{(3)}$	0.11367	$^4I_{(1)}$
2	3474	13/2	–0.98174	^6H	0.15496	$^4I_{(3)}$	0.076379	$^4I_{(1)}$
3	5832	11/2	0.96274	^6H	0.13990	^6F	0.12138	$^4G_{(4)}$
4	7697	9/2	0.96666	^6H	0.16541	$^4G_{(4)}$	0.13186	$^4G_{(1)}$
5	8006	11/2	–0.96684	^6F	0.15824	^6H	0.14203	$^4G_{(3)}$
6	9130	7/2	0.96334	^6H	0.19258	$^4G_{(4)}$	0.15093	$^4G_{(1)}$
7	9355	9/2	–0.93934	^6F	–0.25646	$^4F_{(3)}$	0.14641	$^4F_{(1)}$
8	10192	5/2	–0.96033	^6H	–0.21221	$^4G_{(4)}$	–0.16398	$^4G_{(1)}$
9	11282	7/2	–0.95745	^6F	–0.20681	$^4F_{(3)}$	0.12366	$^4F_{(1)}$
10	12696	5/2	0.96320	^6F	0.14465	$^4F_{(3)}$	–0.12541	$^4D_{(2)}$
11	13498	3/2	–0.95353	^6F	0.17512	$^4D_{(2)}$	–0.16411	$^4D_{(3)}$
12	14050	1/2	0.95504	^6F	–0.19338	$^4D_{(2)}$	0.18127	$^4D_{(3)}$
13	21342	9/2	0.66779	$^4F_{(3)}$	–0.31781	$^4G_{(4)}$	–0.31613	$^4F_{(1)}$
14	21650	15/2	0.65658	$^4I_{(3)}$	0.35015	$^4K_{(1)}$	0.30684	$^4I_{(1)}$
15	23501	11/2	–0.67389	$^4G_{(4)}$	–0.50301	$^4G_{(1)}$	0.34158	$^4G_{(2)}$
16	23900	21/2	0.96483	^4M	–0.25837	^2N	0.04837	^2O

(*Continued*)

Table 6.20 (*Continued*)

	Energy, cm^{-1}	Value of J	Three leading components of the corresponding wave functions (coefficients and LS terms are given). The subscript distinguishes between the terms with the same L and S quantum numbers.					
17	25006	17/2	−0.59202	$^4K_{(1)}$	0.50510	^4L	0.50210	^4M
18	25214	19/2	0.79145	^4M	0.57710	^4L	0.14101	$^2M_{(1)}$
19	25338	13/2	0.75569	$^4I_{(3)}$	0.41052	$^4K_{(1)}$	0.33383	$^4I_{(1)}$
20	26004	7/2	−0.61853	$^4F_{(3)}$	0.32636	$^4G_{(4)}$	−0.28108	$^4F_{(4)}$
21	27603	11/2	0.81318	$^4I_{(3)}$	0.34281	$^4K_{(1)}$	0.31914	$^4I_{(1)}$
22	27623	15/2	0.73768	^4M	0.39469	$^4I_{(3)}$	0.33496	^4L
23	27811	3/2	−0.48568	$^4P_{(2)}$	0.45366	^6P	−0.32338	$^4D_{(3)}$
24	27954	5/2	−0.78480	^6P	0.49850	$^4P_{(2)}$	−0.16953	$^4P_{(1)}$
25	28957	7/2	0.87096	^6P	−0.22503	$^4D_{(1)}$	0.21588	$^4F_{(3)}$
26	29058	17/2	0.76389	^4M	0.59525	$^4K_{(1)}$	−0.23766	$^4K_{(2)}$
27	29366	9/2	−0.85956	$^4I_{(3)}$	−0.30577	$^4I_{(1)}$	−0.24928	$^4I_{(2)}$
28	29890	5/2	−0.56384	$^4F_{(3)}$	0.32002	^6P	0.31863	$^4G_{(4)}$
29	30184	9/2	−0.49245	$^4G_{(4)}$	0.45379	$^4G_{(2)}$	−0.41163	$^4F_{(3)}$
30	30263	19/2	−0.77594	^4L	0.59461	^4M	−0.16781	$^2M_{(1)}$
31	30545	15/2	−0.69493	$^4K_{(1)}$	0.38894	^4L	−0.33441	^4M
32	31222	3/2	0.75035	^6P	−0.31796	$^4D_{(2)}$	0.28895	$^4D_{(3)}$
33	31723	7/2	0.42576	$^4F_{(3)}$	0.42156	$^4G_{(4)}$	−0.42133	$^4G_{(2)}$

	Energy, cm^{-1}	Value of J	Three leading components of the corresponding wave functions (coefficients and LS terms are given). The subscript distinguishes between the terms with the same L and S quantum numbers.					
34	32252	5/2	−0.43160	$^4D_{(2)}$	−0.40282	$^4G_{(4)}$	0.40173	$^4D_{(3)}$
35	32403	1/2	0.47856	$^4D_{(2)}$	−0.47662	$^4D_{(3)}$	−0.46046	$^4P_{(2)}$
36	32630	13/2	−0.58625	$^4K_{(1)}$	0.54561	^4L	0.43830	$^4I_{(3)}$
37	33266	13/2	0.72933	$^4H_{(1)}$	0.56081	$^4H_{(3)}$	−0.19551	^4L
38	33455	17/2	0.83425	^4L	0.33036	$^4K_{(1)}$	−0.26766	^4M
39	33818	3/2	0.77265	$^4F_{(3)}$	−0.34830	$^4F_{(1)}$	0.26855	$^4F_{(4)}$
40	34084	11/2	0.72358	$^4H_{(1)}$	0.48536	$^4H_{(3)}$	0.29396	$^4I_{(2)}$
41	34227	7/2	−0.46550	$^4D_{(3)}$	0.44735	$^4D_{(2)}$	0.37488	^6P
42	34326	9/2	−0.51318	$^4H_{(1)}$	−0.41850	$^4G_{(1)}$	−0.40604	$^4G_{(4)}$
43	34543	5/2	0.57589	$^4F_{(3)}$	−0.36006	$^4G_{(2)}$	−0.27023	$^4D_{(2)}$
44	34873	11/2	−0.67975	$^4G_{(2)}$	0.44713	$^4G_{(3)}$	−0.27450	$^4K_{(1)}$
45	35440	11/2	0.82556	$^4K_{(1)}$	−0.34069	$^4I_{(3)}$	−0.22780	$^4K_{(2)}$
46	35749	15/2	0.74467	^4L	0.45948	$^4K_{(1)}$	−0.33011	^4M
47	35870	13/2	−0.67592	^4L	−0.54996	$^4K_{(1)}$	−0.25318	$^2K_{(1)}$
48	36032	7/2	0.48157	$^4H_{(1)}$	0.37407	$^4G_{(4)}$	0.36777	$^4G_{(1)}$
49	36731	5/2	−0.55910	$^4G_{(2)}$	−0.30405	^6P	−0.28410	$^4P_{(2)}$
50	36739	9/2	0.55695	$^4G_{(2)}$	−0.43996	$^4G_{(3)}$	−0.42845	$^4H_{(3)}$

(Continued)

Table 6.21 Fifty lowest energy levels of free Ho^{3+} ion ($4f^{10}$-configuration) calculated with the Hamiltonian parameters from Table 6.11

	Energy, cm^{-1}	Value of J	Three leading components of the corresponding wave functions (coefficients and LS terms are given). The subscript distinguishes between the terms with the same L and S quantum numbers.					
1	0	8	−0.96631	^5I	0.22371	$^3K_{(2)}$	−0.11778	$^3K_{(1)}$
2	5057	7	−0.98508	^5I	0.14693	$^3K_{(2)}$	−0.07281	$^3K_{(1)}$
3	8580	6	0.97533	^5I	0.13883	$^3H_{(4)}$	−0.08832	$^3H_{(1)}$
4	11158	5	−0.95294	^5I	−0.19760	$^3H_{(4)}$	0.13969	$^3H_{(3)}$
5	13220	4	−0.94726	^5I	−0.22731	$^3H_{(4)}$	0.16351	$^3H_{(3)}$
6	15965	5	−0.89205	^5F	0.32113	$^3G_{(2)}$	0.18460	$^3G_{(1)}$
7	19075	4	−0.95603	^5F	−0.19345	$^3F_{(3)}$	−0.12938	$^3F_{(1)}$
8	19126	2	−0.82396	^5S	−0.37564	$^3P_{(2)}$	−0.26206	^5F
9	20973	8	0.74583	$^3K_{(2)}$	−0.40890	$^3K_{(1)}$	0.32686	^3L
10	21142	3	−0.95076	^5F	−0.19665	$^3D_{(1)}$	−0.14618	$^3F_{(3)}$
11	21653	2	0.78693	^5F	−0.41540	^5S	0.33240	$^3D_{(1)}$
12	22508	6	0.88576	^5G	−0.33936	$^3H_{(4)}$	0.25611	$^3H_{(1)}$
13	22838	1	−0.89235	^5F	−0.38361	$^3D_{(1)}$	−0.20180	$^3D_{(2)}$
14	24376	5	−0.63475	^5G	−0.44578	$^3G_{(2)}$	−0.38609	^5F
15	25821	7	−0.80457	$^3K_{(2)}$	0.43779	$^3K_{(1)}$	−0.31971	^3L
16	26280	4	0.75460	^5G	0.39817	$^3G_{(2)}$	−0.33845	$^3G_{(3)}$

	Energy, cm^{-1}	Value of *J*	Three leading components of the corresponding wave functions (coefficients and LS terms are given). The subscript distinguishes between the terms with the same *L* and *S* quantum numbers.					
17	27923	6	0.60513	$^3H_{(4)}$	−0.50658	$^3H_{(3)}$	0.43261	^5G
18	27982	5	−0.70070	^5G	0.39736	$^3H_{(4)}$	0.30291	$^3G_{(2)}$
19	28349	9	−0.94313	^3L	−0.33243	^3M	0.0000	^1N
20	28850	2	0.45266	^5F	0.42471	^5G	−0.34853	$^3D_{(1)}$
21	29310	3	0.86681	^5G	0.27986	$^3F_{(4)}$	−0.18669	$^3G_{(3)}$
22	29744	6	−0.86356	$^3K_{(2)}$	0.40494	$^3K_{(1)}$	−0.14681	$^3I_{(1)}$
23	30334	4	0.42993	$^3F_{(4)}$	−0.35689	$^3G_{(2)}$	−0.35258	$^3H_{(4)}$
24	31324	2	−0.72611	^5G	−0.32243	$^3D_{(1)}$	−0.29411	$^3F_{(4)}$
25	33195	10	0.98873	^3M	−0.14972	^1N	0.0000	^1N
26	33635	8	0.80222	^3L	0.42149	^3M	−0.36508	$^3K_{(2)}$
27	33775	3	−0.78244	$^3D_{(1)}$	−0.40714	$^3D_{(2)}$	−0.22249	^5G
28	34020	1	0.72989	$^3P_{(2)}$	0.47888	$^3D_{(1)}$	−0.36093	^5F
29	35269	4	−0.53479	^5G	0.46144	^5D	0.35991	$^3G_{(2)}$
30	35705	3	0.52510	$^3G_{(2)}$	−0.44973	$^3F_{(4)}$	−0.37400	$^3F_{(2)}$
31	35721	8	−0.72632	$^1L_{(2)}$	0.47089	^3M	−0.43309	^3L
32	36391	4	−0.51784	^5D	−0.37935	$^3H_{(4)}$	0.34010	$^3H_{(3)}$
33	36817	0	0.93039	$^3P_{(2)}$	0.24398	$^3P_{(3)}$	0.19082	$^3P_{(1)}$

(*Continued*)

Table 6.21 (*Continued*)

	Energy, cm^{-1}	Value of J	Three leading components of the corresponding wave functions (coefficients and LS terms are given). The subscript distinguishes between the terms with the same L and S quantum numbers.					
34	36886	2	0.51405	$^3F_{(2)}$	−0.41765	^5G	0.35080	$^3F_{(4)}$
35	37036	5	0.46364	$^3G_{(2)}$	−0.46195	$^3H_{(4)}$	0.41918	$^3H_{(3)}$
36	37478	7	−0.89615	^3L	0.31038	$^3K_{(2)}$	0.27035	^1K
37	38364	7	0.81549	$^3I_{(1)}$	−0.47088	$^3I_{(2)}$	−0.21769	^1K
38	38383	9	−0.94313	^3M	0.33243	^3L	0.0000	^1N
39	38549	2	−0.67290	$^3P_{(2)}$	−0.39179	$^1D_{(3)}$	0.28686	$^3D_{(1)}$
40	39028	4	−0.73658	$^3F_{(3)}$	−0.41686	$^3F_{(1)}$	0.27025	^5F
41	39482	5	0.73940	$^3I_{(1)}$	0.36362	$^1H_{(1)}$	−0.34357	$^3I_{(2)}$
42	39881	6	−0.80188	$^3I_{(1)}$	0.46632	$^3I_{(2)}$	0.19359	$^3K_{(2)}$
43	40506	3	0.70362	^5D	0.42238	$^3G_{(2)}$	−0.32253	^5G
44	40557	1	−0.54013	^5D	−0.52328	$^3D_{(1)}$	0.50438	$^3P_{(2)}$
45	42001	4	−0.59752	^5D	0.48405	$^3F_{(2)}$	0.40874	$^3F_{(4)}$
46	42510	2	0.88319	^5D	−0.26536	$^3F_{(2)}$	0.24036	$^1D_{(3)}$
47	43181	0	0.95279	^5D	0.18811	$^3P_{(2)}$	−0.17136	$^3P_{(1)}$
48	43360	3	0.55985	$^3F_{(2)}$	−0.54229	^5D	0.32457	$^3G_{(2)}$
49	43637	1	0.81117	^5D	−0.39360	$^3D_{(1)}$	0.31579	$^3P_{(2)}$
50	44664	8	−0.76059	^3M	−0.51867	$^1L_{(2)}$	−0.26672	$^1L_{(1)}$

Table 6.22 Lowest energy levels of free Er^{3+} ion ($4f^{11}$-configuration) calculated with the Hamiltonian parameters from Table 6.11

	Energy, cm^{-1}	Value of J	Three leading components of the corresponding wave functions (coefficients and LS terms are given). The subscript distinguishes between the terms with the same L and S quantum numbers.					
1	0	15/2	−0.98479	^4I	0.17288	^2K	0.01749	^2L
2	6522	13/2	0.99538	^4I	−0.09116	^2K	0.0000	^2L
3	10193	11/2	−0.91108	^4I	−0.37789	$^2H_{(2)}$	−0.06137	^2I
4	12551	9/2	0.76058	^4I	0.42179	$^2H_{(2)}$	0.20944	$^2G_{(2)}$
5	15698	9/2	0.78460	^4F	0.46705	^4I	0.31199	$^2G_{(1)}$
6	19186	3/2	0.82402	^4S	−0.43255	^2P	−0.27944	$^2D_{(1)}$
7	19437	11/2	−0.69868	$^2H_{(2)}$	0.38155	^4I	0.15129	$^2H_{(1)}$
8	21105	7/2	−0.95974	^4F	−0.21512	$^2G_{(1)}$	0.16135	$^2G_{(2)}$
9	22804	5/2	0.91999	^4F	−0.35474	$^2D_{(1)}$	0.13202	$^2D_{(2)}$
10	23176	3/2	0.79031	^4F	−0.45120	$^2D_{(1)}$	0.00256	$^2D_{(2)}$
11	24929	9/2	−0.49818	^4F	0.42700	$^2G_{(1)}$	−0.42581	$^2H_{(2)}$
12	26807	11/2	−0.78792	^4G	0.50337	$^2H_{(2)}$	0.02192	^2I
13	27497	15/2	−0.95582	^2K	−0.23835	^2L	0.0000	^2L

(Continued)

Table 6.22 (*Continued*)

	Energy, cm^{-1}	Value of J	Three leading components of the corresponding wave functions (coefficients and LS terms are given). The subscript distinguishes between the terms with the same L and S quantum numbers.					
14	27865	9/2	0.89718	^4G	0.37001	$^2H_{(2)}$	−0.08577	$^2H_{(1)}$
15	28497	7/2	0.64358	^4G	0.50853	$^2G_{(1)}$	−0.48585	$^2G_{(2)}$
16	32291	3/2	0.60235	^2P	0.45764	$^2D_{(1)}$	0.18558	$^2D_{(2)}$
17	32797	13/2	−0.95074	^2K	0.0000	^2L	0.0000	^2L
18	33849	5/2	0.95694	^4G	−0.18654	$^2F_{(2)}$	−0.08984	$^2D_{(2)}$
19	33923	1/2	0.95947	^2P	0.0000	^2L	0.0000	^2L
20	34466	7/2	−0.73598	^4G	0.52153	$^2G_{(1)}$	−0.39390	$^2G_{(2)}$
21	35525	5/2	−0.76876	$^2D_{(1)}$	0.37630	$^2D_{(2)}$	0.02451	$^2F_{(2)}$
22	36788	9/2	0.56373	$^2H_{(2)}$	−0.39915	$^2G_{(2)}$	−0.28143	$^2H_{(1)}$
23	39439	5/2	−0.66820	^4D	0.51100	$^2D_{(1)}$	0.50647	$^2D_{(2)}$
24	40138	7/2	0.97014	^4D	0.17922	$^2F_{(1)}$	0.10740	$^2F_{(2)}$
25	41206	17/2	1.0000	^2L	0.0000	^2L	0.0000	^2L
26	41285	11/2	0.82617	^2I	0.12378	$^2H_{(2)}$	0.0000	^2L
27	43037	3/2	−0.66192	^4D	−0.58565	$^2D_{(1)}$	0.09351	$^2D_{(2)}$

	Energy, cm^{-1}	Value of *J*	Three leading components of the corresponding wave functions (coefficients and LS terms are given). The subscript distinguishes between the terms with the same *L* and *S* quantum numbers.					
28	43635	13/2	−0.95346	^2I	0.29627	^2K	0.0000	^2L
29	43714	3/2	0.55884	^4D	0.53670	^2P	−0.51900	$^2D_{(2)}$
30	47407	15/2	0.97102	^2L	0.0000	^2L	0.0000	^2L
31	47949	1/2	0.95947	^4D	0.28179	^2P	0.0000	^2L
32	48416	9/2	0.88428	$^2H_{(1)}$	0.42140	$^2H_{(2)}$	0.0000	^2L
33	49772	5/2	0.66634	$^2D_{(2)}$	0.28984	$^2F_{(2)}$	0.06235	$^2D_{(1)}$
34	51371	11/2	0.75808	$^2H_{(1)}$	−0.55672	^2I	0.0000	^2L
35	54926	7/2	0.71916	$^2F_{(2)}$	0.17192	^4G	0.13550	$^2G_{(1)}$
36	56025	3/2	0.82881	$^2D_{(2)}$	−0.24601	$^2D_{(1)}$	0.0000	^2L
37	63709	5/2	0.76646	$^2F_{(2)}$	−0.06984	$^2D_{(1)}$	0.0000	^2L
38	66940	7/2	0.75521	$^2G_{(2)}$	0.02927	$^2F_{(1)}$	0.0000	^2L
39	70965	9/2	0.74809	$^2G_{(2)}$	−0.15257	$^2H_{(1)}$	−0.07521	$^2H_{(2)}$
40	93625	5/2	−0.84596	$^2F_{(1)}$	0.52934	$^2F_{(2)}$	0.02466	$^2D_{(1)}$
41	98241	7/2	−0.75066	$^2F_{(1)}$	0.64220	$^2F_{(2)}$	0.10851	$^2G_{(2)}$

(Continued)

Table 6.23 Energy levels of free Tm^{3+} ion ($4f^{12}$-configuration) calculated with the Hamiltonian parameters from Table 6.11

	Energy, cm^{-1}	Value of J	Three leading components of the corresponding wave functions (coefficients and LS terms are given). The subscript distinguishes between the terms with the same L and S quantum numbers.					
1	0	6	0.99561	^3H	0.093627	^1I	0.0000	^1I
2	5752	4	−0.79025	^3F	−0.54104	^1G	0.28773	^3H
3	8212	5	1.0000	^3H	0.0000	^1I	0.0000	^1I
4	12609	4	0.77511	^3H	0.52396	^3F	−0.35309	^1G
5	14480	3	1.0000	^3F	0.0000	^1I	0.0000	^1I
6	15142	2	0.88201	^3F	−0.44961	^1D	−0.14108	^3P
7	21332	4	0.76329	^1G	0.56250	^3H	−0.31777	^3F
8	27860	2	0.67058	^3P	0.61202	^1D	0.41923	^3F
9	34634	6	0.99561	^1I	−0.093627	^3H	0.0000	^1I
10	34806	0	0.97430	^3P	−0.22525	^1S	0.0000	^1I
11	35599	1	1.0000	^3P	0.0000	^1I	0.0000	^1I
12	37959	2	−0.72830	^3P	0.65061	^1D	0.21516	^3F
13	76415	0	0.97430	^1S	0.22525	^3P	0.0000	^1I

Table 6.24 Energy levels of free Yb^{3+} ion ($4f^{13}$-configuration) calculated with the Hamiltonian parameters from Table 6.11

	Energy, cm^{-1}	Value of J	Leading components of the corresponding wave functions (coefficients and LS terms are given).	
1	0	7/2	1.0000	^2F
2	10248	5/2	1.0000	^2F

Bibliography

1. Liu GK (2005). Electronic energy level structure, in *Spectroscopic Properties of Rare Earths in Optical Materials*. Liu GK, Jacquier B, eds., Tsinghua University Press & Springer-Verlag (Berlin), Heidelberg, pp. 1–84.
2. Svelto O, Longhi S, Della Valle G, et al. (2012). Lasers and coherent light sources, in *Springer Handbook of Lasers and Optics*. Träger F, ed., Springer-Verlag, Berlin, Heidelberg, pp. 641–1170.
3. Henderson B, Bartram RH (2000). *Crystal-Field Engineering of Solid-State Laser Materials*. Cambridge University Press.
4. Kück S (2001). Laser-related spectroscopy of ion-doped crystals for tunable solid-state lasers, *Appl Phys B* **72**(5), 515–562.
5. Bünzli J-C G, Eliseeva SV (2013). Photophysics of lanthanoid coordination compounds, in *Comprehensive Inorganic Chemistry II, From Elements to Applications*. Reedijk J, Poeppelmeier K, eds, vol. 8, 2nd ed, Elsevier, pp. 339–398.
6. Bünzli J-C G, Piguet C (2005). Taking advantage of luminescent lanthanide ions, *Chem Soc Rev* **34**(12) 1048–1077.
7. Bünzli J-C G (2006). Benefiting from the unique properties of lanthanide ions, *Acc Chem Res* **39**(1) 53–61.
8. Wang F, Liu X-G (2009). Recent advances in the chemistry of lanthanide-doped upconversion nanocrystals, *Chem Soc Rev* **38**(4) 976–989.
9. Haschke JM, Stakebake JL (2006). Handling, storage, and disposition of plutonium and uranium, in *The Chemistry of the Actinide and Transactinide Elements*. Morss LR, Edelstein NM, Fuger J, Katz JJ, eds, vol. 5, Springer Publishing Company, Dordrecht, The Netherlands, p. 3199.

10. Natrajan LS (2012). Developments in the photophysics and photochemistry of actinide ions and their coordination compounds, *Chem Soc Rev* **256**(15–16) 1583–1603.

11. Carnall WT, Goodman GL, Rajnak K, Rana RS (1989). A systematic analysis of the spectra of the lanthanides doped into single-crystal LaF_3, *J Chem Phys* **90**(7) 3443–3457.

12. Wybourne BG (1965). *Spectroscopic Properties of Rare Earths.* Interscience Publishers, John Wiley & Sons, Inc., New York.

13. Liu G-K (2005). Systematics of *f*-element crystal-field interaction, *J Solid State Chem* **178**(2) 489–498.

14. Ma C-G, Brik MG, Ryba-Romanowski W, Swart HC, Gusowski MA (2012). Spectroscopy and calculations for $4f^N$->$4f^{N-1}5d$ transitions of lanthanide ions in K_3YF_6, *J Phys Chem A* **116**(36) 9158–9180.

15. Duan CK, Tanner PA (2010). What use are crystal field parameters? A chemist's viewpoint, *J Phys Chem A* **114**(19) 6055–6062.

16. Yeung YY, Tanner P (2013). Parametrization of free ion levels of four isoelectronic $4f^2$ systems: Insights into configuration interaction parameters, *Chem Phys Lett* **590** 46–51.

17. Brik MG, Srivastava AM (2013). Systematic analysis of the spectroscopic characteristics of 3*d* ions in a free state and some cubic crystals, *Opt Mater* **35**(10) 1776–1782.

18. Ma CG, Brik MG (2014). Systematic analysis of spectroscopic characteristics of heavy transition metal ions with $4d^N$ and $5d^N$ (N = 1 ... 10) electronic configurations in a free state, *J Lumin* **145** 402–409.

19. Ma CG, Brik MG, Li QX, Tian Y (2014) Systematic analysis of spectroscopic characteristics of the lanthanide and actinide ions with the $4f^N$ and $5f^N$ (N = 1...14) electronic configurations in a free state. *J Alloys Compds* **599** 93–101.

20. Ma CG, Brik MG, Tian Y, Li QX (2014). Systematic analysis of spectroscopic characteristics of the lanthanide and actinide ions with the $4f^{N-1}5d$ and $5f^{N-1}6d$ electronic configurations in a free state, *J Alloys Compds* **603** 255–267.

21. Morrison CA (1992). *Crystal Fields for Transition-Metal Ions in Laser Host Materials.* Springer-Verlag.

22. Kramida A, Ralchenko Yu, Reader J, NIST ASD Team (2013). NIST Atomic Spectra Database (ver. 5.1), National Institute of Standards and Technology, Gaithersburg, MD. <http://physics.nist.gov/asd> (19.11.13).

23. Elliott JP, Judd BR, Runciman WA (1957). Energy levels in rare-earth ions, *Proc R Soc Lond A* **240**(1223), 509–523.

24. http://www2.phys.canterbury.ac.nz/~mfr24/.

25. Ma C-G, Brik MG, Liu D-X, Feng B, Tian Y, Suchocki A (2016). Energy level schemes of f^N electronic configurations for the di-, tri-, and tetravalent lanthanides and actinides in a free state, *J Lumin* **170** 369–374.

26. Malinowski M, Joubert MF, Mahiou R, Jacquier B (1994). Visible laser-emission of Pr^{3+} in various hosts, *J de Phys IV* **4**(C4) 541–544.

27. Dorenbos P (2013). A review on how lanthanide impurity levels change with chemistry and structure of inorganic compounds, *ECS J Sol State Sci Technol* **2**(2), R3001–R3011.

28. Reid MF, van Pieterson L, Wegh RT, Meijerink A (2000). Spectroscopy and calculations for $4f^N$->$4f^{N-1}5d$ transitions of lanthanide ions in $LiYF_4$, *Phys Rev B* **62**(22) 14744–14749.

29. Wang W, Liu GK, Brik MG, Seijo L, Shi D (2009). 5*f*-6*d* orbital hybridization of trivalent uranium in crystals of hexagonal symmetry: Effects on electronic energy levels and transition intensities, *Phys Rev B* **80**(15) 155120.

30. Reid MF (2005). Transition intensities, in *Spectroscopic Properties of Rare Earths in Optical Materials*. Liu GK, Jacquier B, eds, Tsinghua University Press & Springer-Verlag (Berlin), Heidelberg, pp. 95–129.

31. Burdick GW, Reid MF (2007). $4f^n$–$4f^{(n-1)}5d$ transitions, in *Handbook on the Physics and Chemistry of Rare Earths*. Gschneidner Jr KA, Bünzli JCG, Pecharsky VK, eds, vol. 37, North-Holland Publishing Company (Elsevier), Amsterdam, p. 68.

32. van Pieterson L, Reid MF, Wegh RT, Soverna S, Meijerink A (2002). $4f^n$-> $4f^{n-1}5d$ transitions of the light lanthanides: Experiment and theory, *Phys Rev B* **65**(4) 045113.

33. van Pieterson L, Reid MF, Burdick GW, Meijerink A (2002). $4f^n$-> $4f^{n-1}5d$ transitions of the heavy lanthanides: Experiment and theory, *Phys Rev B* **65**(4) 045114.

34. Reid MF, van Pieterson L, Meijerink A (2002), Trends in parameters for the $4f^N$<-> $4f^{N-1}5d$ spectra of lanthanide ions in crystals, *J Alloys Compds* **344**(1–2) 240–245.

35. Tanner PA, Duan CK, Makhov VN, Kirm M, Khaidukov NM (2009). Vacuum ultraviolet excitation spectra of lanthanide-doped hexafluoroelpasolites, *J Phys Condens Matter* **21**(39) 395504.

36. Duan CK, Tanner PA, Meijerink A, Babin V (2009). Synchrotron excitation, emission and theoretical simulation of lanthanide ions in hexachloroelpasolite crystals, *J Phys: Condens Matter* **21**(39) 395501.

37. Wegh RT, Meijerink A (1999). Spin-allowed and spin-forbidden $4f^N$->$4f^{N-1}5d$ transitions for heavy lanthanides in fluoride hosts, *Phys Rev B* **60**(15) 10820–10830.

38. Ogasawara K, Watanabe S, Sakai Y, et al. (2004). Calculations of complete $4f^n$ and $4f^{n-1}5d^1$ energy level schemes of free trivalent rare-earth ions, *Jpn J Appl Phys* **43**(5) L611–L613.

39. Ogasawara K, Watanabe S, Ishii T, Brik MG (2005), Relativistic calculations of complete $4f^n$ energy level schemes of free trivalent rare-earth ions, *Jpn J Appl Phys* **44**(10) 7488–7490.

40. Peijzel PS, Meijerink A, Wegh RT, Reid MF, Burdick GW (2005), A complete $4f^n$ energy level diagram for all trivalent lanthanide ions, *J Sol State Chem.* **178**(2) 448–453.

Chapter 7

The Group Theory

7.1 Introduction

The group theory emerged from several branches of mathematics: the numbers theory, geometry, and algebraic equations. The foundations of group theory were developed by L. Euler (1707–1783), C.F. Gauss (1777–1855), J.-L. Lagrange (1736–1813), N. H. Abel (1802–1829), E. Galois (1811–1832), and many other prominent mathematicians. At the first glance, the group theory may seem to be a very abstract part of mathematics, which is very far from the real world. However, the area of applications of group theory is surprisingly and unexpectedly wide. It can be used to describe the symmetry properties of various objects. The group theory is very important for combinatorics (permutations and combinations of elements from different sets) and cryptography. It is even applied in the musical set theory, since music—if we forget for the moment about its esthetic component—may be considered as a certain combination of different permutations of the musical notes (well, not every combination of the musical notes can make a pleasant melody though). In chemistry the group theory facilitates description of various properties of molecules and symmetry properties of molecular orbitals. In physics importance of group theory follows from the fact that the symmetry properties of a considered physical system are related to the laws of conservation of certain physical quantities.

Theoretical Spectroscopy of Transition Metal and Rare Earth Ions: From Free State to Crystal Field
Mikhail G. Brik and Chong-Geng Ma

ISBN 978-981-4800-56-3 (Hardcover), 978-0-429-27875-4 (eBook)
www.jennystanford.com

In the present book, we shall focus on how the group theory can be used to describe the energy levels and spectra of impurity ions in crystals. We shall see that the group theory is a very beautiful and powerful tool for this purpose. It allows finding the overall pattern of the degenerated energy levels splitting. In addition, the selection rules for the allowed/forbidden transitions between those split energy levels can also be established in a simple way with the help of the group theory. It should be stressed that the group theory itself does not give any methods to calculate numerically the matrix elements of the transition operators, which quite often can be quite complicated. However, using the symmetry properties of the initial and final states and the transition operator, it is possible to find out whether the matrix element of the corresponding transition between these states is zero or not.

Finally, the group theory provides us with a unified approach to deal with the spectroscopic problems, for example, classification of the energy levels of impurity ions in crystals, selection rules for allowed transitions, analysis of vibrations of molecules, etc.

7.2 Definition and Examples of the Groups

We start with a definition of a group in the mathematical sense of this word. A set G of elements is said to form a group, if the following conditions are fulfilled:

(1) there is a group multiplication rule, and the considered set is closed under this multiplication. In other words, if A and B are the elements of G, then $A \times B = C$, and C is also another element from the same set G. If the group multiplication law is commutative $A \times B = B \times A$, the group is the Abelian group (named after the Norwegian mathematician N. H. Abel);

(2) the group multiplication law is associative: $(A \times B) \times C = A \times (B \times C)$—the order of brackets is of no importance in the group product of three group elements;

(3) there is an identity element E in G, such as $A \times E = E \times A = A$;

(4) for each element A from G there is an inverse element A^{-1}, such as $A \times A^{-1} = A^{-1} \times A = E$.

The nature of the elements forming a group can be arbitrary: These can be the numbers, matrices, functions, etc. By the group multiplication is meant a generalized multiplication, which does not necessarily have to be the multiplication as a mathematical operation. The group multiplication should be understood rather as a correspondence rule, which links together a pair of the group elements A, B with another element C of the same group.

To find out whether a given set is a group in the light of the above-given definition, all four conditions should be checked one by one and all they should hold true simultaneously. If even one out of those four conditions is not fulfilled, the considered set of elements does not form a group.

The property of closeness does not impose any restriction on the number of elements in the group. As a result, the finite and infinite groups in terms of the number of the group elements can be distinguished.

Several examples given below will help to understand the definition of the group and illustrate which sets of elements do or do not form the groups.

Example I. All integer numbers form a group, if the operation of mathematical addition is taken as a group multiplication rule.

Let us check the group properties from the definition:

1. The sum of two integers is an integer number.
2. Addition is an associative operation: $(A + B) + C = A + (B + C)$.
3. Zero is an identity element in such a group: $A + 0 = A$.
4. For each number A, the corresponding negative number "$-A$" is an inverse element: $A + (-A) = 0$.

This is an infinite Abelian group.

Example II. All rational positive numbers form a group, if the operation of mathematical multiplication is taken as a group multiplication rule.

1. The product of two rational numbers is another rational number.
2. Multiplication is an associative operation: $(A \times B) \times C = A \times (B \times C)$.
3. Unity is an identity element: $A \times 1 = A$.

4. For each number A its reciprocal number "$1/A$" is an inverse element: $A \times 1/A = 1$. The word "positive" is very important in the description of the numbers in this example, since it excludes zero from the considered set of numbers and allows each number to have its inverse.

This is another example of an infinite Abelian group.

Example III. A set of square invertible (i.e., with non-zero determinants) matrices of rank n forms an infinite non-Abelian group, if the matrix product is taken as a group multiplication law.

1. The product of two square matrices is a square matrix.
2. Matrix multiplication is an associative operation: $(A \times B) \times C = A \times (B \times C)$.
3. The unity matrix I is an identity element: $A \times I = A$.
4. For each invertible square matrix, there exists an inverse matrix: $A \times A^{-1} = I$.

Example IV. A group can be also defined by the corresponding multiplication table, as in this example. Six elements E, A, B, C, D, F form a group with the following multiplication law (the $A \times B$ product is "analyzed" from the right to the left: The right element is in upper-most row, and the left element is in the left-most column, so $A \times B = D$, $C \times F = B$, etc.).

	E	***A***	***B***	***C***	***D***	***F***
E	*E*	*A*	*B*	*C*	*D*	*F*
A	*A*	*E*	*D*	*F*	*B*	*C*
B	*B*	*F*	*E*	*D*	*C*	*A*
C	*C*	*D*	*F*	*E*	*A*	*B*
D	*D*	*C*	*A*	*B*	*F*	*E*
F	*F*	*B*	*C*	*A*	*E*	*D*

As seen from this table, all four basic properties of a group are fulfilled. It is also easy to find the inverse elements for each member of this group: The E, A, B, C elements are inverse for themselves, whereas the F and D elements are inverse for each other. This is an example of a finite non-Abelian group.

Example V. The natural numbers (0, 1, 2, ...) do <u>not</u> form a group if the mathematical addition is considered as a group multiplication law.

The reason is that we cannot find an inverse element for any member of this set.

Example VI. The natural numbers (0, 1, 2, ...) do <u>not</u> form a group if the mathematical multiplication is considered as a group multiplication law.

Again, as in the previous example an inverse element for any member of this set (except for 1) cannot be found.

Let us list below several important definitions.

The number of elements in a finite group is called its order, or cardinality.

For any element A of any finite group G it is possible to find an integer number m (which is not greater than the number of elements in the group) such that $A \times A \times A ... \times A = A^m = E$. This number m is called the order of the element A in a group G. This follows from the property of closeness of a group and a finite number of elements in it: Taking any element and multiplying it by itself, we sooner or later get the identity element as a result of such multiplication.

It is sometimes possible to select a subset H from the whole group G in such a way, that this subset itself has also all properties of a group. Then such a subset H is called a subgroup H of the group G. Each group has at least two the so-called trivial subgroups: the group itself and the identity element, which alone is a subgroup. If the order of a group G is n, and the order of the subgroup H is m, then $n/m = k$, where k is a positive integer. The last statement is known as the Lagrange's theorem. In principle, a group can have several non-trivial subgroups, and the identity element of the group should simultaneously belong to all these subgroups.

Two elements A and B of a group G are called conjugate to each other, if there exists an element U in G such that $UAU^{-1} = B$. Obviously, an identity element is always conjugate to itself. If A and B are the conjugate elements, and, at the same time, B is conjugate to C, then A and C are also the conjugate elements. All mutually conjugated elements form a class. The class is not a subgroup of a group, but just a set of elements of a group. An important property of all elements in the same class is that they have the same order. The concept of classes is very important for an analysis of the symmetry properties of any object: The symmetry elements, which will be introduced in the next paragraph, can be grouped into the classes in the above-given sense.

If two groups are considered, a group A with n elements and a group B with m elements, it is possible to obtain $n \times m$ new elements of another group, if each element from the group A is multiplied by each element of the group B. This multiplication should be understood as a rule, which establishes a "one-to-one" relation between the pair of elements (one of which belongs to the A group, and the second one belongs to the B group) and an element from the newly formed C group. This new group C is denoted as $A \times B$ and is called a direct product of the groups A and B.

7.3 Symmetry Elements and Group Theory

Each object can be described from the point of view of symmetry of its parts or the symmetry of the whole object. The word "symmetry" comes from the Greek word "συμμετρια," whose meaning is "agreement in appearance, proportion." In the everyday life the term symmetry is related to the way how the objects appear to us—how their parts agree with each other in shape and size and how we perceive those objects. In mathematics the concept of symmetry may be given a more precise definition; it implies that an object remains unchanged (or invariant) after certain symmetry transformations, which are described below.

Let us consider a cube (Fig. 7.1) as an example of a highly symmetric object. The following symmetry elements can be distinguished:

(i) The mirror planes (or the reflection planes). They divide the cube into two equal halves, and each half is a mirror image of its counterpart. The mirror planes shown in Fig. 7.1 are in the xy plane (a horizontal mirror plane), in the zx plane (a vertical mirror plane) and a diagonal mirror plane, which goes through the z axis and diagonals of the cube bases. The standard notations for these mirror planes are σ_h, σ_v, and σ_d, where the horizontal, vertical, and diagonal planes are denoted by the h, v, and d subscripts, correspondingly. Figure 7.2 illustrates the difference between these types of the reflection planes.

(ii) The axes of rotation. The axis of rotation is an axis, which transforms the cube into itself after rotation through a definite angle about this axis. The angle of rotation is $2\pi/n$ radians (n is an integer number). Such an axis is called an n-fold rotation axis and is denoted as C_n. The axes of the fourth, third, and second orders are shown in Fig. 7.1.

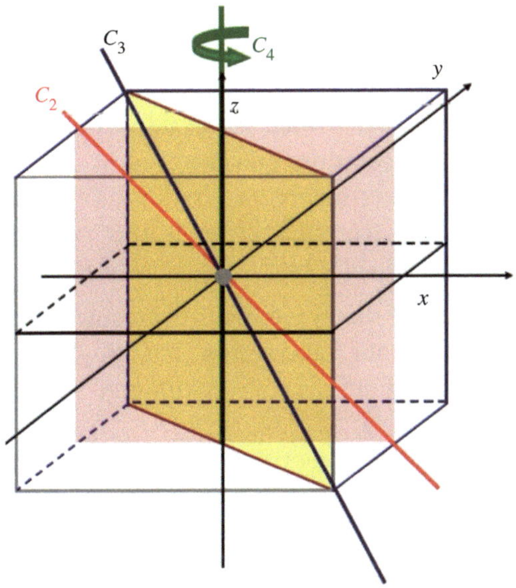

Figure 7.1 Symmetry elements of a cube. See text for detailed explanations.

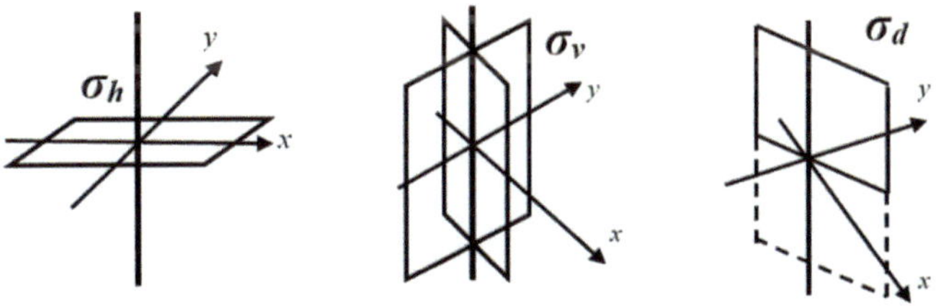

Figure 7.2 Different types of the mirror planes.

(iii) The inversion center. In this case, it is the center of the cube. Each point of the cube becomes another point of the cube at the opposite side after inversion through this center. This symmetry element (denoted as *I*) is of particular importance, since the presence of the inversion center allows to classify the wave functions of the considered system as the odd and even functions.

(iv) Improper rotation. This is a combination of the rotation through $2\pi/n$ radians about an axis with subsequent reflection in the plane perpendicular to this axis (Fig. 7.3). The improper axes are denoted as S_n (the capital letter *S* comes from the word "Spiegel," which means a "mirror" in German). It is easy to see that the order, in which the rotation and reflection are performed, is not important for a final result of the improper rotation. The improper rotation concept allows for representing inversion *I* as an improper rotation of the second order (rotation through 180° and reflection in the plane perpendicular to the axis of rotation): $I = C_2\sigma_h = \sigma_h C_2$.

The set of the symmetry elements of an object forms a group, and this is a very important observation that allows to apply the group theory formalism to describe the symmetry properties of various objects.

So, to sum up, the symmetry elements are the symmetry entities (lines, planes, points), with respect to which the symmetry operations (rotations, reflections, etc.), which transform a considered object into itself, are defined.

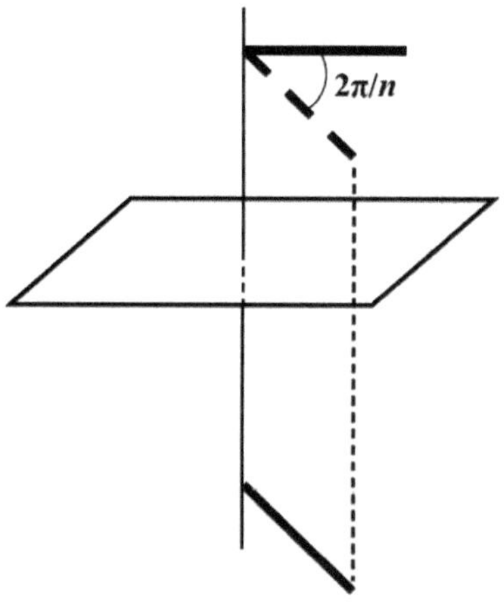

Figure 7.3 Improper rotation.

7.4 An Example of a Symmetry Group: The Group of Rotations of an Equilateral Triangle

In this paragraph we shall consider the set of rotations of an equilateral triangle to see that this set forms a group. Figure 7.4 shows the equilateral triangle *ABC* with its symmetry elements that include only the rotations. There are three rotational axes of the second order (denoted as C_a, C_b, C_c), located in the plane of the triangle, and two axes of rotation through 120° and 240° (counterclockwise), perpendicular to the plane of the triangle (denoted as C_3 and $C_3{}^2$). These coinciding axes are shown by a circle at the center of the triangle. There is also an identity element (rotation through 0° or 360°).

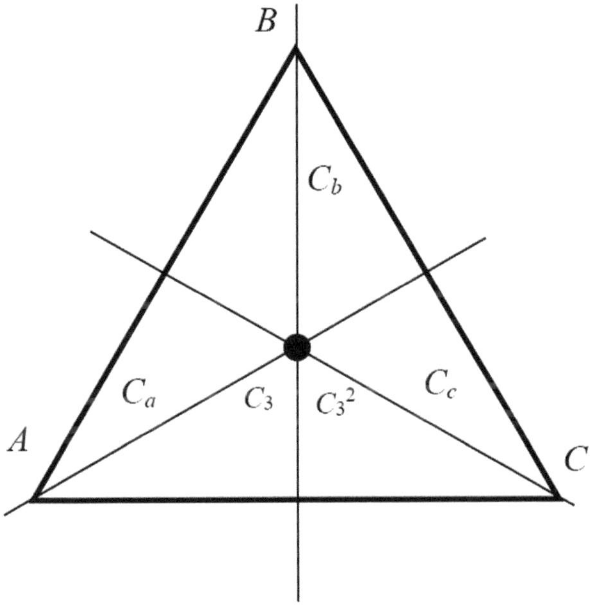

Figure 7.4 Rotations of an equilateral triangle.

As seen from Fig. 7.4, the action of the C_a axis transforms in the triangle *ABC* the vertex *B* into the vertex *C* and vice versa, leaving the vertex *A* at its initial position. In a similar way, the rotation about the C_b axis interchanges the A and C vertices (the vertex B is at the same position), and the rotation about the C_c axis does not alter the position of the C vertex and just swaps the A and B vertices.

Application of the C_3 rotation transforms the triangle vertices as follows: $C \to B$, $B \to A$, $A \to C$, whereas the transformation produced by the C_3^2 rotation is the following: $C \to A$, $B \to C$, $A \to B$.

The group multiplication law can be naturally defined in this case as a successive application of two rotations, for example, the $C_a \times C_3$ product would mean that the C_3 rotation is performed first, and then the C_a rotation is applied to the result of the previous operation. With this convention implemented, it is easy to obtain the following multiplication table for the considered group (Table 7.1).

Table 7.1 Table of multiplication for the symmetry elements of an equilateral triangle

	E	C_3	C_3^2	C_a	C_b	C_c
E	E	C_3	C_3^2	C_a	C_b	C_c
C_3	C_3	C_3^2	E	C_b	C_c	C_a
C_3^2	C_3^2	E	C_3	C_c	C_a	C_b
C_a	C_a	C_c	C_b	E	C_3^2	C_3
C_b	C_b	C_a	C_c	C_3	E	C_3^2
C_c	C_c	C_b	C_a	C_3^2	C_3^2	E

As seen from this table, indeed the set of the equilateral triangle rotations has all the properties of a group. It is also possible to find out from the table, that the second order axes are the inverse elements to themselves (e.g., $C_a \times C_a = E$), and the C_3 and C_3^2 axes are the inverse elements for each other ($C_3 \times C_3^2 = E$).

The order of this group is 6; the order of the identity element E is one, the order of the C_3 and C_3^2 axes is three, and the order of the second order axes C_a, C_b, C_c is two. It is possible to select the following subgroups (apart from two trivial ones): (E, C_3, C_3^2), (E, C_a), (E, C_b), (E, C_c). The following classes of the conjugate elements can be found: (E), (C_3, C_3^2), and (C_a, C_b, C_c) (since $C_3 \times C_a \times C_3^2 = C_c$, $C_3^2 \times C_a \times C_3 = C_b$, $C_a \times C_3 \times C_a = C_3^2$; in all these equations the definition of the conjugate elements and the above shown multiplication Table 7.1 were used).

7.5 Classification of the Point Groups

It was shown in the previous paragraph that a set of the symmetry elements of a certain geometrical object can form a group. If all symmetry operations for a given group are such that at least one point of the considered object is at rest under all symmetry transformations, such a group is called a point group. A short description of all possible point groups that are used for description of the symmetry properties of molecules and crystals (in the ascending order of symmetry, which means in the order of increasing number of the symmetry elements) is given below.

I. The C_n groups.

There is only one symmetry element in these groups—the rotation axis of the n-th order. Each group contains n rotations about this axis through the $2\pi/n$ angle. For example, the C_1 group contains only an identity element. This is the lowest symmetry possible, which can be understood as absence of any symmetry at all. The C_2 group contains an identity element and rotation through 180°, etc.

II. The S_{2n} groups.

These groups have $2n$ elements, each of which is an improper rotation (through π/n angle) around the only axis.

III. The C_{nh} groups.

These groups have the rotational axis of the n-th order and a horizontal mirror plane, which is perpendicular to this axis. There are $2n$ elements in this group: n rotations C_n and n improper rotations (combinations of the rotations C_n and reflection in the horizontal plane perpendicular to the C_n axis).

IV. The C_{nv} groups.

These groups include the axis of the n-th order and n vertical mirror planes, which intersect each other in such a way that the rotational axis is located simultaneously in all these planes (Fig. 7.5). There are $2n$ elements in this group: n rotations C_n and n reflections in n vertical planes.

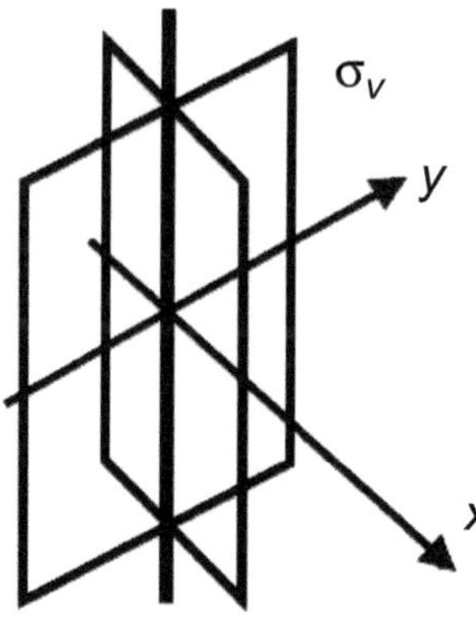

Figure 7.5 The second order rotational axis and two mirror planes in the C_{2v} group.

V. The D_n groups.

If a system has the n-fold rotational axis C_n and n axes of the second order, all of which are in one plane perpendicular to the C_n axis (Fig. 7.6), the symmetry of such a system is described by the D_n group. Note that this plane is not a reflection plane for this group, but just contains all rotational axes of the second order. There are $2n$ elements in this group: n rotations C_n and n rotations C_2.

Figure 7.6 Symmetry axes in the D_3 group.

VI. The D_{nh} groups.

If the plane, in which the second order axes are located (Fig. 7.6), is the mirror plane (this is not the case for the above-described D_n groups!), the number of the symmetry elements in this new D_{nh} group increases, since n vertical mirror planes (each of which contains the vertical axis and one horizontal axis) appear additionally. This group has $4n$ symmetry elements: $2n$ elements from the D_n group, n reflections in the vertical planes, and n improper rotations (combinations of the C_n rotations and reflection in the horizontal plane). Figure 7.7 shows the symmetry elements in the D_{2h} group.

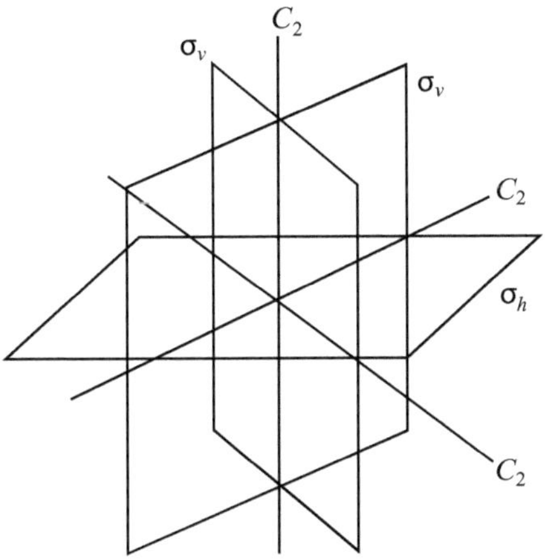

Figure 7.7 Symmetry axes in the D_{2h} group. The vertical mirror planes contain the horizontal rotational axes.

VII. The D_{nd} groups.

These groups are similar to the previous ones, except that the vertical mirror planes are now the diagonal ones (they bisect the angles between the horizontal rotational axes of the second order). For example, for the D_{2d} group the vertical mirror planes will be rotated through 45° with respect to their positions for the D_{2h} group shown in Fig. 7.7. There are $4n$ elements in this group: $2n$ elements from the D_n group, n reflections in the vertical "diagonal" planes, and n C_2 rotations accompanied by the reflection in the vertical "diagonal" planes.

VIII. The T group.

This is the group of all rotational axes of a tetrahedron. There are 12 elements in this group: an identity element E, three C_2 axes, four C_3 axes, and four C_3^2 axes. It is very convenient to draw all these axes, if the tetrahedron is placed into a cube. Then the considered symmetry axes can be easily identified with the symmetry axes of a cube (in Fig. 7.8 one such C_2 axis and one such C_3 axis are shown. The former connects the middle points of the opposite edges of a tetrahedron, and the latter connects the vertex of the tetrahedron with the center of the opposite face).

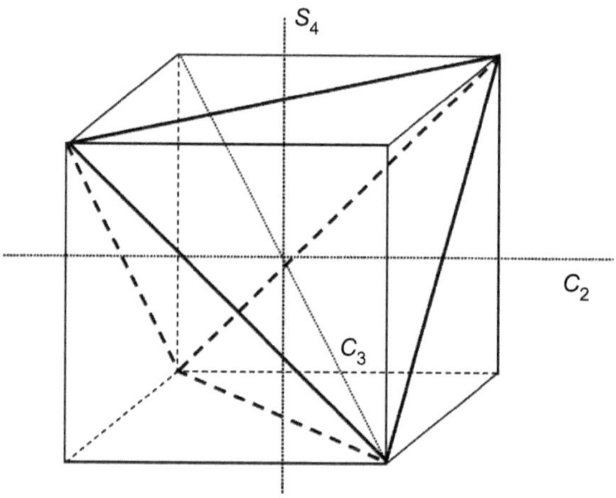

Figure 7.8 The tetrahedron inside a cube. Only one C_2, one C_3 and one S_4 axes are shown for the sake of brevity and simplicity of the figure.

IX. The T_d group.

This group contains not only all rotations from the above-mentioned T group, but also the mirror planes and improper rotations. The group has 24 elements in total: an identity element E, four C_3 axes, four C_3^2 axes, six mirror planes, three S_4 axes, three S_4^2 axes, and three C_2 axes.

X. The T_h group.

The symmetry elements of this group can be obtained from the symmetry elements of the T group by combining each of those elements with an inversion center: $T_h = T \times I$ (it is an example of the product of two elements of a group, which means successive applications of inversion I and then any element of the T group). This converts the C_3 axes into the S_6 axes and adds three mutually perpendicular mirror planes, each of which contains two C_2 axes. There are 24 elements in this group in total.

XI. The O group.

This group contains all symmetry axes of a cube. There are 24 elements in this group: an identity element E, eight C_3 axes, three C_4^2 axes, six C_2 axes, six C_4 axes.

XII. The O_h group.

This group contains all 48 symmetry elements of a cube: an identity element E, three C_2 axes (passing through the centers of the opposite faces), six C_4 axes (passing through the centers of the opposite faces), six C_2 axes (passing through the centers of the opposite edges), eight C_3 axes (coinciding with the cube diagonals), inversion I, three horizontal mirror planes σ_h, eight S_6 axes, six S_4 axes, and six diagonal mirror planes σ_d.

There are altogether 32 point groups, which can be arranged into seven structural types or singonies of crystal lattices (Table 7.2).

Table 7.2 Thirty-two point groups and crystal structures

	Relations between the lattice parameters	Point group	Order
Triclinic	$a \neq b \neq c$, $\alpha \neq \beta \neq \gamma$	C_1	1
		C_i	2
Monoclinic	$a \neq b \neq c$, $\alpha = \beta = \pi/2 \neq \gamma$	C_s	2
		C_2	2
		C_{2h}	4
Orthorhombic	$a \neq b \neq c$, $\alpha = \beta = \gamma = \pi/2$	C_{2v}	4
		D_2	4
		D_{2h}	8
Tetragonal	$a = b \neq c$, $\alpha = \beta = \gamma = \pi/2$	C_4	4
		S_4	4
		C_{4h}	8
		D_{2d}	8
		C_{4v}	8
		D_4	8
		D_{4h}	16

(Continued)

Table 7.2 (*Continued*)

	Relations between the lattice parameters	Point group	Order
Trigonal	$a = b \neq c$, $\alpha = \beta < 2\pi/3$; $\gamma \neq \pi/2$	C_3	3
		S_6	6
		C_{3v}	6
		D_3	6
		D_{3d}	12
Hexagonal	$a = b \neq c$, $\alpha = \beta = \pi/2$; $\gamma = 2\pi/3$;	C_{3h}	6
		C_6	6
		C_{6h}	12
		D_{3h}	12
		C_{6v}	12
		D_6	12
		D_{6h}	24
Cubic	$a = b = c$, $\alpha = \beta = \gamma = \pi/2$	T	12
		T_h	24
		T_d	24
		O	24
		O_h	48

Table 7.2 clearly illustrates the increase of the number of symmetry elements (symmetry ascent), when moving from the C_1 group to the O_h group.

7.6 Representations of the Point Groups

An easy way to introduce the concept of the group representations is to consider in detail an example of the group of rotations of an equilateral triangle (the D_3 group). Let us choose a system of reference x, y, z with the origin at the point of intersection of all symmetry axes. We shall be mainly concerned with how the coordinates x, y, z in the chosen system of reference (Fig. 7.9) will transform into a new set $\{x', y', z'\}$ when all the above-described six symmetry operations of this group are applied.

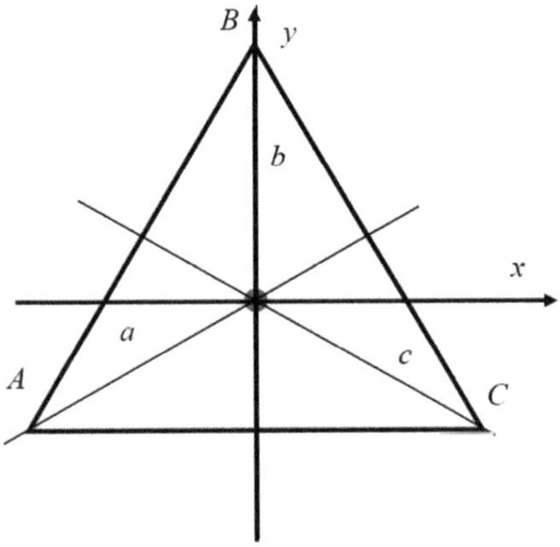

Figure 7.9 An equilateral triangle with the coordinate axes (the z axis is directed out of the plane of figure towards the reader).

An identity element, by its definition, does not produce any change, and we can write immediately that $x' = x$, $y' = y$, $z' = z$. Or, in the matrix form:

$$\begin{pmatrix} x' \\ y' \\ z' \end{pmatrix} = \begin{pmatrix} 1 & 0 & 0 \\ 0 & 1 & 0 \\ 0 & 0 & 1 \end{pmatrix} \begin{pmatrix} x \\ y \\ z \end{pmatrix}. \tag{7.1}$$

To find the coordinate transformation after the C_3 rotation (note that the C_3 axis coincides with the z axis), we employ the following well-known relation from analytical geometry:

$$\begin{aligned} x' &= x\cos\phi + y\sin\phi \\ y' &= -x\sin\phi + y\cos\phi, \end{aligned} \tag{7.2}$$

where ϕ is the angle of rotation around the z axis, and the positive direction of the rotation is counterclockwise (the $+x$ axis is being rotated towards the $+y$ axis). For the C_3 rotation $\phi = 120°$, and the matrix of the coordinate transformation is

$$\begin{pmatrix} x' \\ y' \\ z' \end{pmatrix} = \begin{pmatrix} -\frac{1}{2} & \frac{\sqrt{3}}{2} & 0 \\ -\frac{\sqrt{3}}{2} & -\frac{1}{2} & 0 \\ 0 & 0 & 1 \end{pmatrix} \begin{pmatrix} x \\ y \\ z \end{pmatrix}. \tag{7.3}$$

In a similar way we find that for the C_3 rotation $\varphi = 240°$, and the matrix of the coordinate transformation is

$$\begin{pmatrix} x' \\ y' \\ z' \end{pmatrix} = \begin{pmatrix} -\frac{1}{2} & -\frac{\sqrt{3}}{2} & 0 \\ \frac{\sqrt{3}}{2} & -\frac{1}{2} & 0 \\ 0 & 0 & 1 \end{pmatrix} \begin{pmatrix} x \\ y \\ z \end{pmatrix}. \tag{7.4}$$

Note that both these rotations do not change the z coordinate.

Let us now consider the C_b rotation. It does not change the y coordinate, but just replaces x by $-x$, and z by $-z$, resulting in the following matrix of transformation:

$$\begin{pmatrix} x' \\ y' \\ z' \end{pmatrix} = \begin{pmatrix} -1 & 0 & 0 \\ 0 & 1 & 0 \\ 0 & 0 & -1 \end{pmatrix} \begin{pmatrix} x \\ y \\ z \end{pmatrix}. \tag{7.5}$$

The two remaining rotations C_a and C_c can be treated as follows: We first consider the rotation of the system of reference by 120° counterclockwise (for the C_a rotation) and 240° counterclockwise (for the C_c rotation). These intermediate steps make the y axis to coincide with the axis of rotation and then the transformation (7.5) can be easily applied to the result of this intermediate rotation, which, in turn, can be obtained from Eqs. (7.3) and (7.4). Finally, we arrive at the following transformations:

$$\begin{pmatrix} x' \\ y' \\ z' \end{pmatrix} = \begin{pmatrix} \frac{1}{2} & -\frac{\sqrt{3}}{2} & 0 \\ -\frac{\sqrt{3}}{2} & -\frac{1}{2} & 0 \\ 0 & 0 & -1 \end{pmatrix} \begin{pmatrix} x \\ y \\ z \end{pmatrix} \tag{7.6}$$

for the C_a axis and

$$\begin{pmatrix} x' \\ y' \\ z' \end{pmatrix} = \begin{pmatrix} \frac{1}{2} & \frac{\sqrt{3}}{2} & 0 \\ \frac{\sqrt{3}}{2} & -\frac{1}{2} & 0 \\ 0 & 0 & -1 \end{pmatrix} \begin{pmatrix} x \\ y \\ z \end{pmatrix} \tag{7.7}$$

for the C_c axis.

So, we have found the following "one-to-one" correspondence between the six symmetry operations and six matrices:

$$E \to \begin{pmatrix} 1 & 0 & 0 \\ 0 & 1 & 0 \\ 0 & 0 & 1 \end{pmatrix},\ C_3 \to \begin{pmatrix} -\frac{1}{2} & \frac{\sqrt{3}}{2} & 0 \\ -\frac{\sqrt{3}}{2} & -\frac{1}{2} & 0 \\ 0 & 0 & 1 \end{pmatrix},\ C_3{}^2 \to \begin{pmatrix} -\frac{1}{2} & -\frac{\sqrt{3}}{2} & 0 \\ \frac{\sqrt{3}}{2} & -\frac{1}{2} & 0 \\ 0 & 0 & 1 \end{pmatrix}$$

$$C_a \to \begin{pmatrix} \frac{1}{2} & -\frac{\sqrt{3}}{2} & 0 \\ -\frac{\sqrt{3}}{2} & -\frac{1}{2} & 0 \\ 0 & 0 & -1 \end{pmatrix},\ C_b \to \begin{pmatrix} -1 & 0 & 0 \\ 0 & 1 & 0 \\ 0 & 0 & -1 \end{pmatrix},\ C_c \to \begin{pmatrix} \frac{1}{2} & \frac{\sqrt{3}}{2} & 0 \\ \frac{\sqrt{3}}{2} & -\frac{1}{2} & 0 \\ 0 & 0 & -1 \end{pmatrix}. \tag{7.8}$$

The obtained six matrices form a group, the multiplication table for which is the same as for the group of symmetry operations of an equilateral triangle—this can be easily proved. Moreover, all relations between the constituting elements (including the subgroups, the orders of the elements, their grouping into classes) are essentially the same in both groups. Such groups are called isomorphic. The obtained group of matrices is said to form a representation of the considered D_3 group, since these matrices do indeed represent the symmetry operations in the matrix form.

This is a general algorithm of how to build the representation of a group: first, a set of the linearly independent basis functions should be chosen. It is better to write this basis set as a column vector. Then all symmetry operations of the considered group should be applied to each of the basis functions. The "images" of these basis functions after a symmetry operation from the considered group can be linked to the column vector of initial functions by means of some transformation matrix. The set of the matrices (the number of matrices in a representation is equal to the order of a group) corresponding to the images of these basis functions after transformations form the representation of the group.

The choice of the basis functions is not unique. Therefore, it is possible to have many different representations, depending on the choice of the basis functions. If the following condition is fulfilled between the corresponding matrices (taken for the same symmetry element) of two representations D' and D:

$$D' = U^{-1}DU, \tag{7.9}$$

the representations D' and D are equivalent (physically this means that D' and D are the identical representations with equal traces, or the sums of the matrices' diagonal elements, though they can be expressed in different basis sets), and U is a square invertible matrix.

The rank of the representation matrices is, obviously, equal to the number of functions in a chosen basis set. It may happen, however, that all representation matrices have a so-called block-diagonal form:

$$\begin{pmatrix} 1 & 0 & \cdots & 0 \\ 0 & 1 & \cdots & 0 \\ \vdots & \vdots & \ddots & \vdots \\ 0 & 0 & \cdots & 1 \end{pmatrix}, \begin{pmatrix} D^{(1)}(A) & 0 & \cdots & 0 \\ 0 & D^{(2)}(A) & \cdots & 0 \\ \vdots & \vdots & \ddots & \vdots \\ 0 & 0 & \ldots & D^{(n)}(A) \end{pmatrix}, \\ \ldots, \begin{pmatrix} D^{(1)}(B) & 0 & \cdots & 0 \\ 0 & D^{(2)}(B) & \cdots & 0 \\ \vdots & \vdots & \ddots & \vdots \\ 0 & 0 & \ldots & D^{(n)}(B) \end{pmatrix}, \tag{7.10}$$

where all blocks with the same superscript have the same dimension but different matrix elements, and the A and B letters are used to emphasize that the blocks with the same superscript can contain in general case different matrix elements. The representation given by matrices (7.10) is called a reducible representation, and the diagonal blocks are referred to as the matrices of irreducible representations (their dimensions are smaller than the dimension of matrices in Eq. (7.10)). The procedure of selecting these diagonal blocks is called the decomposition of a reducible representation into irreducible representations. Physically it means that it is possible to select several subsets from the initial set of the basis functions (the number of functions in each subset is equal to the dimension of each block) and the images of the functions from these subsets after applications of the symmetry operations from a given point group are always the linear combinations of the functions from the same subset, never transforming into the wave functions from another subset.

As an example, we consider the matrices in Eq. (7.8). As can be easily seen, they already have a block diagonal form and each matrix can be represented as one 2×2 and one 1×1 block. From the considered $\{x, y, z\}$ basis set we can select two subsets: $\{x, y\}$ and $\{z\}$, which are not mixed up with each other when all symmetry operations are applied.

The matrices of the equivalent representations are related by the transformation (7.9) and, therefore, the sums of the diagonal elements (called the characters and denoted $\chi(A)$, with A being an element of the considered group) of these matrices are equal. Therefore, the character of a representation is a unique common property of all equivalent representations. Besides, since all elements in one class are also connected by the similarity transformation, the characters of all elements in one class are equal. It is also clear that the character of a reducible representation is equal to the sum of the characters of irreducible representations, or traces of the blocks in the block-diagonal form, into which it can be decomposed.

Irreducible representations and their characters have several remarkable properties, which will be briefly mentioned and illustrated below (but without a strict mathematical proof).

Since in any basis set it is possible to choose a symmetric wave function, which will stay invariant under all symmetry transformations (like $x^2 + y^2 + z^2$, for example), each group has a one-dimensional trivial irreducible representation, the characters of which are unity for all elements of a group.

It can be shown that the number of irreducible representations in a group is always equal to the number of classes. The irreducible representations are orthogonal:

$$\sum_{A} G_i(A)_{mk} G_j(A)_{m'k'}^{*} = \frac{n}{\sqrt{l_i l_j}} \delta_{ij} \delta_{mm'} \delta_{kk'}, \qquad (7.11)$$

where the summation is carried over all elements A of the group, i and j denote different irreducible representations with dimensions l_i and l_j, mk, $m'k'$ refer to different matrix elements of these representations, and n is the order of a group.

The sum of the squares of the dimensions l_i of the irreducible representations of a given group is always equal to the order n of this group:

$$\sum_{i=1}^{\text{number of irreps}} l_i^2 = n. \qquad (7.12)$$

Since the point groups of all physical objects include the symmetry operations defined in the three-dimensional space, there can be only one-, two-, and three-dimensional irreducible representations (strictly speaking it is true for the point groups only). A standard practice is to denote the one-dimensional irreducible representations by letters A and B, two-dimensional—by letter E, and three-dimensional—by letter T. The letter "A" is used for the representations, the basis functions of which are symmetric with respect to the rotation about the main axis of the highest order, whereas the letter "B" is used for the representations with the functions, which are antisymmetric with respect to the same rotation. It is also clear that the character of an irreducible representation for an identity element is just the dimension of this irreducible representation. If there are several irreducible representations of the same dimension, they can be distinguished by a subscript, like A_1, T_1, T_2, etc. This is the so-called Mulliken notation. Another notation, where the irreducible representations are denoted by the capital Greek letter Γ with subscripts (like Γ_1, Γ_2, etc.) was introduced by Bethe.

The characters of irreducible representations behave themselves like the orthogonal vectors defined in the k-dimensional space (k stands for the number of irreducible representations):

$$\sum_{A} \chi^{(i)}(A) \chi^{(j)}(A)^{*} = n\delta_{ij}. \qquad (7.13)$$

As follows from Eq. (7.13), the sum of characters of any non-trivial representation is always zero. This can be proved if one of the $\chi^{(i)}$ representations is taken as a unity representation A_1 with all characters equal to unity.

Finally, a very important relation determines, how many times a given i-th irreducible representation occurs in the decomposition of a reducible representation (or, in other words, how many $n^{(i)}$ blocks corresponding to this i-th irreducible representation can be found in a matrix described by Eq. (7.10)):

$$n^{(i)} = \frac{1}{n}\sum_{A}\chi(A)\chi^{(i)}(A)^{*}. \tag{7.14}$$

Again, the sum is running over all elements A of a group. This equation is often used when describing the spectra of impurity ions in solids.

As an example of the decomposition of a reducible representation into the irreducible ones, let us consider the group O of rotations of an ideal octahedron. Table 7.3 shows the characters of five irreducible representations of this group: two one-dimensional representations, one two-dimensional, and two three-dimensional. The sixth representation Γ is shown in the last line. This is a reducible representation, and our aim will be to decompose it into those five irreducible representation of the O group. There are 24 elements in this group, and these elements form five classes.

Table 7.3 Characters of irreducible representations for the O group

	E	$8C_3$	$3C_2$	$6C_2$	$6C_4$
A_1	1	1	1	1	1
A_2	1	1	1	–1	–1
E	2	–1	2	0	0
T_1	3	0	–1	–1	1
T_2	3	0	–1	1	–1
Γ	16	1	8	2	2

We apply Eq. (7.14), carefully performing the summation over all elements of this group (it can be replaced by multiplying a product of two characters of the elements from the same class by the number of the elements in a class). We have:

$$n^{(A_1)} = \frac{1}{24}(16\times1\times1+1\times1\times8+8\times1\times3+2\times1\times6+2\times1\times6)=3,$$

$$n^{(A_2)} = \frac{1}{24}(16\times1\times1+1\times1\times8+8\times1\times3+2 \\ \times(-1)\times6+2\times(-1)\times6)=1,$$

$$n^{(E)} = \frac{1}{24}(16\times2\times1+1\times(-1)\times8+8\times2\times3+2\times0\times6+2\times0\times6)=3,$$

$$n^{(T_1)} = \frac{1}{24}(16\times3\times1+1\times0\times8+8\times(-1) \\ \times3+2\times(-1)\times6+2\times1\times6)=1,$$

$$n^{(T_2)} = \frac{1}{24}(16\times3\times1+1\times0\times8+8\times(-1) \\ \times3+2\times1\times6+2\times(-1)\times6)=1.$$

So, the reducible representation Γ in this example consists of the following irreducible representations of the O group: $\Gamma = 3A_1 + A_2 + 3E + T_1 + T_2$.

A proper decomposition of a reducible representation into the irreducible ones is very important for a qualitative analysis of the degenerated energy level splitting in crystal field and selection rules for various transitions (electric dipole, electric quadrupole, magnetic dipole, etc.) between the pairs of the split levels.

The characters of the irreducible representations of all point groups are given below in Table 7.4 (the x, y, z basis functions, which are the coordinate components of the electric dipole operator, are shown as the basis of some irreducible representations).

It should be also pointed out that the point groups with an inversion center have the so called “gerade” (which means “even” in German) and “ungerade” (“odd” in German) irreducible representations depending on the properties of their basis functions with respect to inversion. They are denoted by the subscripts “g” and “u,” respectively.

7.7 Continuous Groups and Double Groups

A special case of the so-called continuous groups is briefly considered in this paragraph. These are the groups with continuous symmetry operations, describing the axial or spherical symmetry. The simplest continuous group is the C_∞ group, which contains all rotations around an axis through an arbitrary angle ϕ. It can be considered as a limiting case of the C_n group when the order of rotation $n \to \infty$, or the angle of rotation $\phi \to 0$. The linear molecules with all atoms lying on the same straight line belong to this group. If such a molecule has a mirror plane perpendicular to this line, its point group symmetry is $C_{\infty h}$. In a similar way, it is possible to define the $C_{\infty v}$, D_∞, $D_{\infty h}$ groups: All of them describe the symmetry of the axial species. The group of symmetry of a uniform cylinder, for example, is $D_{\infty h}$. It is essential that the continuous groups are the infinite groups, since they have an infinite number of the symmetry elements.

These groups are used for a description of the angular momenta eigenstates and addition of the angular momenta. For the angular momentum, these groups describe the highest (spherical) symmetry of a free atom, which can be reduced if such a free atom is placed into an external electric or magnetic field.

Table 7.4 Characters of irreducible representations of some point groups

C_1	E
A	1

C_2	E	C_2
A (z)	1	1
B(x or y)	1	−1

C_3	E	C_3	C_3^2
A (z)	1	1	1
E ($x \pm i$	1	ε	ε^2
y)	1	ε^2	ε

$$\varepsilon = \exp\left(\frac{2\pi i}{3}\right)$$

C_4	E	C_4	C_2	C_4^3
A (z)	1	1	1	1
B	1	−1	1	−1
E ($x \pm i y$)	1	i	−1	$-i$
	1	$-i$	−1	i

C_6	E	C_6	C_3	C_2	C_3^2	C_6^5
A (z)	1	1	1	1	1	1
B	1	−1	1	−1	1	−1
E_1 ($x \pm i y$)	1	ω^2	$-\omega$	1	ω^2	$-\omega$
	1	$-\omega$	ω^2	1	$-\omega$	ω^2

$E_2\ (x \pm i\,y)$	1	ω	ω^2	−1	$-\omega$	$-\omega^2$
	1	$-\omega^2$	$-\omega$	−1	ω^2	ω

$$\omega = \exp\left(\frac{2\pi i}{6}\right)$$

S_2 (or C_i)	E	I
$A\ (z)$	1	1
$B(x \text{ or } y)$	1	−1

C_S	E	σ
$A\ (x \text{ or } y)$	1	1
$B(z)$	1	−1

S_4	E	S_4	S_2	$S_4{}^3$
A	1	1	1	1
$B(z)$	1	−1	1	−1
$E\ (x \pm i\,y)$	1	i	−1	$-i$
	1	$-i$	−1	i

C_{2h}	E	C_2	σ_h	I
A_g	1	1	1	1
B_g	1	−1	−1	1
$A_u\ (z)$	1	1	−1	−1
$B_u(x \text{ or } y)$	1	−1	1	−1

C_{2v}	E	C_2	σ_v	$\sigma_{v'}$
$A_1\ (z)$	1	1	1	1
$B_1\ (x)$	1	−1	1	−1
A_2	1	1	−1	−1
$B_2\ (y)$	1	−1	−1	1

C_{3v}	E	$2C_3$	$3\sigma_v$
$A_1\ (z)$	1	1	1
A_2	1	1	−1
$E\ (x, y)$	2	−1	0

D_{2d}	E	C_2	$2S_4$	$2C_2$	$2\sigma_d$
A_1	1	1	1	1	1
A_2	1	1	1	−1	−1

B_1	1	1	−1	1	−1
$B_2\ (z)$	1	1	−1	−1	1
$E\ (x, y)$	2	−2	0	0	0

C_{4v}	E	C_2	$2C_4$	$2\sigma_v$	$2\sigma_{v'}$
$A_1\ (z)$	1	1	1	1	1
A_2	1	1	1	−1	−1
B_1	1	1	−1	1	−1
B_2	1	1	−1	−1	1
$E\ (x, y)$	2	−2	0	0	0

D_3	E	$2C_3$	$3C_2$
A_1	1	1	1
$A_2\ (z)$	1	1	−1
$E\ (x, y)$	2	−1	0

D_4	E	C_2	$2C_4$	$2C_2$	$2C'_2$
A_1	1	1	1	1	1
$A_2\ (z)$	1	1	1	−1	−1
B_1	1	1	−1	1	−1
B_2	1	1	−1	−1	1
$E\ (x, y)$	2	−2	0	0	0

D_6	E	C_2	$2C_3$	$2C_6$	$3C_2$	$3C_2'$
A_1	1	1	1	1	1	1
$A_2\ (z)$	1	1	1	1	−1	−1
B_1	1	−1	1	−1	1	−1
B_2	1	−1	1	−1	−1	1
$E_1\ (x, y)$	2	−2	−1	1	0	0
E_2	2	2	−1	−1	0	0

C_{6v}	E	C_2	$2C_3$	$2C_6$	$3\sigma_v$	$3\sigma_{v'}$
$A_1\ (z)$	1	1	1	1	1	1
A_2	1	1	1	1	−1	−1

B_1	1	–1	1	–1	–1	1
B_2	1	–1	1	–1	1	–1
E_1 (x, y)	2	–2	–1	1	0	0
E_2	2	2	–1	–1	0	0

$\boldsymbol{O}$	$\boldsymbol{E}$	$\boldsymbol{8C_3}$	$\boldsymbol{3C_2}$	$\boldsymbol{6C_2}$	$\boldsymbol{6C_4}$
A_1	1	1	1	1	1
A_2	1	1	1	–1	–1
E	2	–1	2	0	0
T_1 (x, y, z)	3	0	–1	–1	1
T_2	3	0	–1	1	–1

$\boldsymbol{T_d}$	$\boldsymbol{E}$	$\boldsymbol{8C_3}$	$\boldsymbol{3C_2}$	$\boldsymbol{6\sigma_d}$	$\boldsymbol{6S_4}$
A_1	1	1	1	1	1
A_2	1	1	1	–1	–1
E	2	–1	2	0	0
T_1	3	0	–1	–1	1
T_2 (x, y, z)	3	0	–1	1	–1

A special case with very important physical applications is the continuous group of spherical symmetry. A free atom in the absence of external fields is described by such symmetry. It contains the group of all three-dimensional rotations as a subgroup. The form of the operator of an infinitesimal rotation about an axis $\mathbf{r}$ through an angle α is $\exp(-i\mathbf{J}\cdot\mathbf{r}\alpha/\hbar)$, where $\mathbf{J}$ is the general angular momentum operator (it can be the orbital momentum $\mathbf{L}$, the spin momentum $\mathbf{S}$ or the total angular momentum $\mathbf{J} = \mathbf{L} + \mathbf{S}$). It is clear then that the properties of this operator should be the same as the properties of the angular momentum. For the sake of simplicity, we restrict our discussion here by the orbital momentum $\mathbf{L}$ only. We already know that the wave functions of free many-electron atoms are expressed in terms of linear combinations of products of the spherical functions $Y_{lm}(\theta,\varphi)$, which make a part of the single-electron wave functions. If we consider the basis function consisting of linear combination of $(2l + 1)$ wave functions with the same quantum number l and different values of m in the range from—l to l (these functions are orthonormalized and form the basis of the $(2l + 1)$-dimensional space), then the rotation through an angle α results in multiplying each function by the $e^{im\alpha}$ factor. Then the matrix representation of such a rotation is diagonal and looks as follows:

$$R = \begin{pmatrix} e^{il\alpha} & 0 & 0 & \cdots & 0 \\ 0 & e^{i(l-1)\alpha} & 0 & \cdots & 0 \\ 0 & \cdots & \ddots & \cdots & 0 \\ 0 & \cdots & \cdots & e^{-i(l-1)\alpha} & 0 \\ 0 & \cdots & \cdots & 0 & e^{-il\alpha} \end{pmatrix}. \tag{7.15}$$

The trace of this matrix or the character of the rotation operator is then

$$\begin{aligned} Tr(R) &= \chi(\alpha) = e^{il\alpha} + e^{i(l-1)\alpha} + \cdots + e^{-il\alpha} \\ &= \frac{e^{i(l+1)\alpha} - e^{-il\alpha}}{e^{i\alpha} - 1} = \frac{\sin\left[\left(l + \frac{1}{2}\right)\alpha\right]}{\sin\left(\frac{1}{2}\alpha\right)}. \end{aligned} \tag{7.16}$$

If the inversion transformation I is considered, the spherical functions $Y_{lm}(\theta,\varphi)$change the sign if l is odd and do not change the sign if l is even. Therefore, the character of the inversion transformation matrix is

$$\chi(I) = \pm(2l + 1). \tag{7.17}$$

The characters of matrices of mirror reflection σ and improper rotation through an angle α can be found if these transformations are represented as a combination of both inversion and rotation: $\sigma = IC_2$, $S(\alpha) = IC(\pi + \alpha)$. The resultant characters will be the characters corresponding to these rotations, taken with the positive/negative sign, depending on the inversion character sign.

Equations (7.16) and (7.17) can be applied in the same way to the states of the orbital momentum L or the total angular momentum J (just by replacing l by L or J in all these equations. For the integer values of J the characters of rotations through the angles α and $(\alpha + 2\pi)$ are equal: $\chi(\alpha + 2\pi) = \chi(\alpha)$ and the 2πrotation acts as an identity element.

The situation becomes different if J takes a half-integer value. Then $\chi(\alpha + 2\pi) = -\chi(\alpha)$, so the rotation by 2π is no longer an identity element, which can be easily checked using Eqs. (7.16)–(7.17). The character of an identity element is $\chi(E) = 2J + 1$ and the character of the 2π rotation (this operation is denoted by R) is $\chi(R) = -(2J + 1)$. The rotation by the 4π angle will become a new identity element, since $\chi(\alpha + 4\pi) = \chi(\alpha)$ in this case of the half-integer J values; the number of symmetry operations in this case increases by considering the products of the operation R with all other elements of a group. These groups are called the double groups. The number of elements in these groups is always doubled with respect to the number of elements in a corresponding "simple" group. The number of classes is also always greater, but not necessarily doubled. The number of irreducible representations increases too. All the above-mentioned properties of the irreducible representations remain valid for the double groups as well.

The half-integer values of J occur when the considered atom or ion has an odd number of electrons in the unfilled shell. The total spin S of such electron configuration is half-integer, and, since the orbital momentum L is an integer number, the total momentum J is half-integer too (all these momenta are in units of $\hbar$). Such atoms and/or ions are called the Kramers ions, and their energy levels always remain doubly degenerated even in the case of very low symmetry. This is because the states with opposite spins in the electrostatic crystal field have the same energy. Only an external magnetic field can split these states and remove this remaining degeneracy.

7.8 Applications of the Group Theory in Spectroscopy: Symmetry Reduction and Splitting of the Energy Levels

It is well known from quantum mechanics that the physical properties of any quantum mechanical system are defined by the solutions of the Schrödinger equation with corresponding Hamiltonian. Since the symmetry transformations of the point groups cannot change the energy of the considered system, the Hamiltonian, its eigenfunctions and eigenvalues should be invariant with respect to the symmetry operations. This means that the eigenfunctions of the Hamiltonian can be classified according to their symmetry: Each energy level is characterized by one or several (if the energy level is degenerated) wave functions, which correspond to definite irreducible representations of this point group. The dimensions of these irreducible representations determine the degeneracy of the energy level: One-dimensional representations correspond to the singlet levels, two-dimensional—to the doublet levels, and three-dimensional—to the triplet levels.

Splitting of the energy levels of a system can be interpreted as a result of a small perturbation (an axial compression or elongation of a crystal, for example). If the symmetry of the perturbation is the same as the symmetry of the system, no further splitting of the energy levels is possible; all energy levels can only undergo a uniform shift upwards or downwards. If, however, the symmetry of the perturbation is lower than the symmetry of the considered system, the degenerated energy levels can be split up into several levels. The number of these newly formed levels can be determined by using a simple group-theoretical consideration.

First, it is clear, that after the low-symmetry perturbation is added, the symmetry of the system becomes that one of the perturbation—this is because if two objects/systems are considered together by placing one into another, only the joint symmetry elements remain in such a new object/system.

Second, the irreducible representations of an initial high-symmetry group become reducible representations in a low-symmetry group that corresponds to the considered perturbation.

Thirdly, the characters of the "old" irreducible representation of a high-symmetry group after lowering the symmetry are simply equal to the characters of the symmetry elements, which are common in both groups.

Let us illustrate these statements by an example. We consider a triplet level, which corresponds to the T_1 irreducible representation of the above considered O point group. If now a perturbation of the D_3 symmetry (a symmetry group of an equilateral triangle) is added, this level should split. The common symmetry elements for these two groups are: E, C_3, and C_2. The characters of the reducible representation Γ (which corresponds to the T_1 irreducible representation in the O group) can be obtained simply eliminating the characters of those elements, which do not "survive" after symmetry lowering. So, the reducible representation Γ is as follows:

	E	$2C_3$	$3C_2$
Γ	3	0	–1

In the D_3 group this representation breaks up into the E and A_2 irreducible representations (we apply Eq. (7.14)): $\Gamma = A_2 + E$. Physically this result means that instead of one triply degenerated level we have one doublet level, whose two wave functions have the symmetry of the two-dimensional irreducible representation E, and one singlet level, whose single wave function has the symmetry of the one-dimensional irreducible representation A_2.

If, however, the T_1 state experiences the action of a perturbation of the C_2 symmetry, then the characters of the reducible representation Γ in this group will be

	E	C_2
Γ	3	–1

and it is decomposed into the three one-dimensional irreducible representations of the C_2 group: $\Gamma = A + 2B$, which corresponds to a complete removal of degeneracy of the triplet level: Three singlet levels are formed instead of one triply degenerated energy level.

This technique is known as the symmetry reduction method. The tables of reduction of all irreducible representations for all groups from the higher to lower symmetries can be found in the literature (see, for example, the book by G. F. Koster et al, which is included in the literature for the further reading at the end of this chapter).

A problem of splitting of the energy levels of free atoms (ions) when they are placed into an external field or into a crystal is of special importance for spectroscopy. It can also be solved by applying the symmetry reduction principle: The symmetry is lowered from the spherical (symmetry of an isolated atom) to a given point group symmetry. The wave functions of a definite LS term of a free ion form a basis of an irreducible representation of the full rotation group, which becomes reducible in a given point group. Application of Eqs. (7.16) and (7.17) allows for finding the characters of this reducible representation, which then should be decomposed into irreducible ones. The number and dimension of these irreducible representations determines the overall pattern of splitting of a given LS term.

It is instructive to consider here a few examples of this kind. The Co^{2+} ion with its $3d^7$ electron configuration has 8 *LS* terms: ^{4}F, ^{4}P, ^{2}P, ^{2}D$_1$, ^{2}D$_2$, ^{2}F, ^{2}G, ^{2}H. In solids, this ion shows a preference to occupy the four-fold coordinated crystallographic sites. If this ion is placed into a crystal field with the T_d symmetry, all free ion terms split, depending on the value of the orbital angular momentum L. The total spin momentum S is not important for this crystal field splitting, since the crystal field can remove partially or completely the orbital degeneracy only. The characters of the reducible representations in the T_d group corresponding to different values of the orbital momentum L are shown in Table 7.5.

Table 7.5 Splitting of *LS* terms of the d^3 (d^7) configuration in T_d group

	E	$8C_3$	$3C_2$	$6\sigma_d$	$6S_4$	T_d **irreducible representations**
$\Gamma(L = 1)$	3	0	−1	−1	1	T_1
$\Gamma(L = 2)$	5	−1	1	1	−1	$T_2 + E$
$\Gamma(L = 3)$	7	1	−1	−1	−1	$A_2 + T_1 + T_2$
$\Gamma(L = 4)$	9	0	1	1	1	$A_1 + E + T_1 + T_2$
$\Gamma(L = 5)$	11	−1	−1	−1	1	$E + 2T_1 + T_2$

The symmetrized combinations of the wave functions of the S, P, D, F, G, H terms (which are characteristic for the transition metal ions) in the cubic crystal field are listed in Appendix III. The crystal field Hamiltonian matrix will have a block-diagonal form in the basis set of those symmetrized wave functions.

The last column of this table shows the decompositions of the reducible representations with the basis sets corresponding to definite values of the orbital momentum in terms of the T_d group irreducible representations. An inspection of this column shows that the *S* term does not split (it can be expected, since this is the orbital singlet level). The *P* term also does not split and remains to be the orbital triplet T_1. The *D* term splits into two levels: the orbital doublet E and the orbital triplet T_2. The *F* term splits into three sublevels: the orbital singlet A_2 and two orbital triplets T_1 and T_2. The *G* term splits into four sublevels: the orbital singlet A_1, the orbital doublet E, two orbital triplets T_1 and T_2. The *H* terms splits into four levels: the orbital doublet E, two orbital triplets T_1 and orbital triplet T_2. The remaining degeneracy of the energy levels can be lifted by a further symmetry lowering, for example, if the tetrahedral complex around the Co^{2+} ion undergoes an axial compression or elongation.

Analysis of the SO splitting of the orbitally degenerated states implies calculations of the direct products of irreducible representations of the "spin" and "orbital" parts and decomposition of the obtained reducible representation into the irreducible once. The double groups should be used for the systems with an odd number of electrons and the "usual" groups are used for the systems with an even number of electrons.

Let us consider an example of the d^7 electron configuration in the tetrahedral crystal field T_d. Table 7.6 gives the irreducible representations of the T_d double group. The two basis functions of the spin-doublet states and the four basis functions of the spin-quartet states transform as the Γ_6 and Γ_8 irreducible representations, respectively. Finding the direct products of these representations with the orbital ones, we arrive at the following pattern of the SO splitting in this case: ${}^4A_1 \rightarrow \Gamma_8 \times A_1 \rightarrow \Gamma_8$, ${}^4A_2 \rightarrow \Gamma_8 \times A_2 \rightarrow \Gamma_8$, ${}^4E \rightarrow \Gamma_8 \times E \rightarrow \Gamma_6 + \Gamma_7 + \Gamma_8$, ${}^4T_1 \rightarrow \Gamma_8 \times T_1 \rightarrow \Gamma_6 + \Gamma_7 + 2\,\Gamma_8$, ${}^4T_2 \rightarrow \Gamma_8 \times T_2 \rightarrow \Gamma_6 + \Gamma_7 + 2\,\Gamma_8$ for the spin-quartet states and ${}^2A_1 \rightarrow \Gamma_6 \times A_1 \rightarrow \Gamma_6$, ${}^2A_2 \rightarrow \Gamma_6 \times A_2 \rightarrow \Gamma_7$, ${}^2E \rightarrow \Gamma_6 \times E \rightarrow \Gamma_8$, ${}^2T_1 \rightarrow \Gamma_6 \times T_1 \rightarrow \Gamma_6 + 2\,\Gamma_8$, ${}^2T_2 \rightarrow \Gamma_6 \times T_2 \rightarrow \Gamma_7 + 2\,\Gamma_8$ for the spin-doublet states.

Table 7.6 Characters of irreducible representations of the T_d double group

Mulliken notation	**Bethe notation**	E	R	$8C_3$	$8RC_3$	$3C_2$ $3RC_2$	$6\sigma_d$ $6R\sigma_d$	$6S_4$	$6RS_4$
A_1	Γ_1	1	1	1	1	1	1	1	1
A_2	Γ_2	1	1	1	1	1	−1	−1	−1
E	Γ_3	2	2	−1	−1	2	0	0	0
T_1	Γ_4	3	3	0	0	−1	−1	1	1
T_2	Γ_5	3	3	0	0	−1	1	−1	−1
	Γ_6	2	−2	1	−1	0	0	$\sqrt{2}$	$-\sqrt{2}$
	Γ_7	2	−2	1	−1	0	0	$-\sqrt{2}$	$\sqrt{2}$
	Γ_8	4	−4	−1	1	0	0	0	0

It can be seen that, for example, the 12-fold degenerated ${}^4T_2$ state (the orbital triplet and spin-quartet, that is why the total degeneracy is 12) is split by the SO interaction into two quartet and two double states. The $^4A_2$ and ^2E states are not split by the SO interaction in the first order of the perturbation theory and remain to be the four-fold degenerated states whose wave functions have the symmetry properties of the Γ_8 irreducible representation.

As for the configuration with even number of electrons, the situation is as follows: The symmetry properties of the basis sets corresponding to the total spin S = 0, 1, 2, 3... are same as the symmetry properties of the orbital moment states with the same values of L. In the T_d group, for example, they will be A_1 for $S(L) = 0$, T_1 for $S(L) = 1$, $E + T_2$ for $S(L) = 2$ and so on. Then the decomposition of the direct product of the irreducible representations corresponding to the orbital and spin parts will give the pattern of the SO splitting. An example of group theory analysis of the SO splitting for the states of the d^2 configuration will be given in the next chapter, when the energy levels of the Cr^{4+} ion in $Y_3Al_5O_{12}$ will be discussed.

It should be kept in mind that the group-theoretical analysis gives the qualitative results only. This means that we can find out how many energy levels will be formed out of a given degenerated level and what are the symmetry properties of their wave functions. At the same time, staying in the group theory framework it is impossible to determine, which energy level is the ground state, what is the order of other energy levels and what are the energy intervals between these split states. To find these quantitative characteristics of the energy levels splitting (i.e., relative positions of the energy levels), the crystal field theory should be used. Its basic ideas will be discussed in Chapter 8.

7.9 Applications of the Group Theory in Spectroscopy: Selection Rules

Before we start discussions on how to use the group theory to analyze the selection rules for the allowed transitions, an idea of the direct product of irreducible representations should be introduced.

Previously we already described the direct product of two groups. Now we consider two systems of functions $\phi_1^{(a)}, \phi_2^{(a)}, ..., \phi_{n_a}^{(a)}$ and $\phi_1^{(b)}, \phi_2^{(b)}, ..., \phi_{n_b}^{(b)}$, which form two basis sets of two irreducible representations $\Gamma^{(a)}$ and $\Gamma^{(b)}$ with dimensions n_a and n_b, respectively.

It is possible to compose a system of $n_a n_b$ functions by arranging them into the products $\phi_i^{(a)}\phi_k^{(b)}$, where $1 \le i \le n_a$, $1 \le k \le n_b$. These functions form the basis of a new representation (reducible, in the general case) of the $n_a \times n_b$ dimension, which is called a direct product of two irreducible representations $\Gamma^{(a)}$ and $\Gamma^{(b)}$. The characters of the direct product of two representations are equal to the products of characters of both involved representations.

Let f be an operator of some physical quantity (electric dipole operator, magnetic dipole operator, etc.). The matrix elements of this operator between two wave functions $\phi_i^{(a)}$ and $\phi_j^{(b)}$ having the symmetry properties of the $\Gamma^{(a)}$ and $\Gamma^{(b)}$ irreducible representations, respectively, are given by the following integral:

$$\int \phi_i^{(a)*} f \phi_j^{(b)} dq \,. \tag{7.18}$$

The symmetry of this operator corresponds to the irreducible representation Γ_f , which is an identity representation, if the operator is a scalar quantity. The whole product in the integral (7.18) corresponds to the direct product of three irreducible representations $\Gamma^{(a)} \times \Gamma_f \times \Gamma^{(b)}$, which forms a reducible representation Γ. If Γ contains an identity representation, the integral (7.18) is not zero and, as a consequence, the transition between these two states $\phi_i^{(a)}$ and $\phi_j^{(b)}$ caused by an operator f is allowed.

This also can be explained from the mathematical point of view: In the general case, the product $\phi_i^{(a)} f \phi_j^{(b)}$ is neither odd nor even function. But if it turns out to be possible to extract an even part from this product, which does not change the sign if the inversion transformation is applied, then the integral (7.18) is not zero. The presence of such an even part in this function is equivalent to the presence of the identity representation in the $\Gamma^{(a)} \times \Gamma_f \times \Gamma^{(b)}$ reducible representation. If there is no identity representation in the reducible representation Γ, then the transition between the two considered states is forbidden. Another (equivalent) formulation of this selection rule is as follows: The transition between two states with the irreducible representations $\Gamma^{(a)}$ and $\Gamma^{(b)}$ is allowed if the direct product $\Gamma^{(a)} \times \Gamma^{(b)}$ contains the irreducible representation Γ_f of the perturbation operator.

So, instead of evaluating the integral (7.18), one can just decompose a direct product of three (or two) irreducible representations. Such a procedure is much easier than the direct calculation of the integrals with the wave functions of the initial and final states. It should be pointed out, however, that the group theory can give only a qualitative answer to the question about the allowed and forbidden transitions. One can say that some transitions are allowed, and some are forbidden, but nothing can be said about the intensities of the allowed ones. It is not possible—using the group theory only—to find a numerical value of probability of the allowed transition. To get a numerical value of the allowed transition probability and see which transitions can be more or less intensive in the spectra, the numerical calculations of the integrals (7.18) will be still required.

As an example, let us derive the symmetry selection rules for the electric dipole transitions in the T_d group (the characters of its irreducible representations are given in Table 7.4). The basis of the T_2 irreducible representation is $\{x, y, z\}$, which exactly corresponds to the components of the radius-vector and, by its definition, electric dipole operator. To find out, whether the electric dipole transition is allowed between the two states $\Gamma^{(a)}$ and $\Gamma^{(b)}$ of the T_d group, we must decompose the direct product $\Gamma^{(a)} \times T_2 \times \Gamma^{(b)}$.

Let us consider the transition from the E state to the T_1 state. The characters of the direct product $E \times T_2 \times T_1$ are as follows:

	E	$8C_3$	$3C_2$	$6\sigma_d$	$6S_4$
$E \times T_2 \times T_1$	18	0	2	0	0

and this 18-dimensional reducible representation can be decomposed into the following irreducible representations of the T_d group: $E \times T_2 \times T_1 = A_1 + A_2 + 2E + 2T_1 + 2T_2$. It contains an identity irreducible representation A_1, which makes the $E \rightarrow T_1$ transition allowed in the electric dipole approximation. All other combinations of the irreducible representation in this group can be considered in a similar way. Table 7.7 shows all allowed (denoted by "+") and forbidden (denoted by "–") electric dipole transitions in the T_d group.

Table 7.8 shows the allowed and forbidden transitions in the same T_d group, if the magnetic dipole operator (whose symmetry is described by the T_1 irreducible representation) is considered as the reason of the transition.

A similar analysis can be performed for any point group and any operator of the transition between two states. It also can help in understanding the polarized spectra of crystals, when intensities of certain transitions are different in different polarizations. In this case, the group selection rules should be analyzed for each polarization separately.

Table 7.7 Allowed and forbidden electric dipole transitions between the states having the symmetry of the irreducible representations of the T_d group

	A_1	A_2	E	T_1	T_2
A_1	–	–	–	–	+
A_2	–	–	–	+	–
E	–	–	–	+	+
T_1	–	+	+	+	+
T_2	+	–	+	+	+

Table 7.8 Allowed and forbidden magnetic dipole transitions between the states having the symmetry of the irreducible representations of the T_d group

	A_1	A_2	E	T_1	T_2
A_1	–	–	–	+	–
A_2	–	–	–	–	+
E	–	–	–	+	+
T_1	+	–	+	+	+
T_2	–	+	+	+	+

Finally, in Table 7.9 we list the irreducible representations of the electric and magnetic dipole operators in various point groups. When the components of the dipole operators are in the brackets, this means that they are transformed according to the double or triple degenerated irreducible representation. When they are listed through the comma, this means that all listed components transform as a given one-dimensional representation. Table 7.9 allows for finding the selection rules for the electric dipole and magnetic dipole transitions in different polarizations, like σ (when the dipole operator components are at the x, y plane) or π (when the dipole operator component is parallel to the z axis).

Table 7.9 Irreducible representations (IR) of the electric (ED, x, y, z) and magnetic (MD, S_x, S_y, S_z) dipole operators in various point groups*

Point group	IR	ED	MD
C_2	A	z	S_z
	B	x, y	S_x, S_y
C_s	A'	x, y	S_z
	A''	z	S_x, S_y
C_i	A_g		S_x, S_y, S_z
	A_u	x, y, z	
C_{2h}	A_g		S_z
	B_g		S_x, S_y
	A_u	z	
	B_u	x, y	
C_{2v}	A_1	z	
	A_2		S_z
	B_1	x	S_y
	B_2	y	S_x
C_3, C_4	A	z	S_z
	E	(x, y)	(S_x, S_y)
C_{3i}, C_{4h}, C_{6h}	A_g		S_z
	E_g		(S_x, S_y)
	A_u	z	
	E_u	(x, y)	
C_{3h}	A'		S_z
	A''	z	
	E'	(x, y)	
	E''		(S_x, S_y)
C_{3v}, C_{4v}, C_{6v}	A_1	z	
	A_2		S_z
	E	(x, y)	(S_x, S_y)

Point group	IR	ED	MD
D_{3d}, D_{4h}	A_{2g}		S_z
	E_g		(S_x, S_y)
	A_{2u}	z	
	E_u	(x, y)	
D_4	A_2	z	S_z
	E	(x, y)	(S_x, S_y)
D_6	A_2	z	S_z
	E_1	(x, y)	(S_x, S_y)
D_{6h}	A_{2g}		S_z
	E_{1g}		(S_x, S_y)
	A_{2u}	z	
	E_{1u}	(x, y)	
T	T	(x, y, z)	(S_x, S_y, S_z)
T_h	T_g		(S_x, S_y, S_z)
	T_u	(x, y, z)	
O	T_1		(S_x, S_y, S_z)
T_d	T_1		(S_x, S_y, S_z)
	T_2	(x, y, z)	
O_h	T_{1g}		(S_x, S_y, S_z)
	T_{1u}	(x, y, z)	

*After Matsumoto M, Chimata K, Koga M (2017). Symmetry analysis of spin-dependent electric dipole and its application to magnetoelectric effects. *J Jpn Phys Soc* **86** 034704.

Point group	IR	ED	MD
C_4, S_4	A	z (C_4)	S_z
	B	z (S_4)	
	E	(x, y)	(S_x, S_y)
C_6	A	z	S_z
	E_1	(x, y)	(S_x, S_y)
D_2	B_1	z	S_z
	B_2	y	S_y
	B_3	x	S_x
D_{2d}	A_2		S_z
	B_2	z	
	E	(x, y)	(S_x, S_y)
D_{2h}	B_{1g}		S_y
	B_{2g}		S_z
	B_{3g}		S_x
	B_{1u}	y	
	B_{2u}	z	
	B_{3u}	x	
D_3	A_2	z	S_z
	E	(x, y)	(S_x, S_y)
D_{3h}	A_2'		S_z
	A_2''	z	
	E'	(x, y)	
	E''		(S_x, S_y)

Bibliography

1. Bishop DM (1973). *Group Theory and Chemistry*. Diver Publications Inc, New York.
2. Ceulemans AJ (2013). *Group Theory Applied to Chemistry*. Springer.
3. Dresselhaus MS, Dresselhaus G, Jorio A (2008). *Group Theory Application to the Physics of Condensed Matter*. Springer.

4. García Solé J, Bausá LE, Jaque D (2005). *An Introduction to the Optical Spectroscopy of Inorganic Solids.* John Wiley & Sons Ltd.

5. Hamermesh M (1964). *Group Theory and Its Applications to Physical Problems,* Addison–Wesley Publishing Co.

6. Henderson B, Bartram RH (2000). *Crystal-Field Engineering of Solid-State Laser Materials.* Cambridge University Press.

7. Henderson B, Imbush GF (2006). *Optical Spectroscopy of Inorganic Solids.* Clarendon Press.

8. Koster GF, Dimmock JO, Wheeler RG, Statz H (1963). *Properties of the Thirty-Two Point Groups.* The MIT Press, Cambridge, Massachusetts.

9. Lifshitz EM, Landau LD, (1958, 1965, 1977). *Quantum Mechanics: Non-relativistic Theory, Vol. 3 (Course of theoretical physics)* (various editions). Elsevier Ltd.

10. Powell RC (1998). *Physics of Solid-State Laser Materials.* Springer.

11. Tsukerblat BS (1994). *Group Theory in Chemistry and Spectroscopy. A Simple Guide to Advanced Usage.* Academic Press, London.

Chapter 8

Basic Postulates of Crystal Field Theory

8.1 Introduction

It was emphasized in the previous chapter that the group theory can give only a qualitative description of the impurity ions energy levels in crystals. The quantitative description, which would include calculations of the split energy levels and energy intervals between these states, is possible only within the framework of the crystal field theory, the main foundation of which was laid by H. Bethe in 1929 and was developed and improved a lot since that time.

When an impurity ion with an unfilled electron shell (most commonly, these are the ions with the *d* and *f* shells) is placed into a crystal lattice, it substitutes for one of the host lattice cations and becomes surrounded by the negatively charged ions. The nearest neighbors of the impurity ion are called ligands. The number of ligands and their geometrical arrangement depend on the crystal lattice symmetry. The crystal lattice ions, both positively and negatively charged, create a crystal field, which acts upon the electrons from the unfilled shell of the impurity ion. The spherical symmetry of an isolated ion is reduced to the point symmetry of the crystal lattice site occupied by the impurity ion. As a result of this symmetry reduction and influence of the crystal field, the impurity ion's LS terms are split into a number of sublevels, with the splitting

Theoretical Spectroscopy of Transition Metal and Rare Earth Ions: From Free State to Crystal Field
Mikhail G. Brik and Chong-Geng Ma

ISBN 978-981-4800-56-3 (Hardcover), 978-0-429-27875-4 (eBook)
www.jennystanford.com

pattern completely determined by the crystal field properties. The methods of calculations of the crystal field splittings of the free ion terms constitute the main content of the crystal field theory, which is discussed in this chapter. We start with the simplest model of point charges and simplest impurity containing only one electron in the unfilled shell and proceed then to the case of the multielectron ions in a crystal field.

8.2 A Single *d*-Electron in a Cubic Crystal Field

In this paragraph, we consider an ion with a single *d*-electron surrounded by six negative point charges $-Ze$, where Z is an integer number, and e is the elementary charge. We assume that this ion (like Ti^{3+}, for example) is at the center of an ideal octahedron formed by those six surrounding ions (Fig. 8.1). The Cartesian and spherical coordinates of all ions from this figure are given in Table 8.1.

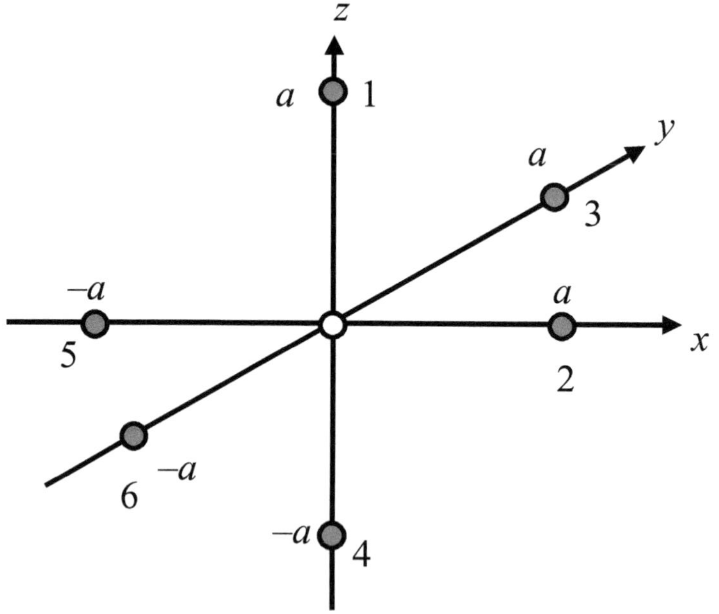

Figure 8.1 A central ion (unfilled circle) surrounded by six point-charges $-Ze$. The distance between each of the six charges and the central ion is a.

Table 8.1 Cartesian and spherical coordinates of ions shown in Fig. 8.1

	Cartesian coordinates			Spherical coordinates		
	x	Y	z	R	θ	ϕ
Central ion	0	0	0	0	0	0
1	0	0	a	a	0	0
2	a	0	0	a	$\pi/2$	0
3	0	a	0	a	$\pi/2$	$\pi/2$
4	0	0	$-a$	a	π	0
5	$-a$	0	0	a	$\pi/2$	π
6	0	$-a$	0	a	$\pi/2$	$3\pi/2$

Note: The numbers of the ions are as in Fig. 8.1.

These surrounding ions create a crystal field of a given symmetry (octahedral in this particular case) around an impurity ion. The main interaction between the central ion and these point charges is the Coulomb interaction, and the potential energy $V(\vec{\mathbf{r}})$ of the single d-electron of the central ion in this crystal field is

$$V(\vec{\mathbf{r}}) = \sum_{i=1}^{6} \frac{Ze^2}{\left|\vec{\mathbf{R}}_i - \vec{\mathbf{r}}\right|}, \tag{8.1}$$

where $\vec{\mathbf{R}}_i\,(i=1,...,6)$ and $\vec{\mathbf{r}}$ are the coordinates of the surrounding point charges and the d-electron of the central ion, respectively. The energy levels E of the considered d-electron will be determined as the solutions of the Schrödinger equation:

$$\left(-\frac{\hbar^2}{2m}\Delta + U(\vec{\mathbf{r}}) + V(\vec{\mathbf{r}})\right)\Psi = E\Psi, \tag{8.2}$$

where $-\frac{\hbar^2}{2m}\Delta$ is the kinetic energy of the electron with mass m, $U(\vec{\mathbf{r}})$ is the potential energy of the electron in a Coulomb field of a nucleus and other core electrons, and $V(\vec{\mathbf{r}})$ is defined by Eq. (8.1). Equation (8.2) is solved using the perturbation theory, assuming that $V(\vec{\mathbf{r}}) << -\frac{\hbar^2}{2m}\Delta + U(\vec{\mathbf{r}})$. Then the free ion wave functions can be taken as the first approximation, and the effects of a small perturbation $V(\vec{\mathbf{r}})$ on these states will be considered. It is already known that there are five wave functions of the d-orbitals (the spin component is neglected at this point), which can be generally written as a product of the radial part $R_{32}(r)$ and the angular part$Y_{2m}(\theta,\varphi)$: $R_{32}(r)$ $Y_{2m}(\theta,\varphi)$, $-2 \le m \le 2$ with the subscripts denoting the corresponding quantum numbers n, l for the radial part and l, m for the angular part, respectively. It is essential that the radial part is the same for all five wave functions. From the quantum mechanical point of view, the problem of calculating new energy levels E, which may arise from the five-fold degenerated d orbital as a result of a small perturbation $V(\vec{\mathbf{r}})$, is reduced to finding the eigenvalues of the following matrix:

	2	**1**	**0**	**−1**	**−2**
2	V_{22}	V_{21}	V_{20}	V_{2-1}	V_{2-2}
1	V_{12}	V_{11}	V_{10}	V_{1-1}	V_{1-2}
0	V_{02}	V_{01}	V_{00}	V_{0-1}	V_{0-2}
−1	V_{-12}	V_{-11}	V_{-10}	V_{-1-1}	V_{-1-2}
−2	V_{-22}	V_{-21}	V_{-20}	V_{-2-1}	V_{-2-2}

(8.3)

In this matrix the numbers in the first row and the first column correspond to the allowed values of the magnetic quantum number m, and the matrix elements $V_{m,m'}$ are calculated as

$$V_{m,m'} = \int (R_{32}(r) Y_{2m}(\theta,\phi))^* V(\vec{\mathbf{r}}) R_{32}(r) Y_{2m'}(\theta,\phi) dq \,. \qquad (8.4)$$

The subscripts m, m' are used to denote the wave functions with the magnetic quantum number m, and integration is carried over all coordinates. The main question now is how to evaluate the integral in Eq. (8.4). The solution is to expand the perturbation operator $V(\vec{\mathbf{r}})$ in terms of the spherical harmonics:

$$\begin{aligned} V(\vec{\mathbf{r}}) &= \sum_{i=1}^{6} \frac{Ze^2}{\left|\vec{\mathbf{R}}_i - \vec{\mathbf{r}}\right|} \\ &= Ze^2 \sum_{i=1}^{6} \sum_{k=0}^{\infty} \frac{r_<^k}{r_>^{k+1}} \frac{4\pi}{2k+1} \sum_{m=-k}^{k} Y_{km}(\theta,\phi) Y_{km}^*(\theta_i,\phi_i) \end{aligned} \quad . \qquad (8.5)$$

Here $r_<$ and $r_<$ are the smallest and the greatest of two distances: r (the electron coordinate) and a (the distance between the central ion and surrounding point charges), (θ, ϕ) and $(\theta_I j_i)$are the spherical angular coordinates corresponding to $\vec{\mathbf{r}}$ and $\vec{\mathbf{R}}_i\,(i = 1,...,6)$, respectively. Assuming that $r < a$, which means that the considered

d-electron is located between the central ion and surrounding point charges, Eq. (8.5) will look as follows:

$$V(\vec{\mathbf{r}}) = \frac{Ze^2}{a}\sum_{i=1}^{6}\sum_{k=0}^{\infty}\left(\frac{r}{a}\right)^k \frac{4\pi}{2k+1}\sum_{m=-k}^{k} Y_{km}(\theta,\phi)Y_{km}^*(\theta_i,\phi_i). \quad (8.6)$$

It is a standard practice to re-write the last expression in new notations:

$$V(\vec{\mathbf{r}}) = \sum_{k=0}^{\infty}\sum_{m=-k}^{k} r^k q_{km} C_m^k(\theta,\phi),$$

where

$$q_{km} = \sqrt{\frac{4\pi}{2k+1}}\frac{Ze^2}{a^{k+1}}\sum_{i=1}^{6} Y_{km}^*(\theta_i,\phi_i) \quad (8.7)$$

and the spherical operators $C_m^k(\theta,\phi)$ are defined as

$$C_m^k(\theta,\phi) = \sqrt{\frac{4\pi}{2k+1}}Y_{km}(\theta,\phi). \quad (8.8)$$

The summation in Eq. (8.6) is from $k = 0$ to $k = \infty$, which implies an infinite number of terms. Luckily, many terms in this infinite sum are vanishing. First, we note that the matrix elements in Eq. (8.4) are proportional to the following integrals from the product of three spherical functions $V_{m,m'} \sim \int_0^{2\pi}\int_0^{\pi} Y_{l_1 m}(\theta,\phi)^* Y_{km}(\theta,\phi)Y_{l_2 m'}(\theta,\phi)\sin\theta d\theta d\phi$ with quantum numbers l_1 and l_2 being different in general (but for the considered case of the d-electron $l_1 = l_2 = 2$). This integral can be expressed in terms of the Wigner 3j-symbols:

$$\int_0^{\pi}\int_0^{2\pi} Y_{l_1 m_1}^* Y_{km} Y_{l_2 m_2} \sin\theta d\phi d\theta$$

$$= (-1)^{m_1}\sqrt{\frac{(2l_1+1)(2k+1)(2l_2+1)}{4\pi}}\begin{pmatrix} l_1 & k & l_2 \\ 0 & 0 & 0 \end{pmatrix}\begin{pmatrix} l_1 & k & l_2 \\ -m_1 & m & m_2 \end{pmatrix}$$

(8.9)

The following conditions should be fulfilled to make the integral (8.9) to be non-zero: (i) $0 \le k \le (l_1 + l_2)$; (ii) k takes only even values from the above specified interval, since $l_1 + l_2 = 4$ (an even number). That is why, for our case of the d-electron configurations only the values $k = 0, 2, 4$ should be considered. If, for example, the atoms with the f electrons in the unfilled shell are analyzed, then only the terms with $k = 0, 2, 4, 6$ would be non-zero.

Using the explicit form of the spherical functions and the spherical coordinates of the ligands from Table 8.1 (inserting them into the $Y_{km}^*(q_i, \phi_i)$ functions), after some algebraic manipulations it is possible to find out the explicit form of the potential $V(\vec{\mathbf{r}})$:

$$V(\vec{\mathbf{r}}) = \frac{6Ze^2}{a} + \frac{7Ze^2}{2a^5} r^4 \left\{ C_0^4(\theta,\phi) + \sqrt{\frac{5}{14}} \left(C_4^4(\theta,\phi) + C_{-4}^4(\theta,\phi) \right) \right\}. \tag{8.10}$$

It is instructive to note that the second order contribution with $k = 2$ vanishes if the surrounding ligands form one of the following ideal polyhedrons: octahedron, tetrahedron, or cube. Such a situation is said to represent a cubic crystal field case. However, the second order contribution is not zero in all other cases of arrangements of ligands around an impurity ion. Even if the coordination polyhedron is a deformed octahedron, tetrahedron, or cube, the terms with $k = 2$ will appear in the expression for $V(\vec{\mathbf{r}})$. In general, the maximal number of the non-zero terms in Eq. (8.10) can be 14 and 27 for the d and f electrons, respectively, in the case of the lowest symmetry possible described by the C_1 point group. The first term in Eq. (8.10) is a constant, which does not depend on the electron coordinates and appears in the same manner in all matrix elements. It does not have any influence on the overall energy level scheme other than a simultaneous and equal shift of all energy levels and, therefore, can be omitted without any lack of generality, if only the relative positions of the calculated energy levels are considered.

Using Eqs. (8.7)–(8.10) and the d-electron wave functions $R_{32}(r)$ $Y_{2m}(\theta, \phi)$, $-2 \leq m \leq 2$ it is possible to evaluate the matrix elements in Eq. (8.3). After straightforward calculations, the following matrix can be obtained:

	2	**1**	**0**	**−1**	**−2**
2	Dq	0	0	0	$5Dq$
1	0	$-4Dq$	0	0	0
0	0	0	$6Dq$	0	0
−1	0	0	0	$-4Dq$	0
−2	$5Dq$	0	0	0	Dq

(8.11)

with

$$D = \frac{35Ze^2}{4a^5}, q = \frac{2}{105}\langle r^4 \rangle_{32}, \langle r^k \rangle_{32} = \int |R_{32}(r)|^2 r^{k+2} dr. \quad (8.12)$$

As seen from the general forms of the D and q expressions, the former characterizes the surrounding ions, whereas the latter depends on the central ion only. Their product Dq has dimension of energy and is called the crystal field strength. It is a combined parameter, which incorporates the properties of both central ion and its environment.

The eigenvalues of the matrix (8.11) can be easily found. The corresponding secular determinant can be written in the following block-diagonal form:

$$\begin{vmatrix} Dq-E & 5Dq & 0 & 0 & 0 \\ 5Dq & Dq-E & 0 & 0 & 0 \\ 0 & 0 & -4Dq-E & 0 & 0 \\ 0 & 0 & 0 & -4Dq-E & 0 \\ 0 & 0 & 0 & 0 & 6Dq-E \end{vmatrix} = 0. \quad (8.13)$$

The calculated eigenvalues are $-4Dq$ (three roots) and $6Dq$ (two roots). So, the five-fold degenerated state of a d-electron (we may also speak about the ^{2}D term of the d^1-configuration) in an octahedral crystal field splits into a triplet, which turns out to be the ground state, and a doublet, which is an excited state, with the energy separation between them being equal to $10Dq$. The barycenter of these two split energy levels is zero: $3\times(-4Dq) + 2\times 6Dq = 0$.

If we consider not an octahedral, but a tetrahedral environment (four ions forming a perfect tetrahedron with an impurity ion at its center), the splitting pattern would be inverted: The ground state would become a doublet, whereas the excited state would be a triplet.

After the energy levels are calculated, it is also possible to find their corresponding eigenfunctions. The functions, which correspond to the doublet level in an octahedral crystal field are: $|20\rangle, \frac{1}{\sqrt{2}}(|22\rangle + |2-2\rangle)$, where the $|lm\rangle$ notation is used and the radial part (which is common for all five functions) is omitted for the sake of brevity. It can be shown by a successive application of all symmetry operations from the O_h group that these functions transform according to the E_g irreducible representation of this group. They are referred to as the d_{e_g} set of wave functions. Similarly, the functions corresponding to the triplet level are $|21\rangle, |2-1\rangle, \frac{-i}{\sqrt{2}}(|22\rangle - |2-2\rangle)$. Their transformation properties are those as of the matrices of the T_{2g} irreducible representation, and they are denoted as the $d_{t_{2g}}$ set of wave functions. In the tetrahedral field these notations are the same, just the subscript "g" should be dropped, since there are no "gerade" ("even") and "ungerade" ("odd") irreducible representations in the T_d group because of absence of the inversion center in the tetrahedron. Figure 8.2 illustrates these both cases.

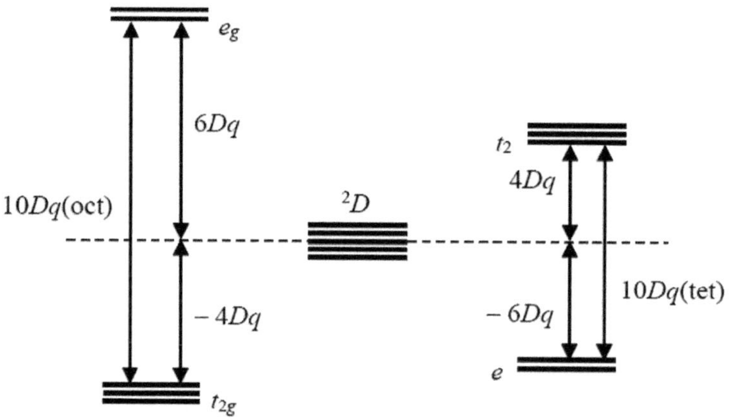

Figure 8.2 Splitting of the ^2D term (center) in an octahedral crystal field (left) and tetrahedral crystal field (right). The barycenter of the energy levels is shown by the dashed line. For a better visualization of the degeneracy, all energy levels are shown by the separate lines (though within each group the energy of all levels is the same). Note the different notations for $10Dq$ in both cases.

It is also instructive to compare the absolute values of the crystal field strength for the ideal octahedral (six ligands), tetrahedral (four ligands) and cubic (eight ligands) environments. We assume that the charges of the surrounding ions are equal in all cases, and interionic distances are chosen in such a way, that the vertices of an octahedron coincide with the centers of the cube faces, and the vertices of the tetrahedron coincide with the vertices of the same cube. It is intuitively clear that $10Dq_{\text{tet}}$ should be lower than $10Dq_{\text{oct}}$—since there are only four ions instead of six and, besides, the distances from the central ion to the ligands are greater in the case of a tetrahedron. Indeed, exact calculations yield

$$Dq_{\text{tet}} = -\frac{4}{9} Dq_{\text{oct}}. \tag{8.14}$$

The negative sign emphasizes the inverse order of the split energy levels in both cases. It is also clear that the Coulomb potential created by eight ions located at the vertices of a cube can be effectively represented as a sum of two potentials from two tetrahedrons, which yields:

$$Dq_{\text{cube}} = 2Dq_{\text{tet}} = -\frac{8}{9} Dq_{\text{oct}}. \tag{8.15}$$

It is also clear that the order of the split energy levels in a cubic environment is the same as in a tetrahedral one.

The order of the doublet and triplet states can also be explained from the point of view of the d-orbitals spatial shapes. Figures 8.3 and 8.4 show the plots of squares of the d orbitals from the e_g and t_{2g} sets. The three orbitals from the t_{2g} set (or the t_2 set for the T_d group) are directed into the space between the negatively charged ligands in the octahedral field (towards the ligands in the cubic and tetrahedral environments). Therefore, the energy of these states is lower in the former case and higher in the latter. On the contrary, two orbitals from the e_g set (or the e set for the T_d group) are oriented towards the negatively charged ligands in the octahedral field (into the space between the ligands in cubic and tetrahedral environments). Therefore, the energy of these states is higher in the former case and lower in the latter.

Equations (8.12) allow for a direct numerical evaluation of the Dq parameter for the ions with a single d-electron. As an example, we consider the Ti^{3+} ion in lithium fluoride LiF. In this material, the Ti^{3+} ions substitute for the Li^+ ion and occupy a position at the center of a slightly distorted octahedron formed by six F^- ions. The octahedron deformation can be caused by the difference in the electric charges between the Li^+ and Ti^{3+} ions. To compensate for the excessive positive charge at the Li position after incorporation of the Ti^{3+} ion, some Li vacancies should be formed. Alternatively, some of the F^- ions may be replaced by the O^{2-} ions. However, we shall neglect these charge compensation mechanisms in the first approximation and take the Ti^{3+}–F^- distance from the LiF structural data as 2.01 Å.[1]

The radial parts of the Ti^{3+} $3d$ wave functions can be approximated by the following combination of the exponential functions[2]:

$$\Psi_{32} = r^2 \begin{Bmatrix} 4.26491\exp(-2.22666r) + 34.47675\exp(-9.19926r) \\ +24.90600\exp(-4.45190r) + 3.59671\exp(-3.45709r) \\ +0.041099\exp(-1.37743r) \end{Bmatrix}. \tag{8.16}$$

The averaged value of $\langle r \rangle^4$ calculated with this wave function is $\langle r \rangle^4$ = 8.34372 a.u.^{4} (where a.u. is the atomic unit of length equal to 0.529177 Å—the radius of the first Bohr's orbit in the hydrogen atom). Since the charge of the F^- is (−1) in units of the proton's charge and the Ti^{3+} – F^- distance is known, the value of Dq can be readily evaluated to be about 340 cm^{-1} only.

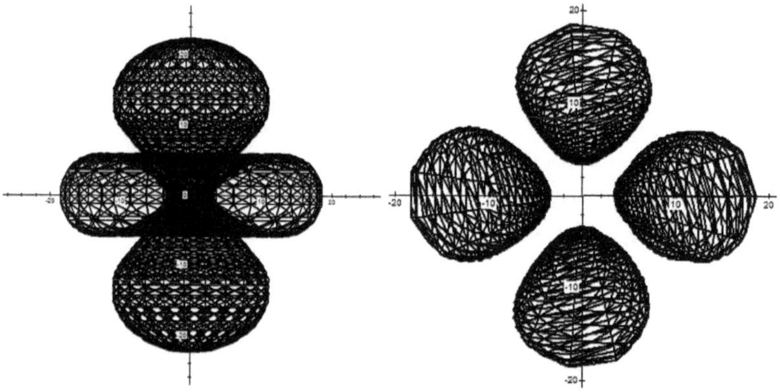

Figure 8.3 Spatial distribution of the d-orbitals from the e_g set.

This value is, however, significantly underestimated in comparison with the experimental results for the LiF:Ti^{3+} material, which are about 1200–1500 cm^{-1}.[3] Such underestimation shows that the point charge model—although it is basically correct—cannot give a precise description of the crystal field splitting. The main reason is that the model treats both central ion and its nearest environment (ligands) as the point charges only. The electronic structure and all quantum mechanical effects arising from it, like modifications of the electron density, formation of the covalent bonds between the crystal lattice ions, etc., as well as deformations of the crystal lattice, which are unavoidable when impurity ions with different electrical charge and radius are incorporated into a crystal, are all neglected.

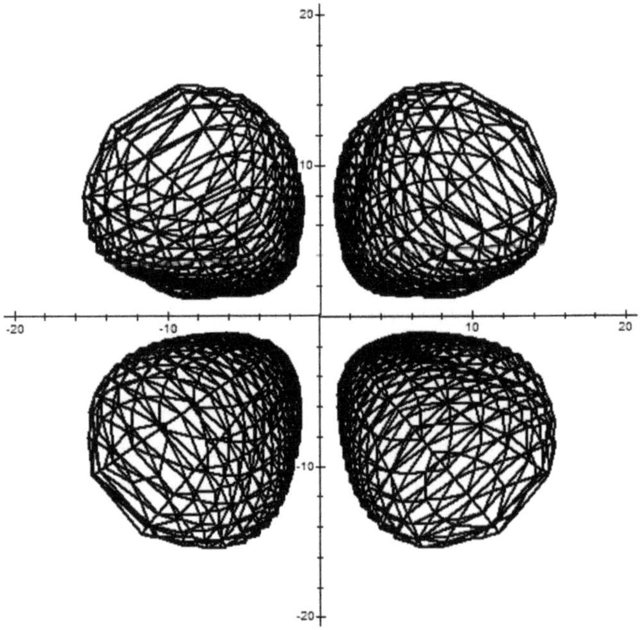

Figure 8.4 Spatial distribution of the *d*-orbitals from the t_{2g} set. Only one of the three orbitals is shown for the sake of brevity; the shapes of these three orbitals are identical. They differ just by the planes, in which their maxima are located (the *xy*, *yz*, and *xz* planes).

To overcome these difficulties, more elaborated models of crystal field have been developed, which include additional interactions to improve agreement between the calculated and observed values.

Nevertheless, even despite these obvious drawbacks, the outlined above point charge model of crystal field was an important step forward in understanding basic properties of the impurity centers in crystals. At least, its relation between the crystal field potential, energy level splitting and the symmetry of the nearest environment created a firm basis for further improvements and developments. Besides, the possibility of expressing the crystal field potential as a linear combination of the spherical operators $C_m^k(\theta, \phi)$ is a key stone for other models of crystal field.

8.3 Many *d*-Electrons in a Crystal Field and Tanabe–Sugano Diagrams

The considered example of an ion with a single *d* electron in a crystal field gives a general idea of how the splitting of the energy levels can be calculated. However, this is the simplest model system, and in most cases of practical interest impurity ions with several electrons should be considered. The Coulomb and SO interactions between these electrons as well as the crystal field effects determine the energy level structure of these ions.

If an impurity ion is placed into a cubic crystal field, three parameters are needed to describe the energy level structure of such a system: the crystal field strength $10Dq$, and two Racah parameters B and C (at this point the SO interaction is neglected). All d^n-configurations were considered by Tanabe and Sugano,[4–8] who calculated the matrices of the Coulomb and crystal field interactions for each d^n-configuration. Since there are three parameters involved - Dq, B, and C, it is impossible to visualize the dependence of the calculated states energy on their values. However, Tanabe and Sugano found an elegant way to overcome this difficulty: They plotted the E/B ratio (where E denotes the energy of a given level) versus the Dq/B ratio for a fixed ratio of C/B. Such diagrams are known as the Tanabe–Sugano (TS) diagrams. They are shown below in Figs. 8.5–8.12 for all configurations of *d* electrons. In all diagrams (except for that one for the d^1 configuration) the horizontal axis shows the Dq/B ratio, and the vertical axis is the E/B ratio. The states with the multiplicity different from that of the ground state are shown by the dashed lines. In all figures, a subscript "g" denoting the even irreducible representations should be added to all labels of energy levels in the O_h group. The C/B ratio is 4.5 for all diagrams.

The TS diagrams can be easily plotted for any C/B ratio and in any range of the Dq/B ratio, using the TS matrices given in Ref. 8. If $Dq/B = 0$, then the free ion energy level scheme is obtained.

The simplest TS diagram is, of course, for the ions with one (or nine) *d*-electron. Quite often it is even considered to be unnecessary to be drawn because of its simplicity. One LS term (^{2}D) of such systems is split into two states $^2T_2$ and ^2E, with the energy separation between them equal to $10Dq$ (Fig. 8.5).

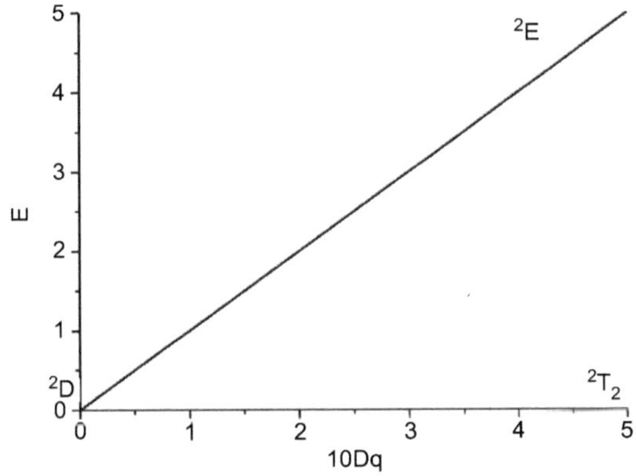

Figure 8.5 TS diagram for the d^1 configuration in an octahedral crystal field and for the d^9 configuration in a tetrahedral crystal field. The order of the energy levels for the d^1 configuration in the T_d symmetry and for the d^9 in the O_h symmetry will be inverted.

For ions with a greater number of the *d* electrons, the number of energy levels in the TS diagrams increases until the *d* shell becomes half-filled, and then decreases again. It should be kept in mind that the TS diagram for an ion with *n d*-electrons in an octahedral crystal field is equivalent to the TS diagram for an ion with (10 – *n*) ions in a tetrahedral crystal field.

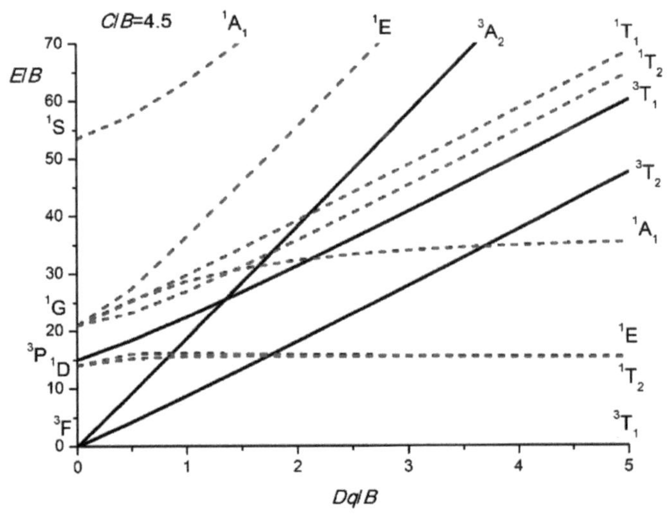

Figure 8.6 TS diagram for the d^2 configuration in an octahedral crystal field and for the d^8 configuration in a tetrahedral crystal field.

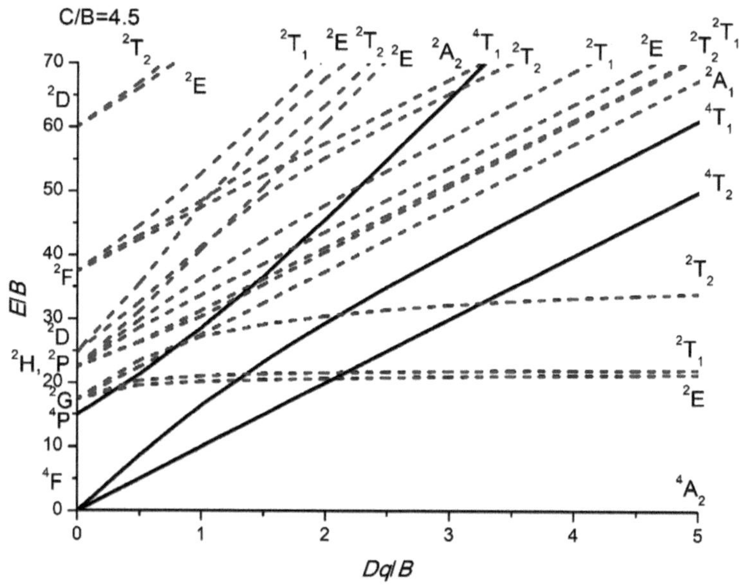

Figure 8.7 TS diagram for the d^3 configuration in an octahedral crystal field and for the d^7 configuration in a tetrahedral crystal.

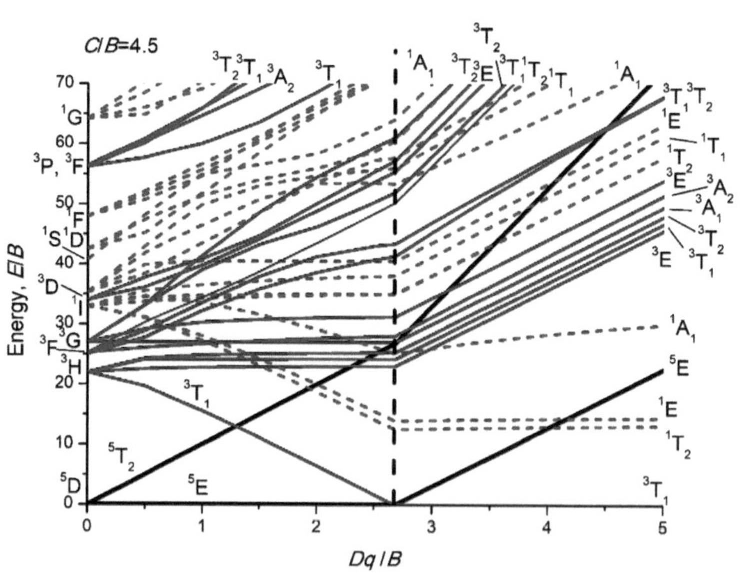

Figure 8.8 TS diagram for the d^4 configuration in an octahedral crystal field and for the d^6 configuration in a tetrahedral crystal field.

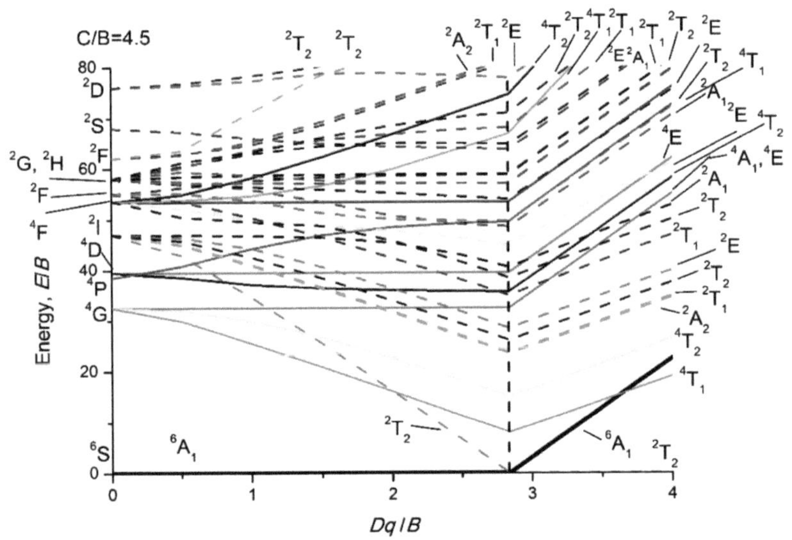

Figure 8.9 TS diagram for the d^5 configuration in an octahedral crystal field.

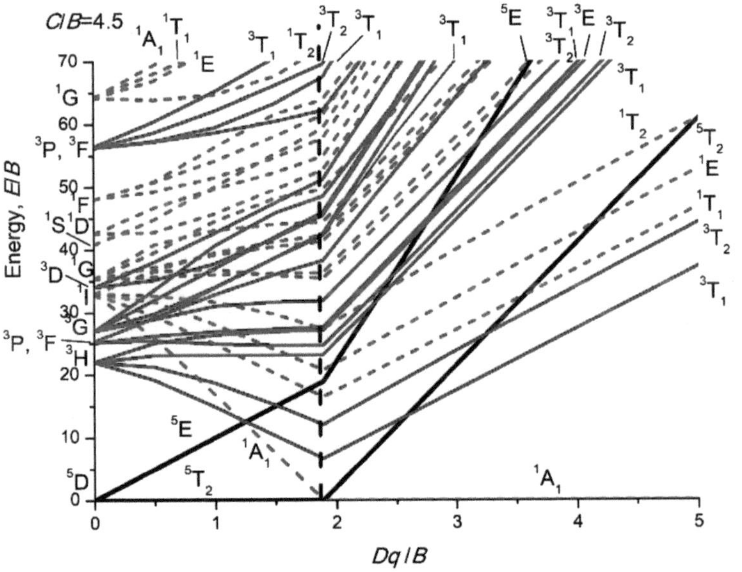

Figure 8.10 TS diagram for the d^6 configuration in an octahedral crystal field and for the d^4 configuration in a tetrahedral crystal field.

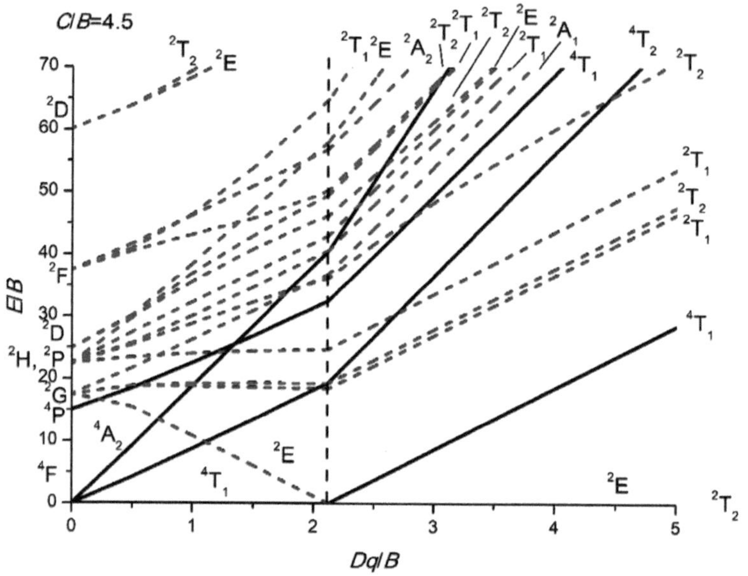

Figure 8.11 TS diagram for the d^7 configuration in an octahedral crystal field and for the d^3 configuration in a tetrahedral crystal field.

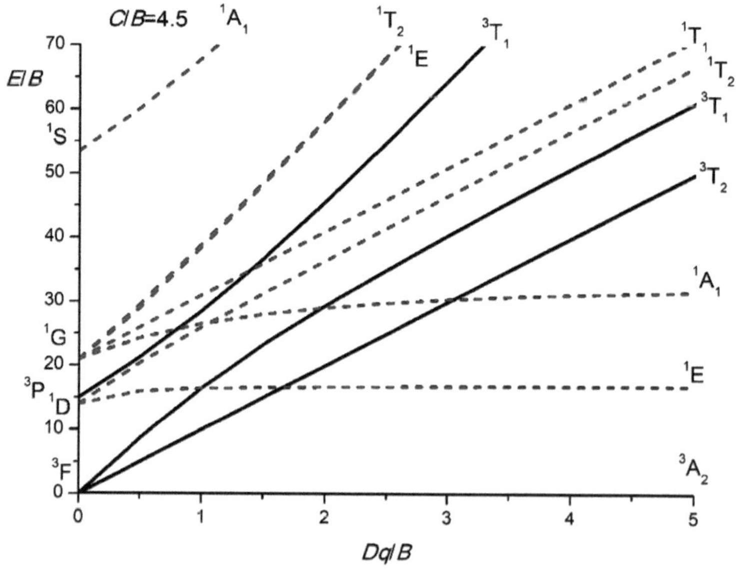

Figure 8.12 TS diagram for the d^8 configuration in an octahedral crystal field and for the d^2 configuration in a tetrahedral crystal field.

Of course, all diagrams are consistent with the group theoretical consideration: The splitting patterns of all LS terms given in Table 7.5 can be easily seen in the TS diagrams.

The TS diagrams are very efficient in analyzing the absorption spectra of *d*-ions. First of all, they always allow to determine a ground state of a considered configuration, and the sequence of the energy levels. Direct comparison of the corresponding TS diagram with an absorption spectrum allows for an assignment of the absorption bands. It should be remembered that the spin-allowed transitions are seen in the absorption spectra as wide bands, whereas the spin-forbidden transitions are revealed as very sharp absorption peaks with much lower intensity because of their forbidden character.

For example, if the absorption spectra of the d^3 ion are considered, then the most prominent bands in the absorption spectrum should correspond—in the order of the increasing energy or decreasing wavelength—to the transitions from the $^4A_2$ ground state to the excited states $^4T_2$, $^4T_1$ (all coming from the ^4F ground term) and to the $^4T_1$ state coming from the ^4P term. The spin-forbidden transitions from the $^4A_2$ ground state to the excited spin-doublets can reveal themselves in sharp features superimposed onto broad bands of the spin-allowed transitions.

In a similar way the absorption spectra of other *d* ions can be analyzed. The only difference is that the TS diagrams for the configurations from 4 to 7 electrons have many energy levels located closely to each other, which sometimes makes the analysis of the absorption spectra somewhat difficult.

As can be easily seen from the TS diagrams, the positions of some energy levels (like ^2E and $^2T_1$ of the d^3-configuration) practically do not depend on the crystal field strength, whereas some other energy levels (like $^4T_2$ from the same configuration) are strongly dependent on the crystal field strength. This observation gives an opportunity of relating the width of the absorption bands to the slope of the energy levels in TS diagram: The greater the slope, the wider the absorption band. The energy levels with the same spin (at least, the closest to each other) arise from the same *LS* term after it is split by the crystal field. The stronger is the crystal field, the greater is the splitting of a particular term. That is why the absorption bands corresponding to the spin-allowed transitions (very often arising from the same *LS* term) are very broad.

For the d^2 configuration in a tetrahedral field the ground state is always $^3A_2$. The first excited state at low values of Dq is $^3T_2$—the state of the same multiplicity as the ground state. Then in the luminescence and absorption spectra the wide bands corresponding to the $^3A_2$ - $^3T_2$ transition are observed. If the Dq value is increased, the spin-singlet state ^1E becomes the first excited state. Then in the absorption and emission spectra very sharp peaks (usually of low intensity) corresponding to the $^3A_2$ - ^1E transition are detected. The former situation is characteristic of a low crystal field, whereas the latter is referred to as a strong crystal field.

A similar observation can be applied to the d^3-configurations in an octahedral field (the only difference will be the multiplicity of the states involved: The ground state is always $^4A_2$, and the first excited state is $^4T_2$ (a weak crystal field; broad emission and absorption bands of the spin-allowed transitions in the experimental spectra) or ^2E (a strong crystal field; very sharp peaks of the spin-forbidden transitions in the experimental spectra).

The TS diagrams for ions with 4–7 d electrons exhibit a particular behavior: The energy levels to not behave continuously and get "broken" at some value of Dq/B. Moreover, the ground state of these configurations is changed after that value. We shall explain such behavior of energy levels for these configurations in the next paragraph.

8.4 The High and Low Spin Complexes

Let us consider the configuration of four d-electrons in an octahedral crystal field. It was shown that the d-orbitals are split into the t_{2g} (the ground state) and e_g (excited state) sets. The former can accommodate up to six, and the latter—up to four electrons. The Coulomb interaction between these electrons gives rise to several degenerated LS terms, which are split in a crystal field and whose energies are shown in the TS diagrams. The first three electrons from these considered four electrons occupy three different t_{2g} states with aligned spins. The fourth electron can be added to these three in two different ways: It can be placed either into one of the t_{2g} states, but with its spin down—to obey the Pauli exclusion principle, or into one of the e_g states with any (up or down) orientation of its spin.

A crucial factor to determine which of the above two situations will be realized is the relation between the spin-pairing energy, which is needed to put two electrons with opposite spins into one orbital, and the crystal field strength or the energy interval between the t_{2g} and e_g states. The spin-pairing energy can be thought of as the energy of interaction between the two magnetic momenta associated with the spin of the electrons. In weak crystal fields the spin-pairing energy exceeds the crystal field strength, and the ground state configuration has three electrons in the t_{2g} states and one electron in the e_g state with all spins parallel to make the total spin $S = 2$. In strong crystal fields, the situation becomes opposite: The spin-pairing energy is smaller than the crystal field strength and adding the fourth electron with an opposite spin into the t_{2g} states is energetically more favorable than placing it into the highly located e_g state. The total spin of such a system is $S = 1$. The first situation is referred to as a "high spin configuration," whereas the second one is a "low spin configuration," with the meaning of the names being directly derived from the total spin values. Figure 8.13 illustrates both cases.

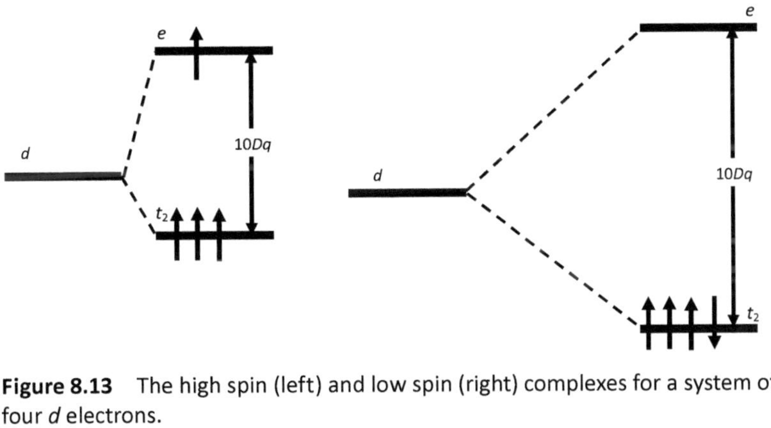

Figure 8.13 The high spin (left) and low spin (right) complexes for a system of four *d* electrons.

The TS diagrams can help in finding out at which crystal field strength the spin-pairing can take place (this is the point at which the ground state is changed, see Figs. 8.8–8.11).

Interesting phenomena can be detected in the systems, which are characterized by the Dq/B values close to those critical ones, when the ground state is changed. It is possible to apply an external hydrostatic (uniform) pressure to a crystal, which initially is characterized by a weak crystal field. The external pressure will decrease distances between ions, which, in turn, will lead to the increase of the Dq value. As a result, at some pressure a weak crystal field can be changed into a strong crystal field. This may have very important practical applications, like a pressure sensor based on the change of the absorption spectra shape.

8.5 Estimations of the Crystal Field Strength *Dq* and Racah Parameters B, C from the Experimental Absorption Spectra. Nephelauxetic Effect

As we have seen, three parameters are needed for a proper description of the energy levels of d-ions in a cubic crystal field: the crystal field strength $10Dq$ (sometimes in the literature both notations—Dq and $10Dq$—are used) and two Racah parameters B and C. The numerical values of these parameters are different for different ions in the same crystal; they also can be different for the same impurity ion in different host materials. These observations suggest that these parameters depend on complicated interrelations between the crystal host and impurity ion—the nature of ligands, interionic separations, peculiar features of the chemical bonds between impurity ion and host lattice ions—all these factors influence the values of Dq, B, and C.

There is a simple way of estimation of these parameters (at least, for the d^2, d^3, d^7, d^8 configurations) using the corresponding TS matrices and experimental absorption spectra.

For example, for the d^2, d^8 configurations it is possible to find the analytical expressions for the energy levels of these ions in a cubic crystal field (Table 8.2). Substituting the experimental energies of the lowest spin-triplet and spin-singlet states into these equations would yield the Dq, B, and C values for a particular system.

Table 8.2 The energy levels of a d^2—configuration as the functions of the crystal field strength $\Delta = 10Dq$ and the Racah parameters B and C

Energy levels	Energy
$^1A_1$	$\Delta + 9B + \frac{9}{2}C \pm \sqrt{\left(\Delta - B - \frac{1}{2}C\right)^2 + 6(2B+C)^2}$
^1E	$\Delta + \frac{1}{2}B + 2C \pm \sqrt{\left(\Delta - \frac{1}{2}B\right)^2 + 12B^2}$
$^1T_2$	$\frac{1}{2}\Delta + \frac{1}{2}B + 2C \pm \sqrt{\frac{1}{4}(\Delta - B)^2 + 12B^2}$
$^3T_1$	$\frac{1}{2}\Delta - \frac{1}{2}B \pm \sqrt{\frac{1}{4}(\Delta + 9B)^2 + 36B^2}$
$^1T_1$	$\Delta + 4B + 2C$
$^3T_2$	$\Delta - 8B$
$^3A_2$	$2\Delta - 8B$

Note: The energy levels of a d^8 – configuration can be obtained by replacing Δ by $-\Delta$.

The situation becomes more complicated for the d^3, d^7 configurations. The reason is that some of the corresponding TS matrices have a greater than two dimension in this case, and analytical expressions for the energy levels can be obtained only approximately.

The position of the first broad spin-allowed band in the spectra of the ions with the d^3 configuration in the octahedral crystal field (which corresponds to the $^4A_2 \rightarrow {}^4T_2$ absorption transition) immediately gives the value of $10Dq$. Then, if ΔE is the energy difference between the maxima of two lowest broad spin-allowed absorption bands ($^4A_2 \rightarrow {}^4T_2$ and $^4A_2 \rightarrow {}^4T_1$ transitions), the following approximate equation can be used to find the value of the Racah parameter B:

$$\frac{B}{Dq}=\frac{\left(\frac{\Delta E}{Dq}\right)^2-10\left(\frac{\Delta E}{Dq}\right)}{15\left(\frac{\Delta E}{Dq}-8\right)}. \tag{8.17}$$

Finally, the position of the first sharp absorption peak (which corresponds to the $^4A_2 \rightarrow {}^2E$ spin-forbidden transition) is needed to find the value of *C* from the equation:

$$\frac{E(^2E)}{B}=3.05\frac{C}{B}+7.90-1.80\frac{B}{Dq}. \tag{8.18}$$

This transition is very weak in the absorption spectra, and it can be easier to use the position of the sharp spin-forbidden emission line of the $^2E \rightarrow {}^4A_2$ transition.

An example of applications of these equations is given below. According to the literature data,[9] the lowest energy levels of Cr^{3+} ion in Cs_2NaYF_6 crystal are as follows (all in cm^{-1}): $^4A_2$ – 0; ^2E – 15309; $^4T_2$ – 16130; $^4T_1$ – 23254. From the position of the $^4T_2$ level we get immediately that $Dq = 1613$ cm^{-1}. Consequent application of the two last equations gives the values of *B* and *C* (in cm^{-1}) as 740 and 3303, respectively.

Comparison of these values with those for the free Cr^{3+} ions (Table 4.9 and corresponding data from Chapter 5) shows a significant reduction of the Racah parameters in crystals. This is the manifestation of the so-called nephelauxetic effect, which is due to the covalent bond formation between the impurity ion and ligands. Because of the overlap of the impurity ion and ligands wave functions, the impurity ion's electrons become delocalized and spend some time at the ligand's orbitals, which leads to a decrease of the Coulomb interaction between the impurity ion's electrons. As a consequence, the Racah parameters values are decreased, and amount of such reduction depends on the impurity ion and the host material. Some time back this effect was described by the nephelauxetic ratio

$$\beta=\frac{B_{\text{complex}}}{B_{\text{free ion}}}, \tag{8.19}$$

where B_{complex} and $B_{\text{free ion}}$ stand for the Racah parameters of a considered ion in a crystal and in a free state. Due to the nephelauxetic effect, the Racah parameter B for some ions in crystals can be up to two times smaller than in a free state. This is particularly true for crystals with dominating covalent bonds. However, as it can be seen from Eq. (8.19), the Racah parameter C is completely ignored. This is because for a long time it was assumed that the C/B ratio for the ions in crystals is equal to (or varies very little from) the same ratio for free ions. However, it is not the case for many materials. Recently, a new nephelauxetic ratio β_1 was introduced[10–14] defined as follows:

$$\beta_1 = \sqrt{\left(\frac{B}{B_0}\right)^2 + \left(\frac{C}{C_0}\right)^2}, \tag{8.20}$$

where B, C (B_0, C_0) are the Racah parameters of the considered $3d$ ion in a crystal (free state), respectively). This parameter describes the nephelauxetic effect much better and allows one to find a linear relation between the energy of the spin-forbidden transitions and β_1 value. An interested reader is advised to look into the above-cited original publications for more details.

8.6 Crystal Field of Lower Than Cubic Symmetries and Exchange Charge Model of Crystal Field

The previous discussion of the d ions energy levels in crystals was based on the assumption that the crystal field around an impurity ion has a perfect cubic symmetry, which means that the ligands form an ideal octahedron or a tetrahedron. Although quite interesting and important results can be obtained in this way—like the energy level splitting, symmetry of the split states, assignment of the main features in the experimental spectra, qualitative understanding of difference in the intensities of the spin-allowed and spin-forbidden transitions located according to the TS diagram, still one question remains open: What would happen if the crystal field symmetry is lower than cubic? Practical importance of such question follows from that fact that in most doped solids used in optical applications, the point symmetry of the crystal lattice sites occupied by the impurities is indeed lower than cubic.

The first qualitative answer to that question comes from the group theory, which says that any symmetry lowering is accompanied by the further splitting of the orbitally degenerated energy levels, if the symmetry is low enough to produce such splitting. So, the orbital triplet and doublet states denoted in the TS diagrams by the T and E letters, respectively, may split to remove partially or completely their cubic symmetry degeneracy. The splitting pattern depends on the actual symmetry. As for the quantitative answer to the raised question, still decomposition of the Coulomb potential into the spherical harmonics series would be required. As a final result, the crystal field Hamiltonian will contain more terms, with different combinations of the p and k indices.

The theory, which allows one to describe the crystal field effects on the energy levels of an impurity ion located at the crystal lattice site with arbitrary point symmetry, is based on the possibility to represent the resulting crystal field Hamiltonian in the following form:

$$H = \sum_{p=0,2,4,6} \sum_{k=-p}^{p} B_p^k O_p^k, \tag{8.21}$$

where O_p^k are the linear combinations of irreducible tensor operators (they will be defined below) acting on the angular parts of the impurity ion wave functions. Several definitions and normalizations of operators O_p^k exist in the literature.[15-17] To be consistent, in what follows we shall use the Stevens' normalization,[18-20] if not stated otherwise explicitly. The first term B_p^k stands for the crystal field parameters (CFPs). They have dimension of energy, and all information about the geometrical structure of an impurity center is "hidden" in the CFPs.

The general recipe of calculating the energy levels of an impurity ion in a crystal is as follows: If the B_p^k values are known and the O_p^k operators are defined, the eigenvalues of the operator (8.21) in the basis set spanned by all wave functions of a considered multielectron configuration will give the required energy level scheme. Of course, to do that, the wave functions of the considered multielectron configuration should be built up properly. The rank of the CFPs is defined by p; the range of p variation covers all even values from 0 to 4 for the d-electrons and from 0 to 6 for the f-electrons.

As for the ways of determination of the B_p^k values, there are two possible methods:

(i) An assumption can be made on what is the point symmetry of the crystal lattice site occupied by the impurity. Then the number of non-zero CFPs is determined (see Table 8.3). If the experimental energy levels are known from the absorption spectra, an initial starting set of B_p^k can be taken to start the fitting procedure, in which all B_p^k are treated as the fitting parameters and the fit continues until the root mean square (rms) deviation between the calculated and experimental energy levels is minimized. A strong feature of such approach is that the rms deviation can be very small, which means a very good correlation between the theoretical and experimental energy level schemes. A drawback of this method is a very large number of fitting parameters (in the case of the C_1 symmetry there are 14 CFPs for the *d*-ions and 27 for the *f*-ions); it also looks like merely a mathematical rather than a physical problem. Besides, the number of iteration steps may depend on the initial set of parameters, and the competition between the local and global minima when performing the fitting can obscure the final result.

(ii) Taking the structure of the crystalline hosts that accommodates a given impurity ion (the structure can be obtained from the experimental X-ray diffraction crystallographic data), it is possible—in the framework of different crystal field models—to calculate the B_p^k values. Then the crystal field Hamiltonian can be diagonalized and the calculated results can be compared with the corresponding experimental data. Although this approach also may be based upon some fitting parameters, their number is considerably smaller than in the previous approach. Such approach is more physical, since it is possible to distinguish between different interactions that contribute to the CFPs values.

Below we give a detailed description of the so-called exchange charge model (ECM) of crystal field developed by B. Z. Malkin.[21] The main ECM idea is to represent the CFPs as a sum of two terms:

$$B_p^k = B_{p,q}^k + B_{p,S}^k \quad (8.22)$$

The first term $B^k_{p,q}$ describes the electrostatic interaction an impurity ion and ions of crystal lattice, which are treated as the point charges with the electrical charges equal to the formal valence of the ions derived from the crystal's chemical formula. These charges are placed at the crystal lattice sites.

Table 8.3 Non-zero terms in the crystal field Hamiltonian (Eq. 8.21)

Symmetry	Point group	Non-zero CFPs
Cubic	T, T_d, T_h, O, O_h	$B_4^0, B_4^4, B_6^0, B_6^4$
Hexagonal	$C_{6v}, D_6, D_{6h}, D_{3h}$	$B_2^0, B_4^0, B_6^0, B_6^6$
	C_6, C_{3h}, C_{6h}	$B_2^0, B_4^0, B_6^0, B_6^6, B_6^{-6}$
Trigonal	C_{3v}, D_3, D_{3d}	$B_2^0, B_4^0, B_4^3, B_6^0, B_6^3, B_6^6$
	C_3, S_6	$B_2^0, B_4^0, B_4^3, B_6^0, B_6^3, B_6^6, B_4^{-3}, B_6^{-3}, B_6^{-6}$
Tetragonal	$C_{4v}, D_{2d}, D_{4h}, D_4$	$B_2^0, B_4^0, B_4^4, B_6^0, B_6^4$
	S_4, C_4, C_{4h}	$B_2^0, B_4^0, B_4^4, B_6^0, B_6^4, B_4^{-4}, B_6^{-4}$
Orthorhombic	C_{2v}, D_2, D_{2h}	$B_2^0, B_2^2, B_4^0, B_4^2, B_4^4, B_6^0, B_6^2, B_6^4, B_6^6$
Monoclinic	C_2, C_{2h}, Cs,	$B_2^0, B_2^2, B_4^0, B_4^2, B_4^4, B_6^0, B_6^2, B_6^4, B_6^6, B_2^{-2}, B_4^{-2}, B_4^{-4}, B_6^{-2}, B_6^{-4}, B_6^{-6}$
Triclinic	C_1, C_i	$B_2^0, B_2^1, B_2^2, B_4^0, B_4^1, B_4^2, B_4^3, B_4^4, B_6^0, B_6^1, B_6^2, B_6^3, B_6^4, B_6^5, B_6^6, B_2^{-1}, B_2^{-2}, B_4^{-1}, B_4^{-2}, B_4^{-3}, B_4^{-4}, B_6^{-1}, B_6^{-2}, B_6^{-3}, B_6^{-4}, B_6^{-5}, B_6^{-6}$

The second term $B^k_{p,S}$ is proportional to the overlap integrals between the wave functions of an impurity ion and ligands. Its proper calculation implies a quantum-mechanical treatment of both central ion and ligands, to distinguish between the overlap integrals corresponding to the electron orbitals with different l and m quantum numbers. This term models the effects of the chemical (covalent, since the overlap integrals are involved) bond formation; its inclusion considerably improves and develops the original point charge model of crystal field. As a result, agreement between the ECM calculated and experimentally observed energy levels of the impurity ions is greatly improved. The analytical expressions for calculating both terms are as follows:

$$B_{p,q}^{k} = -K_{p}^{k}(1-\sigma_{p})e^{2}\left\langle r^{p}\right\rangle \sum_{i} q_{i}\frac{V_{p}^{k}(\theta_{i},\phi_{i})}{R_{i}^{p+1}}, \tag{8.23}$$

$$B_{p,S}^{k} = K_{p}^{k}e^{2}\frac{2(2p+1)}{2l+1} \sum_{i}\left(G_{s}S(s)_{i}^{2}+G_{\sigma}S(\sigma)_{i}^{2}+\gamma_{p}G_{\pi}S(\pi)_{i}^{2}\right)\frac{V_{p}^{k}(\theta_{i},\phi_{i})}{R_{i}}. \tag{8.24}$$

The summation in both equations is performed over the crystal lattice ions denoted by index i with charges q_i. The R_i, θ_i, ϕ_i entries denote the spherical coordinates of the i-th ion of crystal lattice in the coordinate system, whose origin coincides with the impurity ion position. The averaged values $\left\langle r^{p}\right\rangle$ of p-th power of the impurity ion electrons coordinate can be found in the literature.[22] Alternatively, they can be calculated numerically using the corresponding ions' radial wave functions (see also Chapters 5 and 6). The values of the numerical factors K_p^k and expressions for the polynomials V_p^k are given below in Tables 8.4 and 8.5, respectively.[21] The value of l in Eq. (8.24) is 2 (3) for the d- (f-) electrons, respectively.

Table 8.4 Numerical factors K_p^k from Eqs. (8.22) and (8.23)

P	k	a_{pk}	b_{pk}	p	k	a_{pk}	b_{pk}
0	0	1	1	6	0	1/16	1
1	0	1	1		1	$\sqrt{42}/8$	7
	1	$\sqrt{2}$	1		2	$\sqrt{105}/16$	7
2	0	1/2	1		3	$\sqrt{105}/8$	21
	1	$\sqrt{6}$	3		4	$\sqrt{126}/16$	21
	2	$\sqrt{6}/2$	3		5	$\sqrt{693}/8$	231

(Continued)

Table 8.4 *(Continued)*

3	0	1/2	1	6	6	$\sqrt{231}/16$	231
	1	$\sqrt{3}/2$	1	7	0	1/16	1
	2	$\sqrt{30}/2$	5		1	$\sqrt{14}/32$	1
	3	$\sqrt{5}/2$	5		2	$\sqrt{21}/16$	3
4	0	1/8	1		3	$\sqrt{42}/32$	3
	1	$\sqrt{5}/2$	5		4	$\sqrt{462}/16$	33
	2	$\sqrt{40}/8$	5		5	$\sqrt{462}/32$	33
	3	$\sqrt{35}/2$	35		6	$\sqrt{3003}/16$	429
	4	$\sqrt{70}/8$	35		7	$\sqrt{858}/32$	429
5	0	1/8	1	8	0	1/128	1
	1	$\sqrt{30}/8$	3		1	$\sqrt{2}/32$	3
	2	$\sqrt{210}/4$	21		2	$\sqrt{315}/32$	45
	3	$\sqrt{35}/8$	7		3	$\sqrt{2310}/32$	165
	4	$\sqrt{630}/8$	63		4	$\sqrt{1386}/64$	99
	5	$\sqrt{63}/8$	63		5	$\sqrt{18018}/32$	1287
					6	$\sqrt{429}/32$	429
					7	$\sqrt{12870}/32$	6435
					8	$\sqrt{12870}/128$	6435

Note: $a_{pk} = a_{p\text{-}k}$ (for even values of k), $a_{pk} = -\ a_{p\text{-}k}$ (for odd vales of k); $K_p^0 = a_{p0}^2, K_p^k = 0.5a_{pk}^2 / b_{pk} (k \neq 0)$.

The shielding constant σ_p is required only for the *f* ions, to take into account the screening effects of the filled 5*s* and 5*p* orbitals; it is identically zero for the *d*-electrons, since the *d*-shells are the outer

Table 8.5 Explicit expressions for the V_p^k polynomials

Second rank polynomials	**Third rank polynomials**	**Fourth rank polynomials**
$V_2^0 = 3\gamma^2 - 1$	$V_3^0 = \gamma(5\gamma^2 - 3)$	$V_4^0 = 35\gamma^4 - 30\gamma^2 + 3$
$V_2^1 = 3\alpha\gamma$	$V_3^1 = \alpha(5\gamma^2 - 1)$	$V_4^1 = 5\alpha\gamma(7\gamma^2 - 3)$
$V_2^{-1} = 3\beta\gamma$	$V_3^{-1} = \beta(5\gamma^2 - 1)$	$V_4^{-1} = 5\beta\gamma(7\gamma^2 - 3)$
$V_2^2 = 3(\alpha^2 - \beta^2)$	$V_3^2 = 5\gamma(\alpha^2 - \beta^2)$	$V_4^2 = 5(\alpha^2 - \beta^2)(7\gamma^2 - 1)$
$V_2{}^2 = 6\alpha\beta$	$V_3^{-2} = 10\alpha\beta\gamma$	$V_4^{-2} = 10\alpha\beta(7\gamma^2 - 1)$
	$V_3^3 = 5\alpha(\alpha^2 - 3\beta^2)$	$V_4^3 = 35\alpha\gamma(\alpha^2 - 3\beta^2)$
	$V_3^{-3} = 5\beta(3\alpha^2 - \beta^2)$	$V_4^{-3} = 35\beta\gamma(3\alpha^2 - \beta^2)$
		$V_4^4 = 35(\alpha^4 - 6\alpha^2\beta^2 + \beta^4)$
		$V_4^{-4} = 140\alpha\beta(\alpha^2 - \beta^2)$
Fifth rank polynomials	**Sixth rank polynomials**	
$V_5^0 = \gamma(63\gamma^4 - 70\gamma^2 + 15)$	$V_6^0 = 231\gamma^6 - 315\gamma^4 + 105\gamma^2 - 5$	
$V_5^1 = 3\alpha(21\gamma^4 - 14\gamma^2 + 1)$	$V_6^1 = 7\alpha\gamma(33\gamma^4 - 30\gamma^2 + 5)$	
$V_5^{-1} = 3\beta(21\gamma^4 - 14\gamma^2 + 1)$	$V_6^{-1} = 7\beta\gamma(33\gamma^4 - 30\gamma^2 + 5)$	
$V_5^2 = 21\gamma(\alpha^2 - \beta^2)(3\gamma^2 - 1)$	$V_6^2 = 7(\alpha^2 - \beta^2)(33\gamma^4 - 18\gamma^2 + 1)$	
$V_5^{-2} = 42\alpha\beta\gamma(3\gamma^2 - 1)$	$V_6^{-2} = 14\alpha\beta(33\gamma^4 - 18\gamma^2 + 1)$	
$V_5^3 = 7\alpha(\alpha^2 - 3\beta^2)(9\gamma^2 - 1)$	$V_6^3 = 21\alpha\gamma(\alpha^2 - 3\beta^2)(11\gamma^2 - 3)$	
$V_5^{-3} = 7\beta(3\alpha^2 - \beta^2)(9\gamma^2 - 1)$	$V_6^{-3} = 21\beta\gamma(3\alpha^2 - \beta^2)(11\gamma^2 - 3)$	
$V_5^4 = 63\gamma(\alpha^4 - 6\alpha^2\beta^2 + \beta^4)$	$V_6^4 = 21(\alpha^4 - 6\alpha^2\beta^2 + \beta^4)(11\gamma^2 - 1)$	
$V_5^{-4} = 252\alpha\beta\gamma(\alpha^2 - \beta^2)$	$V_6^{-5} = 84\alpha\beta(\alpha^4 - \beta^2)(11\gamma^2 - 1)$	
$V_5^5 = 63\alpha(\alpha^4 - 10\alpha^2\beta^2 + 5\beta^4)$	$V_6^5 = 231\alpha\gamma(\alpha^4 - 10\alpha^2\beta^2 + 5\beta^4)$	
$V_5^{-5} = 63\beta(5\alpha^4 - 10\alpha^2\beta^2 + \beta^4)$	$V_6^{-5} = 231\beta\gamma(5\alpha^4 - 10\alpha^2\beta^2 + \beta^4)$	
	$V_6^6 = 231(\alpha^2 - \beta^2)(\alpha^4 - 14\alpha^2\beta^2 + \beta^4)$	
	$V_6^{-6} = 462\alpha\beta(3\alpha^4 - 10\alpha^2\beta^2 + 3\beta^4)$	

Note: In all equations, $\alpha = \chi/r$, $\beta = y/r$, $\gamma = z/r$ (x, y, z are the Cartesian coordinates of ligands in a system of reference centered at the impurity ion, $r = \sqrt{x^2 + y^2 + z^2}$).

ones. The overlap integrals between the d-functions of the central ion and p-, s-functions of the ligands are denoted by S_s, S_σ, S_π. If the $\langle lm|l'm'\rangle$ notation) is used; then $S_s = \langle d0|s0\rangle$, $S_\sigma = \langle d0|p0\rangle$, $S_\pi = \langle d1|p1\rangle$. These integrals can be calculated only numerically; the main mathematical difficulty is that the wave functions involved are centered on different ions and a coordinate transformation should be made to express all coordinates in one system of reference.[23] The ECM dimensionless adjustable parameters are denoted by G_s, G_σ, G_π. Their numerical values are determined by matching the positions of the three lowest experimentally observed spectral bands to the corresponding calculated energy levels. Very often they can be taken equal: $G_s = G_\sigma = G_\pi = G$. In this case the number of the ECM fitting parameters is reduced to one only, and only the lowest in energy absorption band is required to determine the value of G by adjusting the corresponding calculated energy level to its experimental counterpart. The numerical factors γ_p are $\gamma_2 = 1$, $\gamma_4 = -4/3$ for the d-electrons and $\gamma_2 = 3/2$, $\gamma_4 = 1/3$, $\gamma_6 = -3/2$ for the f-electrons.

The operators O_p^k are the Stevens homogeneous spherical polynomials. They are expressed in terms of the spherical operators $C_p^k = \sqrt{\frac{4\pi}{2p+1}} Y_{pk}$ (Y_{pk} are the spherical functions):

$$O_p^k = \frac{1}{a_{pk}} Z_p^k, \tag{8.25}$$

and Z_p^k are the real tesseral harmonics[24]

$$Z_p^k = \begin{cases} C_p^{-k} + (-1)^k C_p^k, & k > 0 \\ C_p^k, & k = 0 \\ -i\left(C_p^{-k} - (-1)^k C_p^k\right), & k < 0 \end{cases}. \tag{8.26}$$

As it was already mentioned, the maximum number of terms in Eq. (8.21) is 14 for the d-ions, and 27 for the f-ions (in this case thirteen terms with $p = 6$ are added). However, the number of non-zero CFPs can be reduced significantly if the symmetry of the considered complex is high enough. In the case of perfect cubic symmetry, only two parameters survive in Eq. (8.21): B_4^0 and $B_4^4 = 5B_4^0$, with this relation holding true for the above given definitions of the operators and CFPs. It should be emphasized that to ensure the simplest structure of the crystal field Hamiltonian that is determined by the symmetry, the z axis of the system of reference centered at the impurity ion should be parallel to the highest symmetry axis of the impurity center (this symmetry axis may not always coincide with one of the crystallographic axes). This significantly simplifies diagonalization of the crystal field Hamiltonian, which can be even done analytically in some cases.

For the computational purposes it is more convenient to express the O_p^k operators in terms of the unit tensor operators u_p^k. To this end, we have to introduce new definitions. The convenience of the above-mentioned representation can be explained as follows: Calculations of the matrix elements of the crystal field Hamiltonian are reduced to the integrals of the form $\int_0^{\pi}\int_0^{2\pi} Y_{l_1m_1}{}^{*} Y_{pk} Y_{l_2m_2} \sin\theta\, d\phi\, d\theta$ with the values of l_1, l_2, and p fixed, and m_1, m_2, k varied. The l_1, l_2 numbers correspond to the orbital quantum numbers of the impurity ion electrons, m_1, m_2 are their magnetic quantum numbers, p is the CFPs rank, and k is from $-p$ to p. This integration can be done in a routine way if the Wigner–Eckart theorem is used. The mathematical expression of this theorem is as follows: For the operator T_{kq}(with $-k \le q \le k$), which is referred to as an irreducible tensor operator of rank k, the matrix element between two functions $\langle JM|$ and $\langle J'M'|$ is equal to

$$\langle JM|T_{kq}|J'M'\rangle = (-1)^{J-M}\langle J\|T_k\|J'\rangle \begin{pmatrix} J & k & J' \\ -M & q & M' \end{pmatrix}. \qquad (8.27)$$

The pre-factor $\langle J\|T_k\|J'\rangle$, which does not depend on M, M' and q, is called a "reduced matrix element," and the last term in the right-hand side is the Wigner $3j$-symbol. The huge advantage of the Wigner–Eckart theorem is that if the reduced matrix element between the states with the J and J' values is known, any integral for any values of M, M', and k can be readily evaluated without integration, but just by a simple multiplication of the reduced matrix element by the corresponding $3j$-symbol. Note that to emphasize the difference between the matrix elements and reduced matrix elements (which are not equal), two vertical bars from both sides are used for the latter.

The unit tensor operator u_p^k is defined in such a way that its reduced matrix element between the one-electron wave functions with the quantum numbers of orbital momentum l_1 and l_2 is determined by the following expression:

$$\langle l_1 \| u_k \| l_2 \rangle = \delta_{l_1 l_2}, \quad \delta_{l_1 l_2} = \begin{cases} 1, & l_1 = l_2 \\ 0, & l_1 \neq l_2 \end{cases}. \tag{8.28}$$

Then the spherical operators C_p^k can be written as

$$C_p^k = \langle l \| C_p \| l' \rangle u_p^k, \tag{8.29}$$

and the reduced matrix elements in the last equation are

$$\langle l \| C_p \| l' \rangle = (-1)^l \sqrt{(2l+1)(2l'+1)} \begin{pmatrix} l' & p & l \\ 0 & 0 & 0 \end{pmatrix}. \tag{8.30}$$

The last 3*j*-symbol is non-zero, if $(l + l' + p)/2$ is an integer number. Again, if we consider the case of equivalent electrons $l = l'$, we see that only even numbers of p from the interval from 0 to $2l$ give non-zero contributions. In the case of the *d*-electrons ($l = l' = 2$) we have

$$\langle 2 \| C_2 \| 2 \rangle = -\frac{1}{7}\sqrt{70}, \; \langle 2 \| C_4 \| 2 \rangle = \frac{1}{7}\sqrt{70}. \tag{8.31}$$

And in another important case of the *f*-electrons ($l = l' = 3$) we have

$$\langle 3 \| C_2 \| 3 \rangle = -\frac{2}{15}\sqrt{105}, \; \langle 3 \| C_4 \| 3 \rangle = \frac{1}{11}\sqrt{154}, \; \langle 3 \| C_6 \| 3 \rangle = -\frac{10}{429}\sqrt{3003}. \tag{8.32}$$

Finally, coming back to our intention to express the operators from Eq. (8.25) in terms of the unit tensor operators u_p^k, we can write

$$\begin{aligned} O_p^0 &= \frac{1}{a_{p0}} \langle l \| C_p \| l' \rangle u_p^0, \\ O_p^k &= \frac{1}{a_{pk}} \langle l \| C_p \| l' \rangle \left(u_p^{-k} + (-1)^k u_p^k \right), \; k > 0, \\ O_p^k &= \frac{(-i)}{a_{pk}} \langle l \| C_p \| l' \rangle \left(u_p^{-k} - (-1)^k u_p^k \right), \; k < 0. \end{aligned} \tag{8.33}$$

The matrix elements of the u_p^k operators between all LS terms of all electronic configurations of practical interest (including those with the *d*- and *f*-electrons) are tabulated in the book by C. W. Nielsen and G. F. Koster.[25] Together with the equations given above, they can be used for a computer generation of the matrix elements of the O_p^k operators for any ion with the *d*- or *f*-electron configurations.

Table 8.6 Numerical approximations of the overlap integrals between some 3*d* ions and ligands

	O^{2-}	F^{-}	S^{2-}
Ti^{3+}	$-0.41047 + 0.08221R$ [a] $0.34999 - 0.06621R$ $1.6847\exp(-0.84645R)$		
V^{4+}	$-0.31129 + 0.06631R$ [a] $0.29370 - 0.06006R$ $1.68240\exp(-0.97694R)$		
Cr^{5+}	$-0.98756\exp(-0.83927R)$ [a] $0.83581\exp(-0.75558R)$ $1.55630\exp(-1.07040R)$		
Mn^{6+}	$-0.91299\exp(-0.92478R)$ [a] $0.86054\exp(-0.85516R)$ $1.38650\exp(-1.14700R)$		
V^{3+}	$-0.89227\exp(-0.68515R)$ [b] $0.67342\exp(-0.53979R)$ $1.59400\exp(-0.88172R)$		$2.4687\exp(-0.8454R)$ [e] $0.97599\exp(-0.55674R)$ $2.4069\exp(-0.88214R)$
Cr^{4+}	$0.01293 + 4.1973\exp(-1.28143R)$ [c] $0.22132 - 0.04368R$ $0.01413 + 4.63085\exp(-1.42861R)$		

(*Continued*)

Table 8.6 *(Continued)*

	O^{2-}	F^{-}	S^{2-}
Mn^{5+}	$-0.82303 \exp(-0.82905\,R)$ [d] $0.70370 \exp(-0.74130\,R)$ $1.49570 \exp(-1.10540\,R)$		
Cr^{3+}	$-1.5941 \exp(-0.74497\,R)$ [e] $1.3496 \exp(-0.76490\,R)$ $1.2614 \exp(-0.86530\,R)$	$2.7775 \exp(-1.0277\,R)$ [f] $1.1674 \exp(-0.7105\,R)$ $2.0930 \exp(-1.0235\,R)$	
Mn^{4+}	$-0.92125 \exp(-0.69807\,R)$ [f] $0.35114 \exp(-0.74151\,R)$ $1.46440 \exp(-1.10230\,R)$	$2.9781 \exp(-1.1738\,R)$ [f] $1.2177 \exp(-0.82329\,R)$ $1.9908 \exp(-1.1314\,R)$	
Fe^{5+}	$1.00200 \exp(-0.81119\,R)$ [c] $0.90092 \exp(-0.84777\,R)$ $1.27430 \exp(-1.09140\,R)$		
Ni^{2+}	$-0.99799 \exp(-0.73145\,R)$ [g] $0.84696 \exp(-0.68325\,R)$ $1.13280 \exp(-0.86486\,R)$	$2.2782 \exp(-1.0188\,R)$ [h] $1.0744 \exp(-0.72669\,R)$ $1.5892 \exp(-0.98151\,R)$	
Cu^{2+}	$-0.97125 \exp(-0.75587\,R)$ [g] $0.80480 \exp(-0.69594\,R)$ $1.16820 \exp(-0.90598\,R)$		

	O^{2-}	F^-	S^{2-}
Co^{2+}		$2.8041 \exp(-1.0393\ R)$ [h] $1.2050 \exp(-0.73591\ R)$ $1.7064 \exp(-0.9792\ R)$	$1.9413 \exp(-0.83737\ R)$ [e] $0.20781 - 0.03168\ R$ $1.9413 \exp(-0.87809\ R)$
Cr^{2+}			$1.7815 \exp(-0.72552\ R)$ [e] $0.25776 - 0.041053\ R$ $2.1967 \exp(-0.81329\ R)$
V^{2+}		$2.5658 \exp(-0.86398\ R)$ [g] $1.1444 \exp(-0.5957\ R)$ $2.0825 \exp(-0.88627\ R)$	$1.8673 \exp(-0.69949\ R)$ [i] $0.30378 - 0.04383\ R$ $2.4000 \exp(-0.79664\ R)$

Note: For each pair of ions in the table, the expressions from the top to the bottom correspond to the $S_s = \langle d0|s0\rangle$, $S_\sigma = \langle d0|p0\rangle$, $S_\pi = \langle d1|p1\rangle$ overlap integrals. R is expressed in atomic units.

[a]$2.817 < R < 3.817$; [b]$2.73 < R < 3.73$; [c]$3 < R < 4$; [d]$2.62 < R < 3.62$; [e]$3.75 < R < 4.75$; [f]$2.93 < R < 3.93$; [g]$3.3 < R < 4.3$; [h]$3.317 < R < 4.317$; [i]$4 < R < 5$.

Some explanations regarding the calculations of the overlap integrals $S_s = \langle d0|s0\rangle, S_\sigma = \langle d0|p0\rangle, S_\pi = \langle d1|p1\rangle$ should also be added. They can be computed numerically using the tabulated in the literature the wave functions of various ions (see, for example, a famous paper by E. Clementi and C. Roetti, for example Ref. 2). Usually these integrals are calculated for 10–20 different interionic distances r and the obtained sets of the calculated values are generally approximated by the exponential functions of the form $A\exp(-ar)$, where A and a are constants, or by some other decreasing function of distance.

Such approximating functions of the overlap integrals between typical ligands and many 3*d* ions are given in Table 8.6.

One more practical point has to be mentioned: If the charges of the crystal lattice are expressed as the integer numbers of the elementary charge (e.g., –1 for F^-, +3 for Al^{3+}, etc.) and all distances are expressed in Å, the calculated CFP values will be in cm^{-1}, if the factor 1.16×10^5 will be placed before the summation signs in Eqs. (8.23) and (8.24). This multiplier appears after converting the unit of energy erg of the CGS system into cm^{-1} (note that the elementary charge in the CGS system is 4.8×10^{-10} electrostatic units and the distance is measured in cm).

The ECM has been successfully used for analysis of the optical absorption spectra of both 3*d* and 4*f* ions in crystals. Several examples of its application will be shown in the next paragraph.

8.7 Application of the Exchange Charge Model of Crystal Field to the Calculations of Energy Levels of the 3*d* Ions in Crystals

In this paragraph, we illustrate how the ECM can be used to calculate the energy levels of the 3*d* ions in crystals. The considered example is the Cr^{4+} ion in the yttrium–aluminum garnet, $Y_3Al_5O_{12}$.

The four-valent Cr ions have been a subject of thorough investigations since the late 80s of the previous century, because of tunable laser generation in the infrared range,[26–48]; although this list of references is quite extensive and covers both experimental and theoretical studies, it serves for the representative purposes only and does not reflect the whole content published on this topic.

The yttrium–aluminum garnet $Y_3Al_5O_{12}$ (YAG) is just one member from a large class of the garnet crystals. Many of them are used for lasing, lighting, scintillating applications,[49–51] etc. We shall focus on the energy level scheme of the Cr^{4+} ions in YAG: how it can be obtained starting from the structure of the host and how the obtained results can be compared with the corresponding experimental data.

The YAG structure is cubic and is described by the Ia-3d space group (No. 230) with the lattice constant a = 12.0062 Å.[52] There are two Al sites in this structure: four- and six-fold coordinated. The Y ions are eight-fold coordinated. Comparing the ionic radii of the host cations (Al^{3+}—0.39 Å for the four-fold coordination and 0.535 Å for the six-fold coordination, Y^{3+}—1.019 Å for the eight-fold coordination) with that one for the Cr^{4+} ions (0.41 Å for the four-fold coordination and 0.55 Å for the six-fold coordination)[53] and taking into account preference of the Cr^{4+} ions to occupy the tetrahedral sites, it is clear that the four-fold coordinated aluminum ions will be replaced by the dopants. Of course, the dopant concentration is very small—not more than 1–2% only. If only the nearest four ligands around the tetrahedral Al ion are considered, then the local site symmetry is D_{2d}.[39] Table 8.7 collects the Cartesian coordinates of the oxygen ions in the AlO_4 tetrahedral cluster obtained from Ref. 46, whereas Fig. 8.14 shows the nearest environment around the Cr^{4+} ion in the system of reference corresponding to the data from Table 8.7 (the z axis is directed from the plane of figure towards the reader). All Cr–O distances in this complex are 1.75378 Å.

Table 8.7 Cartesian coordinates (in Å) of all ions in the AlO_4 complex in $Y_3Al_5O_{12}$

	x	y	z
Central ion (Al or Cr)	0	0	0
O1	1.3504	0	1.1190
O2	−1.3504	0	1.1190
O3	0	1.3504	−1.1190
O4	0	−1.3504	−1.1190

Since the electric charges of the substituting and substituted ions are different, some charge compensation should be realized. One of such possible mechanisms is co-doping with the Mg^{2+} ions, like

$Al^{3+} + Al^{3+} \rightarrow Cr^{4+} + Mg^{2+}$. Since such charge compensation defects are not in the first coordination sphere, we shall not take them into account.

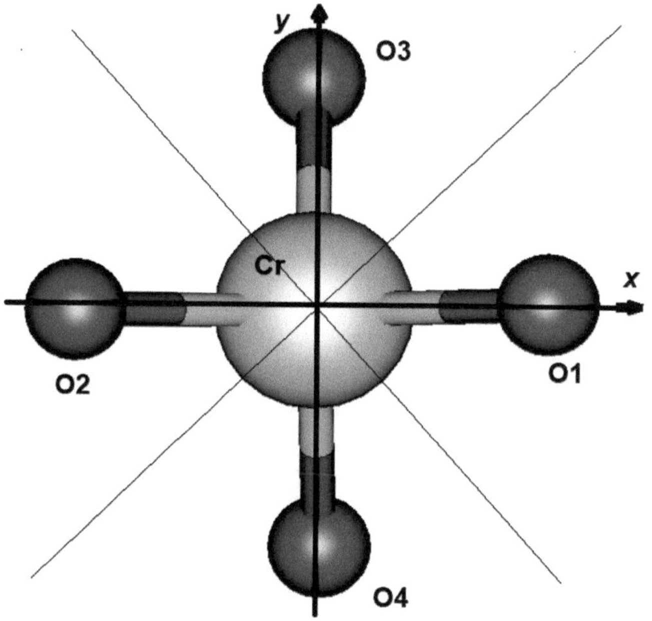

Figure 8.14 Tetrahedral CrO_4 cluster in $Y_3Al_5O_{12}$. The O1 and O2 ions are located above the central ion (positive *z* coordinate), the O3 and O4 ions are located below the central ion (negative *z* coordinate). The *z* axis goes out of the figure plane and is the S_4 axis and C_2 axis of the D_{2d} group. Two other C_2 axes are shown by the thin lines. Dihedral reflection planes σ_d coincide with the coordination planes *xz* and *yz*. Such choice of the reference system simplifies the ligands coordinates by placing them into the *xz* and *yz* planes. Drawn with VESTA.[54]

Deviation from the ideal tetrahedral symmetry is illustrated by the angles between the chemical bonds in the considered complex. In the ideal tetrahedron this angle is 109.47°, but in the complex shown in Fig. 8.14 the O1-Cr-O2 and O3-Cr-O4 angles are 100.71°, whereas the O3-Cr-O2 and O1-Cr-O4 angles are 114.02°, which corresponds to the tetrahedron elongation along the *z* axis.

Using all relevant equations of the ECM (Eqs. (8.22) and (8.23), Tables (8.5)–(8.7)) with all numerical factors, polynomials and overlap integrals, the CFPs values were calculated and they are shown in Table 8.8, with separated point charge and exchange charge contributions. The structure of the crystal field Hamiltonian (its non-zero terms) agrees with the data from Table 8.3.

Table 8.8 Calculated point charge $B_{p,q}^{k}$ and exchange charge $B_{p,S}^{k}$ CFPs (in cm^{-1}) for the Cr^{4+} ion in the CrO_4 complex in $Y_3Al_5O_{12}$

	$B_{p,q}^{k}$	$B_{p,S}^{k}$	B_{p}^{k}
B_2^0	3267.6	2492.5	5760.1
B_4^0	−678.7	−2188.4	−2867.1
B_4^4	2447.1	7890.7	10337.8

Note: The ECM parameter $G = 6.0$ (see text for more details).

As seen from Table 8.8, the exchange charge contributions to the total CFPs are several times greater than (for the fourth rank parameters) or very close to (the second rank parameters) the point charge contributions. This circumstance is an additional illustration of inability of the point charge model to reproduce the crystal field splittings properly. From the other side, it also demonstrates advanced character of the ECM, which considerably extends the point charges model by considering the covalent and overlap effects.

The matrix elements of the crystal field Hamiltonian were calculated using Eqs. (8.25)–(8.33). The value of the ECM parameter G was determined to match the first calculated spin-allowed transition from the $^3A_2$ ground state to one of the components of the $^3T_2$ states (here the irreducible representations of the T_d group are used for simplicity) to its experimental value at about 7812 cm^{-1}.[55] By changing G and calculating the energy levels for each G value, we arrived at $G = 6.0$. When constructing the crystal field Hamiltonian matrix, it is necessary to add to all diagonal elements arising from the same LS term the energy of the corresponding free ion terms expressed in Racah parameters B and C. If there are several identical LS terms in a particular electron configuration, then the Coulomb interaction will mix them up and non-diagonal matrix elements will also appear. The reader may find all electrostatic matrices in the book by Nielsen and Koster.[25]

In the considered case of the Cr^{4+} ion in $Y_3Al_5O_{12}$ the values of the Racah parameters B and C for were taken as 500 cm^{-1} and 3370 cm^{-1}, respectively. All calculated energy levels are collected in Table 8.9, in comparison with the experimental data from Ref. 55. Since the actual symmetry of the considered cluster is not ideally tetrahedral, the symmetry reduction effects should be considered. According to the group theory,[56] the following reduction of the T_d irreducible representation in the D_{2d} point group occurs: $A_1 \rightarrow A_1$, $A_2 \rightarrow B_1$, $E \rightarrow A_1 + B_1$, $T_1 \rightarrow A_2 + E$, $T_2 \rightarrow B_2 + E$.

Table 8.9 Calculated energy levels (in cm^{-1}) of the Cr^{4+} ion in the CrO_4 complex in $Y_3Al_5O_{12}$

D_{2d} point group with the Td point group parent states and free ion LS terms	Calculated without SO interaction (this work)	Calculated with SO interaction (this work)	Experimental data[55]
$^3B_1$ ($^3A_2$ (^3F))	0	0*, 0.7	0
$^3B_2$ ($^3T_2$ (^3F))	7876	7865, 7884*	7812
$^1A_1$ (^1E (^1D))	8185	7977	8292
$^3A_2$ ($^3T_1$ (^3F))	8282	8282*, 8474	8264
^3E ($^3T_2$ (^3F))	9948	9801, 9847, 9963*, 10058, 10086	9823
$^1B_1$ (^1E (^1D))	10616	10675	
$^1A_2$ ($^1T_1$ (^1G))	15556	15551	
^3E ($^3T_1$ (^3F))	16219	16187, 16196, 16226*, 16247, 16291	~ 16000 (broad band)
$^1B_2$ ($^1T_2$ (^1D))	18434	18431	
$^1A_1$ ($^1A_1$ (^1G))	18545	18575	
^1E ($^1T_2$ (^1D))	20083	20081*	
^3E ($^3T_1$ (^3P))	22099	22045, 22048, 22120*, 22171, 22185	
$^3A_2$ ($^3T_1$ (^3P))	25648	25637*, 25659	
$^1A_1$ (^1E (^1G))	26231	26247	

D_{2d} point group with the Td point group parent states and free ion LS terms	Calculated without SO interaction (this work)	Calculated with SO interaction (this work)	Experimental data[55]
$^1E\ (^1T_1\ (^1G))$	26514	26552*	
$^1E\ (^1T_2\ (^1G))$	28889	28901*	
$^1B_1\ (^1E\ (^1G))$	34977	34990	
$^1B_2\ (^1T_2\ (^1G))$	35036	35049	
$^1A_1\ (^1A_1\ (^1S))$	50655	50670	

Note: The doublet states remaining after SO splitting are denoted with an asterisk.

As seen from Table 8.9, good agreement between the calculated and experimental energy levels was obtained, what demonstrates the high potential of ECM and its applicability to handling the spectra of the 3*d* ions in crystals.

Figure 8.15 shows the calculated energy levels (in a real scale, but without the spin-singlet states) with allowed electric and magnetic dipole transitions. The group selection rules from Chapter 7 were used to identify the allowed transitions, which are shown by the vertical arrows.

The left part of Fig. 8.15 shows the spin-triplet states of the d^2 electron configuration in the ideal tetrahedral environment. The $^3A_2 \rightarrow {}^3T_2$ transition is forbidden in the electric dipole approximation, whereas the $^3A_2 \rightarrow {}^3T_1$ transitions are allowed. That is why the former always has a lower intensity in the experimental absorption spectra than the latter ones.

Table 8.9 also lists the calculated energy levels of the Cr^{4+} ions in the YAG crystal with SO interaction taken into account. The matrix elements of the SO interaction between different terms of the l^n configurations can be calculated as follows[57]:

$$l^n SLJM|\sum_i a(r_i) l_i s_i | l^n S'L'J'M'$$

$$= \delta_{JJ'}\delta_{MM'}(-1)^{S'+L+J} \times \xi_{nl}\sqrt{l(l+1)(2l+1)}\left(l^n SL\left\|V^{11}\right\|l^n S'L'\right)\begin{Bmatrix} S & L & J \\ L' & S' & 1 \end{Bmatrix}, \tag{8.34}$$

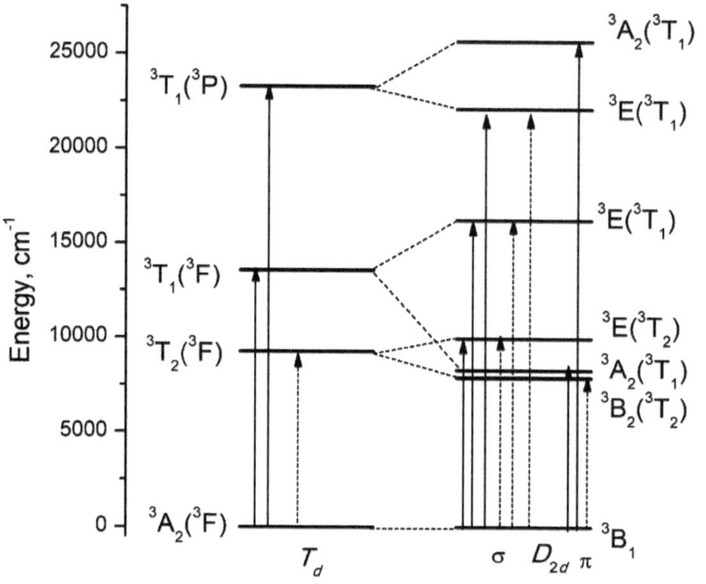

Figure 8.15 Calculated energy levels of the Cr^{4+} ions in $Y_3Al_5O_{12}$. The allowed electric and magnetic dipole transitions are shown by the solid and dashed lines, respectively. The cases of **π**- and **σ**-polarizations for the D_{2d} group are shown separately.

where $a(r_i)$ is the radial part of the SO interaction, ξ_{nl} is the SO interaction constant (taken as 180 cm^{-1} in this case), the matrix elements $\left(l^n SL\left\|V^{11}\right\|l^n S'L'\right)$ are tabulated in Ref. 25, and $\begin{Bmatrix} S & L & J \\ L' & S' & 1 \end{Bmatrix}$ is the Wigner 6*j*-symbol. In the D_{2d} group the spin-triplet functions transform as A_2+E, and the spin-singlet functions—as A_1. Therefore, to find, for example, the SO splitting of the ^3E state in this case, it is necessary to find the direct product (A_2+E) × E and decompose it into the irreducible representations of the D_{2d} group. With this method, the following results on the SO splitting of the d ^{2} state in the D_{2d} symmetry can be obtained: $^3A_1 \rightarrow A_2+E$, $^3A_2 \rightarrow A_1+E$, $^3B_1 \rightarrow B_2+E$, $^3B_2 \rightarrow B_1+E$, $^3E \rightarrow A_1+A_2+B_1+B_2+E$; $^1A_1 \rightarrow A_1$, $^1A_2 \rightarrow A_2$, $^1B_1 \rightarrow B_1$, $^1B_2 \rightarrow B_2$, $^1E \rightarrow E$. Some states remain doubly degenerated after SO splitting, and the results of the group theory analysis are fully consistent with the data from Table 8.9. It can also be noticed that the magnitude of the SO splitting is just a few tens or a few hundred wave numbers. This is the common feature of the 3*d* ions and experimental detection of the SO split states in their spectra can be quite difficult, since they are always "hidden" by the broad spectral bands.

One important note ought to be made regarding the principal possibility to observe and use the *d-d* transitions. According to the selection rules, these transitions should be forbidden as the transitions between the states of the same parity. This strict parity selection rule is partially lifted because of slight (but still non-zero!) mixture of the 3*d* states with the opposite parity high-lying states of the same impurity ion, like the 4*p* states. That small contribution does allow to detect the *d-d* transitions in the experimental absorption and emission spectra. In addition, admixture of the ligands *p* states to the central ion's *d* wave functions also contributes to the partial removal of the forbidden character of the *d-d* transitions.

More examples of application of the ECM to the 3*d* ions in crystals, including energy levels calculations, analysis of the allowed and forbidden transitions in the polarized spectra, studies of the microscopic crystal field effects, etc., can be found in many publications.[58–82]

8.8 A Single *f*-Electron in a Cubic Crystal Field

A good model system to study the crystal field splitting of a single *f* electron states in a cubic crystal field is BaF_2, which crystallizes in the cubic Fm-3m space group with the lattice constant 6.1964 Å.[83] One unit cell of BaF_2 is depicted in Fig. 8.16. Doped with the Ce^{3+} ions, it is used for scintillating applications because of intensive 5*d*-4*f* emission transitions of the Ce^{3+} ions.[84] However, in this paragraph only the lowest energy levels coming from the $4f^1$ ground state electron configuration of the Ce^{3+} ions will be considered. In the free state, the SO coupling splits the only ^{2}F term of the $4f^1$ electron configuration into two states only: $^2F_{5/2}$ (six-fold degenerated ground state) and $^2F_{7/2}$ (eight-fold degenerated excited state).

Since the fluorine ions form an ideal cube around the Ba ions (which will be substituted by the Ce ions), only B_4^0, B_4^4, B_6^0, B_6^4 crystal field parameters (Table 8.3) are not zero. Taking the structural data from Ref. 85, it is easy to obtain the following Cartesian coordinates of nearest eight fluorine ions around Ce^{3+} ion (Table 8.10).

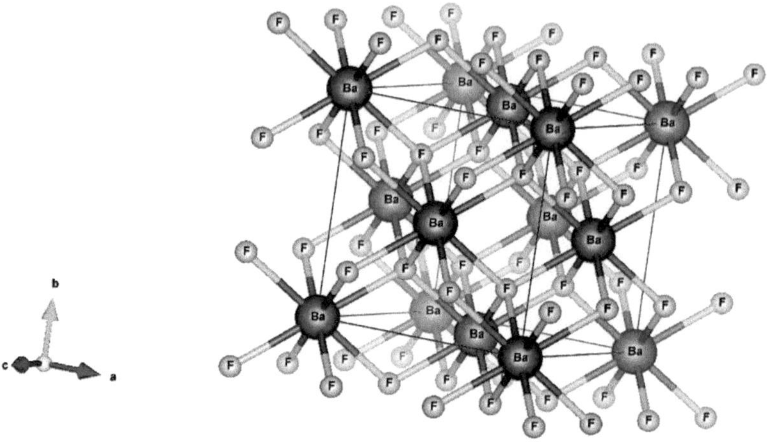

Figure 8.16 Crystal structure of BaF_2. Eight fluorine ions around one barium ion form an ideal cube. Drawn with VESTA.[54]

Table 8.10 Cartesian coordinates (in Å) all ions in CeF_8 cluster in BaF_2

Ion	x	y	Z	Distance from the Ce^{3+} ion
Ce^{3+}	0	0	0	2.6847
F^-	1.5500	−1.5500	−1.5500	2.6847
F^-	−1.5500	1.5500	1.5500	2.6847
F^-	−1.5500	1.5500	−1.5500	2.6847
F^-	−1.5500	−1.5500	1.5500	2.6847
F^-	−1.5500	−1.5500	−1.5500	2.6847
F^-	1.5500	1.5500	1.5500	2.6847
F^-	1.5500	1.5500	−1.5500	2.6847
F^-	1.5500	−1.5500	1.5500	2.6847

The values of the electron density moments for the Ce^{3+} ion are (in atomic units): $\langle r^2 \rangle = 1.2$, $\langle r^4 \rangle = 3.455$, $\langle r^6 \rangle = 21.225$.[86] Since for the rare-earth ions the magnitude of the SO interaction exceeds that one of the crystal field, the latter should be considered as a small perturbation to the states produced as a result of the SO splitting of the free ions LS terms. From the point of view of calculations of the crystal field Hamiltonian matrix elements this requires the change of the basis set from the $|LM\rangle$ functions that are characteristic for the $3d$ ions to the $|JM\rangle$ functions. The CFPs calculated with the help of Eqs. (8.23) and (8.24) correspond to the so-called Stevens normalization, which can be directly used for the calculations of the energy levels of the d-ions in the $|LM\rangle$ basis set. The Wybourne normalization should be used if the energy levels of the f ions are calculated in the $|JM\rangle$ basis set. The relations between these the CFPs sets are as follows[87]:

$$B_{2,W}^{\pm2} = \frac{2}{\sqrt{6}} B_{2,S}^{\pm2}, \ B_{2,W}^{\pm1} = -\frac{1}{\sqrt{6}} B_{2,S}^{\pm1}, \ B_{2,W}^{0} = 2B_{2,S}^{0},$$

$$B_{4,W}^{\pm4} = \frac{8}{\sqrt{70}} B_{4,S}^{\pm4}, \ B_{4,W}^{\pm3} = -\frac{2}{\sqrt{35}} B_{4,S}^{\pm3}, \ B_{4,W}^{\pm2} = \frac{4}{\sqrt{10}} B_{4,S}^{\pm2},$$

$$B_{4,W}^{\pm1} = -\frac{2}{\sqrt{5}} B_{4,S}^{\pm1}, \ B_{4,W}^{0} = 8B_{4,S}^{0}, \ B_{6,W}^{\pm6} = \frac{16}{\sqrt{231}} B_{4,S}^{\pm6},$$

$$B_{6,W}^{\pm5} = -\frac{8}{3\sqrt{77}} B_{6,S}^{\pm5}, \ B_{6,W}^{\pm4} = \frac{16}{3\sqrt{14}} B_{6,S}^{\pm4}, \ B_{6,W}^{\pm3} = -\frac{8}{\sqrt{105}} B_{6,S}^{\pm3},$$

$$B_{6,W}^{\pm2} = \frac{16}{\sqrt{105}} B_{6,S}^{\pm2}, \ B_{6,W}^{\pm1} = -\frac{8}{\sqrt{42}} B_{6,S}^{\pm1}, \ B_{6,W}^{0} = 16B_{6,S}^{0}. \quad (8.35)$$

Using the ionic coordinates from Table 8.10, the ECM Eq. (8.23) for the point charge contribution to the CFPs only and the conversion relations (8.35), the following non-zero CFPs for Ce^{3+} ions in BaF_2 were obtained (all values in cm^{-1}): $B_4^0 = -701.0$, $B_4^4 = -418.9$, $B_6^0 = 95.6$, $B_6^4 = -178.9$. It can easily be seen that in the case of the cubic crystal field the following relations are true for the Wybourne CFPs: $B_4^4 = \frac{5}{\sqrt{70}} B_4^0$, $B_6^4 = -\sqrt{\frac{7}{2}} B_6^0$.[88] The group theoretical analysis shows that the $^2F_{5/2}$ state in a cubic crystal field splits into a doublet Γ_7 and a quartet Γ_8, whereas the $^2F_{7/2}$ state splits into two doublets Γ_6, Γ_7 and a quartet Γ_8. Since the Ce^{3+} ion is a Kramers ion, the double groups should be used to describe its energy levels splitting, and each energy level remains at least doubly degenerated.

The next step is to diagonalize the crystal field Hamiltonian in the space of the fourteen functions of the Ce^{3+} $4f^1$ electron configuration. To perform the energy level assignment, the symmetry adapted combinations of the $|JM\rangle$ wave functions were taken from Ref. 89. These functions are as follows:

$$\Psi_1\left(\Gamma_7, {}^2F_{5/2}\right) = \sqrt{\frac{1}{6}}\left|\frac{5}{2}\frac{5}{2}\right\rangle - \sqrt{\frac{5}{6}}\left|\frac{5}{2} - \frac{3}{2}\right\rangle,$$

$$\Psi_2\left(\Gamma_7, {}^2F_{5/2}\right) = -\sqrt{\frac{5}{6}}\left|\frac{5}{2}\frac{3}{2}\right\rangle + \sqrt{\frac{1}{6}}\left|\frac{5}{2} - \frac{5}{2}\right\rangle,$$

$$\Psi_1\left(\Gamma_8, {}^2F_{5/2}\right) = -\sqrt{\frac{1}{6}}\left|\frac{5}{2}\frac{3}{2}\right\rangle - \sqrt{\frac{5}{6}}\left|\frac{5}{2} - \frac{5}{2}\right\rangle, \tag{8.36}$$

$$\Psi_2\left(\Gamma_8, {}^2F_{5/2}\right) = \left|\frac{5}{2}\frac{1}{2}\right\rangle,$$

$$\Psi_3\left(\Gamma_8, {}^2F_{5/2}\right) = \left|\frac{5}{2} - \frac{1}{2}\right\rangle,$$

$$\Psi_4\left(\Gamma_8, {}^2F_{5/2}\right) = \sqrt{\frac{5}{6}}\left|\frac{5}{2}\frac{5}{2}\right\rangle + \sqrt{\frac{1}{6}}\left|\frac{5}{2} - \frac{3}{2}\right\rangle$$

for the Γ_7, Γ_8 states of the ${}^2F_{5/2}$ level, and

$$\Psi_1\left(\Gamma_6, {}^2F_{7/2}\right) = -\sqrt{\frac{7}{12}}\left|\frac{7}{2}\frac{1}{2}\right\rangle - \sqrt{\frac{5}{12}}\left|\frac{7}{2} - \frac{7}{2}\right\rangle,$$

$$\Psi_2\left(\Gamma_6, {}^2F_{7/2}\right) = \sqrt{\frac{5}{12}}\left|\frac{7}{2}\frac{7}{2}\right\rangle + \sqrt{\frac{7}{12}}\left|\frac{7}{2} - \frac{1}{2}\right\rangle,$$

$$\Psi_1\left(\Gamma_7, {}^2F_{7/2}\right) = \sqrt{\frac{3}{4}}\left|\frac{7}{2}\frac{5}{2}\right\rangle - \sqrt{\frac{1}{4}}\left|\frac{7}{2} - \frac{3}{2}\right\rangle,$$

$$\Psi_2\left(\Gamma_7, {}^2F_{7/2}\right) = \sqrt{\frac{1}{4}}\left|\frac{7}{2}\frac{3}{2}\right\rangle - \sqrt{\frac{3}{4}}\left|\frac{7}{2} - \frac{5}{2}\right\rangle,$$

$$\Psi_1\left(\Gamma_8, {}^2F_{7/2}\right) = \sqrt{\frac{3}{4}}\left|\frac{7}{2}\frac{3}{2}\right\rangle + \sqrt{\frac{1}{4}}\left|\frac{7}{2} - \frac{5}{2}\right\rangle, \tag{8.37}$$

$$\Psi_2\left(\Gamma_8, {}^2F_{7/2}\right) = -\sqrt{\frac{5}{12}}\left|\frac{7}{2}\frac{1}{2}\right\rangle + \sqrt{\frac{7}{12}}\left|\frac{7}{2} - \frac{7}{2}\right\rangle,$$

$$\Psi_3\left(\Gamma_8, {}^2F_{7/2}\right) = \sqrt{\frac{7}{12}}\left|\frac{7}{2}\frac{7}{2}\right\rangle - \sqrt{\frac{5}{12}}\left|\frac{7}{2} - \frac{1}{2}\right\rangle,$$

$$\Psi_4\left(\Gamma_8, {}^2F_{7/2}\right) = \sqrt{\frac{1}{4}}\left|\frac{7}{2}\frac{5}{2}\right\rangle + \sqrt{\frac{3}{4}}\left|\frac{7}{2} - \frac{3}{2}\right\rangle$$

for the Γ_6, Γ_7, Γ_8 states of the $^2F_{7/2}$ level (the subscript enumerates the wave functions of the same symmetry in the same irreducible representation). The crystal field Hamiltonian will have a block-diagonal form in this symmetrized basis set.

With the Ce^{3+} SO constant of 647 cm^{-1} [90] and the basis set defined by Eqs. (8.36) and (8.37), the following energy levels were obtained (in cm^{-1}, the energy of the ground state is taken as zero): 0 (a quartet Γ_8), 192 (a doublet Γ_7), 2227 (a doublet Γ_6), 2297 (a quartet Γ_8), 2519 (a doublet Γ_7), as illustrated in Fig. 8.17.

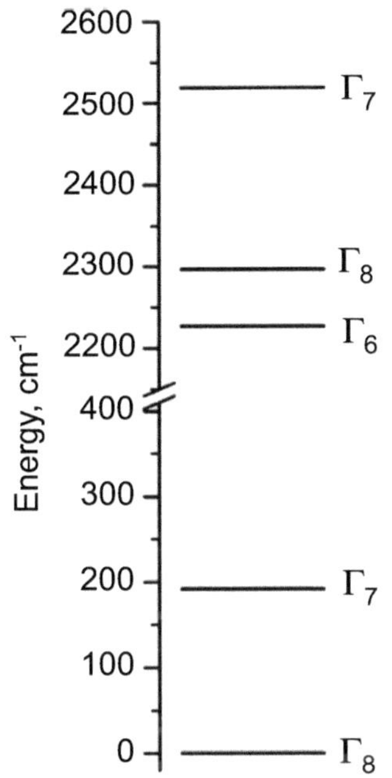

Figure 8.17 Calculated 4*f* energy levels splitting for the Ce^{3+} ions in BaF_2.

Since the point charge model only was used in this example, the overall splitting of the Ce 4*f* is underestimated. It has been shown, both experimentally and theoretically[91] that the Ce 4*f* levels extend to about 3200 cm^{-1} in crystals. Agreement between the calculated and observed 4*f* energy levels can be improved if more elaborated crystal field models, like the ECM described above, are used. This is discussed in the next paragraph.

8.9 Application of the Exchange Charge Model of Crystal Field to the Analysis of the 4*f* Ions Spectra in Crystals

Application of the ECM requires calculations of the overlap integrals between the wave functions of an impurity ion and ligands, as Eq. (8.24) implies. For example, these overlap integrals for the Ce^{3+} $4f$ functions and O^{2-} $2s$, $2p$ functions were calculated numerically for different interionic separations R and then they were approximated by the following functions (R is measured in the atomic units and is in the range from 0 to 12 a.u.)[92,93]:

$$S_s = \langle f0|s0\rangle = 1.24783(R-1.16708)\exp(-1.90106R)R^{2.16972},$$

$$S_\sigma = \langle f0|p0\rangle = -0.75305(R-1.180304)\exp(-1.59584R)R^{1.79939},$$

$$S_\pi = \langle f\pm1|p\pm1\rangle = 1.13834(R-0.10336)\exp(-1.45377R)R^{0.59915}.$$

(8.38)

Figure 8.18 visualizes the calculated dependences from Eq. (8.38).

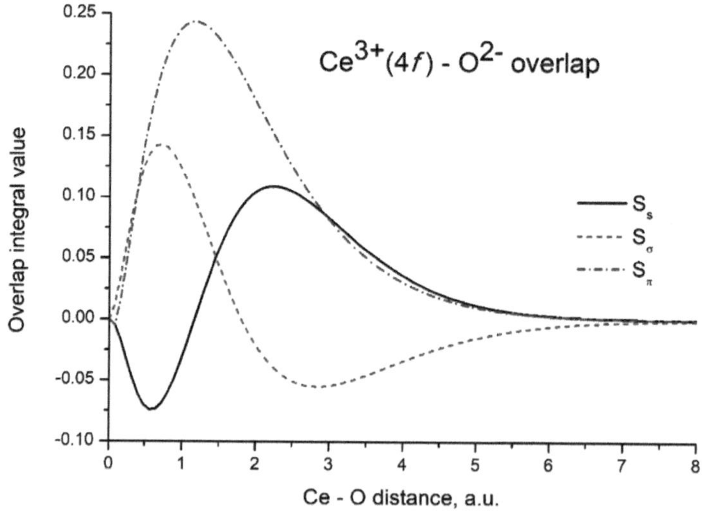

Figure 8.18 Calculated overlap integrals between the Ce^{3+} $4f$ states and O^{2-} $2s$, $2p$ states as a function of the interionic distance.

The overlap integrals from Eqs. (8.38) were used to calculate the CFPs and $4f$ energy levels of the Ce^{3+} ion in $YAlO_3$ (YAP).[85] After doping, the Ce^{3+} ions occupy the Y^{3+} sites with the C_{1h} symmetry, and the crystal field Hamiltonian for the $4f$ states of cerium is defined as follows:

$$\begin{aligned}\mathbf{H}(4f^1) = {} & E_{avg} + \zeta_{4f}\mathbf{s}_f \cdot \mathbf{l}_f + B_0^2\mathbf{C}_0^{(2)} \\ & + B_2^2(\mathbf{C}_2^{(2)} + \mathbf{C}_{-2}^{(2)}) + B_0^4\mathbf{C}_0^{(4)} + B_2^4(\mathbf{C}_2^{(4)} + \mathbf{C}_{-2}^{(4)}) \\ & + B_4^4(\mathbf{C}_4^{(4)} + \mathbf{C}_{-4}^{(4)}) + B_0^6 C_0^{(6)} + B_2^6(\mathbf{C}_2^{(6)} + \mathbf{C}_{-2}^{(6)}) \\ & + B_4^6(\mathbf{C}_4^{(6)} + \mathbf{C}_{-4}^{(6)}),\end{aligned} \tag{8.39}$$

where the first term stands for the average energy of the Ce^{3+} $4f^1$ configuration, second term describes the SO coupling, and all the remaining terms describe the crystal field. Using the YAP structural data, the following CFPs values were calculated (in cm^{-1}, Wybourne notation): $B_0^2 = -30$, $B_1^2 = -680$, $B_2^2 = -60$, $B_0^4 = -369$, $B_1^4 = 1397$, $B_2^4 = -199$, $B_3^4 = 759$, $B_4^4 = -777$, $B_0^6 = 1319$, $B_1^6 = -160$, $B_2^6 = 390$, $B_3^6 = -350$, $B_4^6 = -891$, $B_5^6 = -110$, $B_6^6 = 488$. Besides, $E_{avg} = 1654$ cm^{-1}, $\zeta_{4f} = 617$ cm^{-1}. Diagonalization of the crystal field Hamiltonian in the basis set spanned by the wave functions of the $^2F_{5/2}$ state (six functions) and $^2F_{7/2}$ state (eight functions) resulted in getting the following energy level scheme (Table 8.11).

Table 8.11 Calculated and experimental $4f$ energy levels of the Ce^{3+} ions in the YAM crystal[91]

Level No.	^2L_J	E_{calc}	E_{exp}
1	$^2F_{5/2}$	0	0
2		334	—
3		722	—
4	$^2F_{7/2}$	2107	2085
5		2432	2485
6		2772	2699
7		3209	3257

Comparison of the calculated and observed energy levels (Table 8.11) yields good agreement between the two data sets, which confirms the validity and applicability of the ECM to the calculations of the 4*f* ions energy levels in solids. More examples of calculations of the 4*f* and 5*f* ions energy levels, transition intensities, paramagnetic susceptibilities, etc., can be found in other references.[21,91–113]

8.10 Fundamental Difference in the Energy Levels Formation of the 3*d* and 4*f* Ions in Crystals

An attentive reader might have noticed that different basis sets are used to calculate the crystal field splittings of the 3*d* and 4*f* ions. In the former case, the $|LM\rangle$ functions form the basis set, whereas in the latter case the $|JM\rangle$ functions are used to diagonalize the crystal field Hamiltonian. The physical foundations for such choice of the basis sets become after consideration of Figs. 8.19 and 8.20.

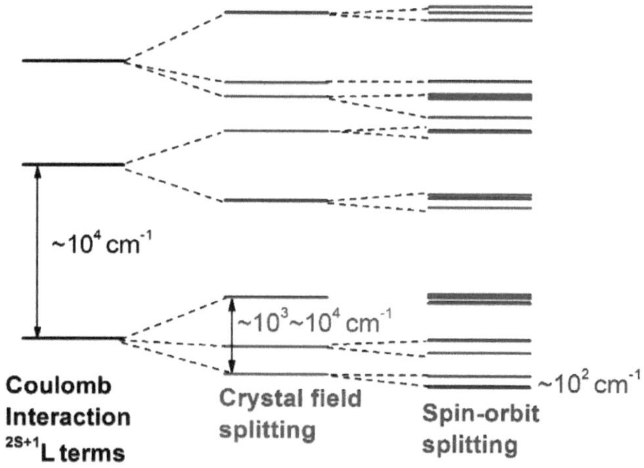

Figure 8.19 The general scheme of formation of energy levels of the 3*d* ions.

An initial step in the energy levels formation is the same in both cases: The Coulomb interaction among the electrons in both *d*-and *f*-shells unfilled shells produces the ^{2S+1}L terms, with the energy separation between the nearest terms of the order of 10^4 cm^{-1}. Since the 3*d* ions are located in the first half of the Periodic table, the SO splitting (or the fine structure) of these terms is not very large. The crystal field can split the free ion's ^{2S+1}L terms into a number of sublevels, and the overall magnitude of each ^{2S+1}L term splitting can reach 10^3–10^4 cm^{-1}, which is comparable with the energy separation between the free ion terms. And only then one can take into account the SO interaction, which can split the crystal field levels into not more than $2S+1$ states (if the number of electrons in the d shell is even) or $S+1/2$ states (if the number of electrons in the d shell is odd) with the overall magnitude of this splitting to be of ~10^2 cm^{-1} only (Fig. 8.19).

However, in the case of the $4f$ ions (the $5f$ ions also follow the same pattern) the SO interaction is much stronger due to considerably greater atomic numbers of these elements. Then it comes immediately after the Coulomb interaction to produce the $^{2S+1}L_J$ manifolds, with the overall SO splitting of each free ion ^{2S+1}L term to be about 10^3–10^4 cm^{-1}. Afterwards, the crystal field produces small crystal field splittings of the $^{2S+1}L_J$ manifolds, of about ~10^2 cm^{-1} only (Fig. 8.20).

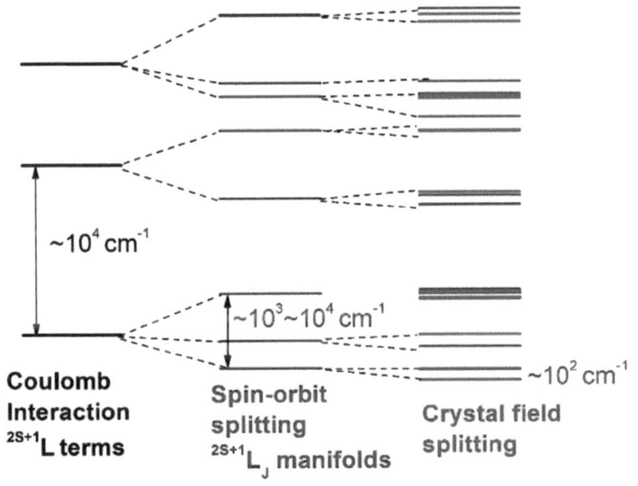

Figure 8.20 The general scheme of formation of energy levels of the $4f$ ions.

To sum up, the order of SO and crystal field interactions for the $3d$ and $4f$ ions are interchanged. The examples of the $3d$ and $4f$ ions energy levels calculations given in this chapter confirm this relation between the magnitudes of these interactions.

Bibliography

1. Davies JJ (1974). EPR and ENDOR of titanium-doped fluoride, *J Phys C: Solid State Phys* **7** (3) 599–609.
2. Clementi E, Roetti C (1974). Roothaan-Hartree-Fock atomic wavefunctions: Basis functions and their coefficients for the ground and excited states of neutral and ionized atoms, $Z \leq 54$, *Atom Data Nucl Data Tables* **14** (3–4) 177–478.
3. Wu SY, Gao XY, Yan WZ, Wei WH (2004). Theoretical studies of the local structure for the trigonal Ti^{3+} center in LiF cristal, *Spectrochim Acta A* **60**(11) 2531–2535.
4. Tanabe Y, Sugano S (1954). On the absorption spectra of complex ions. I, *J Phys Soc Jpn* **9**(5), 753–766.
5. Tanabe Y, Sugano S (1954). On the absorption spectra of complex ions. II, *J Phys Soc Jpn* **9**(5), 766–779.
6. Tanabe Y, Sugano S (1956). On the absorption spectra of complex ions. III the calculation of the crystalline field strength, *J Phys Soc Jpn* **11**(8), 864–877.
7. Tanabe Y, Kamimura H (1958). On the absorption spectra of complex ions IV. The effect of the spin–orbit interaction and the field of lower symmetry on *d* electrons in cubic field, *J Phys Soc Jpn* **13**(4), 394–411.
8. Sugano S, Tanabe Y, Kamimura H (1970). *Multiplets of Transition-Metal Ions in Crystals*, New York, Acad. Press.
9. Rudowicz C, Brik MG, Avram NM, Yeung YY, Gnutek P (2006). Crystal field analysis of the energy level structure of $Cs_2NaAlF_6{:}Cr^{3+}$, *J Phys: Condens Matter* **18**(22) 5221–5234.
10. Brik MG, Srivastava AM (2013). Electronic energy levels of the Mn^{4+} ion in the perovskite, $CaZrO_3$, *ECS J Sol State Sci Technol* **2**(7) R148–R152.
11. Brik MG, Srivastava AM, Avram NM, Suchocki A (2014). Empirical relation between covalence and the energy position of the Ni^{2+} ^1E state in octahedral complexes, *J Lumin* **148** 338–341.
12. Brik MG, Camardello SJ, Srivastava AM (2015). Influence of covalency on the Mn^{4+} $^2E_g \rightarrow ^4A_{2g}$ emission energy in crystals, *ECS J Sol State Sci Technol* **4**(3) R39–R43.
13. Brik MG, Camardello SJ, Srivastava AM, Avram NM, Suchocki A (2016). Spin-forbidden transitions in the spectra of transition metal ions and nephelauxetic effect, *ECS J Sol State Sci Technol* **5**(1) R3067–R3077.

14. Ma CG, Wang Y, Liu DX et al. (2018). Origin of the β_1 parameter describing the nephelauxetic effect in transition metal ions with spin-forbidden emissions, *J Lumin* **197** 142–146.

15. Rudowicz C (1987). Concept of spin Hamiltonian, forms of zero-field splitting and electronic Zeeman Hamiltonians and relations between parameters used in EPR. A critical review, *Magn Res Rev* **13** 1–89.

16. Rudowicz C (1988). Corrigendum to: Concept of spin Hamiltonian, forms of zero-field splitting and electronic Zeeman Hamiltonians and relations between parameters used in EPR. A critical review, *Magn Res Rev* **13** 335.

17. Rudowicz C, Misra SK (2001). Spin-Hamiltonian formalisms in electron magnetic resonance (EMR) and related spectroscopies, *App. Spectrosc Rev* **36**(1) 11–63.

18. Stevens KWH (1952). Matrix elements and operator equivalents connected with the magnetic properties of rare earth ions. *Proc Phys Soc A* **65** 209–215.

19. Rudowicz C (1985). Transformation relations for the conventional O_k^q and normalized O'^q_k Stevens operator equivalents with k = 1 to 6 and $-k \leq q \leq k$, *J Phys C: Solid State Phys* **18**(7) 1415–1430.

20. Rudowicz C (1985). Corrigendum to: Transformation relations for the conventional O_k^q and normalized O'^q_k Stevens operator equivalents with k = 1 to 6 and $-k \leq q \leq k$, *J Phys C: Solid State Phys* **18**(19) 3837.

21. Malkin BZ (1987). Crystal field and electron-phonon interaction in rare-earth ionic paramagnets, in: *Spectroscopy of Solids Containing Rare-Earth Ions*. Kaplyanskii AA, Macfarlane BM, eds, Amsterdam: North Holland, pp. 13–50.

22. Abragam AG, Bleaney B (1970). *Electron Paramagnetic Resonance of Transition Ions*. Oxford, Clarendon, Chapter 7.

23. Brik MG, Avram NM (2009). Numerical calculations of the overlap integrals between the wave functions of 3d ions and ligands, *AIP Conf Proc*, **1131** 86–91.

24. Whittaker ET, Watson GN (1990). *A Course in Modern Analysis*, 4th ed., Cambridge, England, Cambridge University Press.

25. Nielsen CW, Koster GF (1963). *Spectroscopic Coefficients for the p^n, d^n, and f^n Configurations*, MIT Press, Cambridge, Massachusetts.

26. Petričević V, Gayen SK, Alfano RR, Yamagishi K, Anzai H, Yamaguchi Y (1988). Laser action in chromium-doped forsterite, *Appl Phys Lett* **52**(13) 1040–1042.

27. Petričević V, Gayen SK, Alfano RR (1988). Laser action in chromium-activated forsterite for near-infrared excitation: Is Cr^{4+} the lasing ion? *Appl Phys Lett* **53**(26) 2590–2592.

28. Angert NB, Borodin NI, Garmash VM, Zhitnyuk VA, Okhrimchuk AG, Siyuchenko OG, Shestakov AV (1988). Lasing due to impurity color centers in yttrium aluminum garnet crystals at wavelengths in the range 1.35–1.45 μm, *Sov J Quant Electron* **18**(1) 73–74.

29. Eilers H, Dennis WM, Yen WM, Kück S, Petermann K, Huber G, Jia W (1993). Performance of a Cr:YAG laser, *IEEE J Quantum Electron* 29(9) 2508–2512.

30. Deka C, Bass M, Chai BHT, Shimony Y (1993). Optical spectroscopy of Cr^{4+}:Y_2SiO_5, *J Opt Soc Am B* **10**(9) 1499–1507.

31. Avanesov AG, Denker BI, Galagan BI, Osiko VV, Shestakov AV, Sverchkov SE (1994). Temperature stimulated emission from Cr(IV)-activated yttrium orthosilicate, *Kvantovaya Elektronika* **21**(3) 216–216.

32. Sennaroglu A, Pollock CR, Nathel H (1995). Efficient continuous-wave chromium-doped YAG laser, *J Opt Soc Am B* **12**(5) 930–937.

33. Avanesov AG, Dvornikova VG, Zhorin VV, Kaneeva AF, Lebedev VA, Pisarenko VF, Shestakov AV (1995). Monocrystalline silicates as a new class of materials for tunable lasers, *Izv Akad Nauk Ser Fiz* **59**(6) 10–16.

34. Kück S, Petermann K, Pohlmann U, Huber G (1996). Electronic and vibronic transitions of the Cr^{4+}-doped garnets $Lu_3Al_5O_{12}$, $Y_3Al_5O_{12}$, $Y_3Ga_5O_{12}$ and $Gd_3Ga_5O_{12}$, *J Lumin* **68**(1) 1–14.

35. Avanesov AG, Lebedev VA, Zhorin VV, Okhrimchuk AG, Shestakov AV (1997). Silicate crystals for infra-red lasers, *J Lumin* **72–74** 155–156.

36. Brik MG, Zhorin VV (1997). Energy levels of Cr^{4+} in YAG. *J Lumin* **72–74** 149–151.

37. Brik MG, Shchekoldin DG (1998). Electronic structure of Cr^{4+}:$Y_3Al_5O_{12}$. *Opt Spectr* **84**(5) 683–684.

38. Okhrimchuk AG, Shestakov AV (2000). Absorption saturation mechanism for YAG:Cr^{4+} crystals, *Phys Rev B* **61**(2) 988–995.

39. Kück S (2001). Laser-related spectroscopy of ion-doped crystals for tunable solid-state lasers, *Appl Phys B* **72** 515–562.

40. Ishii T, Ogasawara K, Adachi H, Tanaka I (2001). Theoretical calculation for the multiplet structure of the tetrahedrally coordinated Cr^{4+} in $Y_3Al_5O_{12}$, *J Chem Phys* **115**(1) 492–508.

41. Ishii T, Fujimura K, Ogasawara K, Adachi H, Tanaka I (2001). Theoretical calculation for the multiplet structures of tetrahedrally coordinated Cr^{4+} in silicate crystals, *J Phys: Condens Matter* **13**(25) 5757–5784.

42. Brik MG, Avram NM, Avram CN, Tanaka I (2001). Effects of the spin-triplet states mixture and electron-phonon coupling in $Y_3Al_5O_{12}$:Cr^{4+}, *Eur Phys J: Appl Phys* **29**(3) 239–245.

43. Brik MG, Ogasawara K (2007). Comparative study of the absorption spectrum of Li_2CaSiO_4:Cr^{4+}: First-principles fully relativistic and crystal field calculations, *Opt Mater* **30**(3) 399–406.

44. Sharonov My, Bykov AB, Owen S, Petričević V, Alfano RR, Beall GH, Borrelli N (2004). Spectroscopic study of transparent forsterite nanocrystalline glass-ceramics doped with chromium, *J Opt Soc Am B* **21**(11) 2046–2052.

45. Li C, Xu JL, Liu W, Zheng DY, Zhang SL, Zhang Y, Lin H, Liu L, Liu JH, Zeng FM (2015). Synthesis and characterization of Cr^{4+}-doped Ca_2GeO_4 tunable crystal, *J Alloys Compd* **636** 211–215.

46. Sato Y, Taira T (2017). Model for the polarization dependence of the saturable absorption in Cr^{4+}:YAG, *Opt Mater Express* **7**(2) 577–586.

47. Slavkina VV, Mitrokhin VP, Subbotin KA, Lis DA, Lis ON, Zharikov EV (2017). First tunable and mode-locked laser action at Cr^4+ and Sc^{3+} co-doped forsterite Cr,Sc:Mg_2SiO_4 crystals. *Conference on Lasers and Electro-Optics Europe / European Quantum Electronics Conference* (CLEO/Europe-EQEC), Munich, Germany, June 25–29.

48. Sanina VV, Mitrokhin VP, Subbotin KA, Lis DA, Lis ON, Ivanov AA, Zharikov EV (2018). Tunable and mode-locked laser action of Cr^{4+} in codoped forsterite Cr, Sc:Mg_2SiO_4, *Laser Phys* **28**(1) 015803.

49. Asami K, Ueda J, Kitaura M, Tanabe S (2018). Investigation of luminescence quenching and persistent luminescence in Ce^{3+} doped $(Gd,Y)_3(Al,Ga)_5O_{12}$ garnet using vacuum referred binding energy diagram, *J Lumin* **198** 418–426.

50. Xu J, Ueda J, Tanabe S (2017). Toward tunable and bright deep-red persistent luminescence of Cr^{3+} in garnets, *J Am Ceram Soc* **100**(9) 4033–4044.

51. Ueda J, Katayama M, Asama K, Xu J, Inada Y, Tanabe S (2017). Evidence of valence state change of Ce^{3+} and Cr^{3+} during UV charging process in $Y_3Al_2Ga_3O_{12}$ persistent phosphors, *Opt Mater Express* **7**(7) 2471–2476.

52. Nakatsuka A, Yoshiasa A, Yamanaka T (1999). Cation distribution and crystal chemistry of $Y_3Al_{5-x}Ga_xO_{12}$ ($0 \le x \le 5$) garnet solid solutions, *Acta Cryst B* **55**(3) 266–272.

53. Shannon RD (1976). Revised effective ionic radii and systematic studies of interatomic distances in halides and chalcogenides, *Acta Cryst A* **32** 751–767.

54. Momma K, Izumi F (2011). VESTA 3 for three-dimensional visualization of crystal, volumetric and morphology data, *J Appl Crystallogr* **44** 1272–1276.

55. Eilers H, Hömmerich U, Jacobsen SM, Yen WM, Hoffman KR, Jia W (1994). Spectroscopy and dynamics of

Cr^{4+}:$Y_3Al_5O1_{12}$, *Phys Rev B* **49**(22) 15505–15513.

56. Koster GF, Dimmock JO, Wheeler RG, Statz H (1963). *Properties of the Thirty-Two Point Groups.* M.I.T. Press, Cambridge, Massachusetts.

57. Sobelman II (1992). *Atomic Spectra and Radiative Transitions.* Springer series on atoms + plasmas, 12.

58. Brik MG, Avram NM (eds) (2013). *Optical Properties of 3d-Ions in Crystals: Spectroscopy and Crystal Field Analysis.* Springer and Tsinghua University Press.

59. Jousseaume C, Vivien D, Kahn-Harare A, Malkin BZ (2002). Long-lifetime fluorescence and crystal field calculation on Cr^{4+}-doped Li_2MSiO_4, M-Mg, Zn, *Opt Mater* **24**(1–2) 143–150.

60. Brik MG, Avram NM, Avram CN (2004). Crystal field analysis of energy level structure of the Cr_2O_3 antiferromagnet, *Solid State Commun* **132**(12) 831–835.

61. El-Korashy A, Brik MG (2005) Crystal growth, spectroscopic and crystal field studies of $[N(CH_3)_4]_2MnCl_4$ and $[N(CH_3)_4]_2CoCl_4$ single crystals in the paraelectric phase, *Solid State Commun* **135**(5) 298–303.

62. Brik MG, Avram NM, Avram CN (2006). Comparative crystal field study of Ni^{2+} energy levels in $NiCl_2$, $NiBr_2$, and NiI_2 crystals, *Physica B* **371**(1) 43–49.

63. Brik MG (2006). Crystal field analysis of the absorption spectra and electron-phonon interaction in $Ca_3Sc_2Ge_3O_{12}$:Ni^{2+}, *J Phys Chem Solids* **67**(4) 738–744.

64. Rudowicz C, Brik MG, Avram NM, Yeung YY, Gnutek P (2006). Crystal field analysis of the energy level structure of Cs_2NaAlF_6:Cr^{3+}, *J Phys: Condens Matter* **18**(22) 5221–5234.

65. Brik MG, Yeung YY (2008). Semi-ab initio calculations of superposition model and crystal field parameters for Co^{2+} ions using the exchange charge model, *J Phys Chem Solids* **69**(10) 2401–2410.

66. Brik MG, Avram NM (2009). Microscopic analysis of the crystal field strength and electron-vibrational interaction in cubic $SrTiO_3$ doped with Cr^{3+}, Mn^{4+} and Fe^{5+} ions, *J Phys: Condens Matter* **21**(15) 155502.

67. Krsmanović RM, Antić Z, Mitrić M, Dramićanin MD, Brik MG (2011). Structural, spectroscopic and crystal field analyses of Ni^{2+} and Co^{2+} doped Zn_2SiO_4, *Appl Phys A* **104**(1) 483–492.

68. Brik MG, Avram CN (2011). Exchange charge model and analysis of the microscopic crystal field effects in $KAl(MoO_4)_2$:Cr^{3+}, *J Lumin* **131**(12) 2642–2645.

69. Brik MG, Srivastava AM, Avram NM (2011). Comparative analysis of crystal field effects and optical spectroscopy of six-coordinated Mn^{4+} ion in the $Y_2Ti_2O_7$ and $Y_2Sn_2O_7$ pyrochlores, *Opt Mater* **33**(11) 1671–1676.

70. Srivastava AM, Brik MG (2012). Ab initio and crystal field studies of the Mn^{4+}-doped Ba_2LaNbO_6 double-perovskite, *J Lumin* **132**(3) 579–584.

71. Brik MG, Srivastava AM (2013). On the optical properties of the Mn^{4+} ion in solids. *J Lumin* **133** 69–72.

72. Brik MG, Srivastava AM (2013). Comparative crystal field analysis of energy level schemes and nephelauxetic effect for Cr^{4+}, Cr^{3+}, and Mn^{4+} ions in $Y_2Sn_2O_7$ pyrochlore, *Opt Mater* **35**(6) 1251–1256.

73. Peng MY, Yin XW, Tanner PA, Brik MG, Li PF (2015). Site occupancy preference, enhancement mechanism, and thermal resistance of Mn^{4+} red luminescence in $Sr_4Al_{14}O_{25}$:Mn^{4+} for warm WLEDs, *Chem Mater* **27**(8) 2938–2945.

74. Anghel S, Brik MG, Ma CG, Sushkevich K, Kulyuk L (2015). On the excitation spectra of Cr^{3+}/Cr^{2+} and V^{3+} co-doped $ZnAl_2S_4$ single crystals, *J Lumin* **166** 282–288.

75. Medić MM, Brik MG, Dražić G, Antić Ž, Lojpur VM, Dramićanin MD (2015). Deep-red emitting Mn^{4+} doped Mg_2TiO_4 nanoparticles, *J Phys Chem C* **119**(1) 724–730.

76. Brik MG, Srivastava AM (2016). A computation study of site occupancy in the commercial $Mg_{28}Ge_{7.55}F_{15.04}$:$Mn^{4+}$ phosphor, *Opt Mater* **54** 245–251.

77. Brik MG, Papan J, Jovanović DJ, Dramićanin MD (2016). Luminescence of Cr^{3+} ions in $ZnAl_2O_4$ and $MgAl_2O_4$ spinels: correlation between experimental spectroscopic studies and crystal field calculations, *J Lumin* **177** 145–151.

78. Li PF, Brik MG, Li LJ, Han J, Li XM, Peng MY (2016). Prediction on Mn^{4+}-doped germanate phosphor by crystal field calculation on basis of

exchange charge model: A case study of $K_2Ge_4O_9$:Mn^{4+}, *J Am Ceram Soc* **99**(7) 2388–2394.

79. Srivastava AM, Brik MG (2017). The nature of Mn^{4+} luminescence in the orthorhombic perovskite, $GdAlO_3$, *Opt Mater* **63** 207–212.
80. Wu WL, Fang MH, Zhou WL, Lesniewski T, Mahlik S, Grinberg M, Brik MG, Sheu HS, Cheng BM, Wang J, Liu RS (2017). High color rendering index of Rb_2GeF_6:Mn^{4+} for light-emitting diodes, *Chem Mater* **29**(3) 935–939.
81. Đordević V, Brik MG, Srivastava AM, Medić M, Vulić P, Glais E, Viana B, Dramićanin MD (2017). Luminescence of Mn^{4+} ions in $CaTiO_3$ and $MgTiO_3$ perovskites: Relationship of experimental spectroscopic data and crystal field calculations, *Opt Mater* **74** 46–51.
82. Zhu YW, Chen DQ, Huang L, Liu Y, Brik MG, Zhong JS, Wang J. (2018). Phase-transition-induced giant enhancement of red emission in Mn^{4+}-doped fluoride elpasolite phosphors, *J Mater Chem C* **6**(15) 3951–3960.
83. Radtke AS, Brown GE (1974). Frankdicksonite, BaF_2, a new mineral from Nevada, *Am Mineral* **59**, 885–888.
84. Visser R, Dorenbos P, van Ejk CWE, Hollander RW, Schotanus P (1991). Scintillation properties of Ce^{3+} doped BaF_2 crystals, *IEEE Trans Nucl Sci* **38**(2) 178–183.
85. Wyckoff RWG (1963). *Crystal Structures*, Interscience Publishers, New York, New York.
86. Abraham A, Bleaney B (1970). *Electron Paramagnetic Resonance of Transition Ions*. Oxford, England: Oxford University Press.
87. Newman D, Ng B (2000). *Crystal Field Handbook*, Cambridge University Press.
88. Liu GK (2005). *Spectroscopic Properties of Rare Earth in Optical Materials*, in Liu GK, Jacquier B, eds, Springer, New York.
89. Leushin AM (1968). Tables of the functions transforming according to the irreducible representations of the crystallographic point groups (in Russian: Таблицы функций, преобразующихся по неприводимым представлениям кристаллографических точечных групп), Moscow, Nauka.
90. Carnall WT, Goodman GL, Rajnak K, Rana RS (1988). A systematic analysis of the spectra of the lanthanides doped into single crystal LaF_3, *J Chem Phys* **90**(7) 3443–3457.
91. Wang YJ, Włodarczyk D, Li LA et al. (2017). Electronic structure of Ce^{3+} in yttrium and lutetium orthoaluminate crystals and single crystal layers, *J Alloys Compds* **723** 157–163.

92. Ma CG, Trevisani M, Piccinelli F, Ivanovskikh KV, Bettinelli M, Brik MG (2013). Analysis of vacuum ultraviolet electronic spectra of Ce^{3+} and Pr^{3+} ions in $Ca_9Lu(PO_4)_7$: Crystal-field calculations and simulation of optical spectra, *J Phys: Condens Matter* **25**(16) 165503.
93. Przybylińska H, Ma CG, Brik MG, et al. (2013). Evidence of multicenter structure of cerium ions in gadolinium gallium garnet crystals studied by infrared absorption spectroscopy, *Phys Rev B* **87** 045114.
94. Avanesov AG, Zhorin VV, Malkin BZ, Pisarenko VF (1992). Crystal field in heterodesmic compounds, *Fizika Tverdogo Tela* **34**(9) 2899–2906.
95. Popova MN, Klimin SA, Malkin BZ, Kasatkina LA, Cao G, Crow J (1996). Crystal field and spectrum of Pr^{4+} in $BaPrO_3$, *Phys Lett A* **223**(4) 308–312.
96. Macfarlane RM, Meltzer RS, Malkin BZ (1998). Optical measurement of the isotope shifts and hyperfine and superhyperfine interactions of Nd in the solid state, *Phys Rev B* **58**(9) 5692–5700.
97. Zhorin VV, Liu GK (1998). Modelling crystal-field interactions for *f*-elements in $LaCl_3$, *J Alloys Compds* **275** 137–141.
98. Liu GK, Li ST, Zhorin VV, Loong CK, Abraham MM, Boatner LA (1998). Crystal-field splitting, magnetic interaction, and vibronic excitations of $Cm_{244}{}^{3+}$ inYPO_4 and $LuPO_4$, *J Chem Phys* **109**(16) 6800–6808.
99. Liu GK, Zhorin VV, Li ST, Beitz JV (2000). Crystal field analysis and Monte Carlo simulation of lattice disordering for Cm^{3+} inYPO_4 and $LuPO_4$, *J Chem Phys* **112**(1) 373–382.
100. Liu GK, Zhorin VV, Antonio MR, Li ST, Williams CW, Soderholm L (2000). Studies of local structure of Cm^{3+} in borosilicate glass using laser and x-ray spectroscopic methods and computational modeling, *J Chem Phys* **112**(3) 1489–1496.
101. Malkin BZ, Leushin AM, Iskhakova AI et al. (2000). EPR and optical spectra of Yb^{3+} in $CsCdBr_3$: Charge-transfer effects on the energy-level structure of Yb^{3+} in the symmetrical pair centers, *Phys Rev B* **62**(11) 7063–7070.
102. Malkin BZ, Zakirov AR, Popova MN, et al. (2004). Optical spectroscopy of $Yb_2Ti_2O_7$ and $Y_2Ti_2O_7$: Yb^{3+} and crystal-field parameters in rare-earth titanate pyrochlores, *Phys Rev B* **70**(7) 075112.
103. Lago J, Zivkovic I, Malkin BZ, et al. (2010). $CdEr_2Se_4$: A New Erbium Spin Ice System in a Spinel Structure, *Phys Rev Lett* **104**(24) 247203.
104. Malkin BZ, Lummen TTA, van Loosdrecht, Dhalenne G, Zakirov AR (2010). Static magnetic susceptibility, crystal field and exchange interactions in rare earth titanate pyrochlores, *J Phys: Condens Matter* **22**(27) 276003.

105. Malkin BZ, Pytalev DS, Popova MN, Baibekov EI, Falin ML, Gerasimov KI, Khaidukov NM (2012). Random lattice deformations in rare-earth-doped cubic hexafluoroelpasolites: High-resolution optical spectroscopy and theoretical studies, *Phys Rev B* **86**(13) 134110.

106. Klekovkina VV, Malkin BZ (2014). Crystal field and magnetoelastic interactions in $Tb_2Ti_2O_7$, *Opt Spectr* **116**(6) 849–857.

107. Malkin BZ, Nikitin SI, Mumdzhi IE et al. (2015). Magnetic and spectral properties of the multisublattice oxides SrY_2O_4:Er^{3+} and $SrEr_2O_4$, *Phys Rev B* **92**(9) 094415.

108. Malkin BZ, Abishev NM, Baibekov EI, Pytalev DS, Boldyrev KN, Popova MN, Bettinelli M (2017). Distribution function of random strains in an elastically anisotropic continuum and defect strengths of Tm^{3+} impurity ions in crystals with zircon structure, *Phys Rev B* **96**(1) 014116.

109. Ma CG, Brik MG, Ryba-Romanowski W, Swart HC, Gusowski MA (2012). Spectroscopy and calculations for $4f^N$ -> $4f^{N-1}5d$ transitions of lanthanide ions in K_3YF_6, *J Phys Chem A* **116**(36) 9158–9180.

110. Ma CG, Brik MG (2013). First-principles calculations of structural and electronic properties of pure and Tm^{2+}-doped $SrCl_2$, *Phys Stat Solidi A* **250**(4) 858–863.

111. Przybylińska H, Ma CG, Brik MG et al. (2013). Electronic structure of Ce^{3+} multicenters in yttrium aluminum garnets, *Appl Phys Lett B* **102**(24) 241112.

112. Syvorotka II, Sugak D, Wierzbicka A, et al. (2015). Optical properties of pure and Ce^{3+} doped gadolinium gallium garnet crystals and epitaxial la, *J Lumin* **164** 31–37.

113. Wittlin A, Przybylińska H, Berkowski M, et al. (2015). Ambient and high pressure spectroscopy of Ce^{3+} doped yttrium gallium garnet, *Opt Mater Express* **5**(8) 1868–1880.

114. Bersuker IB (2010). *Electronic Structure and Properties of Transition Metal Compounds: Introduction to the Theory*. 2nd edition, John Wiley & Sons.

115. Brik MG, Avram NM (eds) (2013). *Optical Properties of 3d-Ions in Crystals: Spectroscopy and Crystal Field Analysis*. Springer and Tsinghua University Press.

116. Cohen-Tannoudji C, Diu B, Laloe F (1977). *Quantum Mechanics*. John-Wiley, Paris.

117. Cowan RD (1981). *The Theory of Atomic Structure and Spectra*. University of California Press, Berkeley.

118. Edmonds AR (1957). *Angular Momentum in Quantum Mechanics.* Princeton University Press.

119. Figgis BN, Hitchman MA (2000). *Ligand Field Theory and Its Applications.* Wiley-VCH.

120. Griffiths DJ (1995). *Introduction to Quantum Mechanics.* Prentice-Hall, New Jersey.

121. Henderson B, Imbush GF (1989 or 2006). *Optical Spectroscopy of Inorganic Solids.* Clarendon Press.

122. Henderson B, Bartram RH (2000). *Crystal-Field Engineering of Solid-State Laser Materials.* Cambridge University Press.

123. Landau LD, Lifshitz EM (1977). *Quantum Mechanics.* Pergamon Press, Oxford.

124. Powell RC (1998). *Physics of Solid-State Laser Materials.* Springer.

Chapter 9

Electron-Vibrational Interaction and Its Manifestation in the Experimental Absorption and Emission Spectra of Impurity Ions in Crystals

9.1 Introduction

This chapter will be concerned with the simplified models for the analysis of the electronic and optical properties of impurity ions is solids. It will start with the necessary explanations of why certain simplifications are strongly needed and will be continued with their subsequent introduction and description. The most important effects of interaction of the impurity centers with the crystal lattice vibrations and their manifestation in the optical spectra will be explained as well.

9.2 Adiabatic Approximation and Configurational Coordinate Model

The Schrödinger equation can be solved exactly only for the simplest systems, such as the hydrogen-like atoms. In the case of the multielectron atoms, the Coulomb interaction between the electrons

Theoretical Spectroscopy of Transition Metal and Rare Earth Ions: From Free State to Crystal Field
Mikhail G. Brik and Chong-Geng Ma

ISBN 978-981-4800-56-3 (Hardcover), 978-0-429-27875-4 (eBook)
www.jennystanford.com

is not spherical, and an effective potential should be introduced to model this non-spherical part of the Coulomb interaction. The choice of such a potential is an approximation already.

The situation becomes more complicated in molecules and in crystals, since the number of particles—both nuclei and electrons—increases dramatically. In one mole of a solid, there are $\sim 10^{23}$ atoms, and each atom has many electrons. Therefore, the number of equations to be solved for a complete description of a solid is beyond even the best computational facilities available nowadays. To overcome these difficulties, the so-called adiabatic (or Born–Oppenheimer) approximation has been developed. Its main ideas are described below.

Let us consider a molecule, which consists of N nuclei with masses M_α, electrical charges Z_α, coordinates R_α (α = 1...N) and n electrons with coordinates r_i (i = 1...n). In the general case, the nuclei can be different. The total Hamiltonian of this molecule is as follows:

$$\begin{aligned} H = -\frac{\hbar^2}{2m}\sum_{i=1}^{n}\Delta_i - \frac{\hbar^2}{2}\sum_{\alpha=1}^{N}\frac{\Delta_\alpha}{M_\alpha} + \frac{1}{2}\sum_{i=1}^{n}\sum_{j=1}^{n}\frac{e^2}{|r_i - r_j|} \\ + \frac{1}{2}\sum_{\alpha=1}^{N}\sum_{\beta=1}^{N}\frac{Z_\alpha Z_\beta e^2}{|R_\alpha - R_\beta|} - \sum_{\alpha=1}^{N}\sum_{i=1}^{n}\frac{Z_\alpha e^2}{|R_\alpha - r_i|} \end{aligned}. \quad (9.1)$$

The first term describes the kinetic energy of n electrons (m is the electron's mass), the second term is the kinetic energy of N nuclei, the third and the fourth terms are the potential energies of the Coulomb repulsion between all electrons and nuclei, respectively (a pre-factor 1/2 is needed in order not to take twice into account interaction between each pair of electrons or nuclei), and the fifth term describes the potential energy of the Coulomb attraction between electrons and nuclei. For the sake of brevity, we introduce a notation

$$V(r,R) = \frac{1}{2}\sum_{i=1}^{n}\sum_{j=1}^{n}\frac{e^2}{|r_i - r_j|} + \frac{1}{2}\sum_{\alpha=1}^{N}\sum_{\beta=1}^{N}\frac{Z_\alpha Z_\beta e^2}{|R_\alpha - R_\beta|} - \sum_{\alpha=1}^{N}\sum_{i=1}^{n}\frac{Z_\alpha e^2}{|R_\alpha - r_i|}. \quad (9.2)$$

The stationary Schrödinger equation for this molecule will be written as follows:

$$\left\{-\frac{\hbar^2}{2m}\sum_{i=1}^{n}\Delta_i - \frac{\hbar^2}{2}\sum_{\alpha=1}^{N}\frac{\Delta_\alpha}{M_\alpha} + V(r,R)\right\}\Psi(r,R) = E\Psi(r,R), \quad (9.3)$$

where E is the energy eigenvalue, and $\Psi(r, R)$ is the corresponding wave function. Equation (9.3) can be simplified considerably, if the $\Psi(r, R)$ wave function will be represented as a product of two terms, one of which depends on the nuclear coordinates R only, and another one depends on both electron and nuclear coordinates r and R:

$$\Psi(r, R) = \phi(R)\Phi(r, R). \tag{9.4}$$

Such assumption is based on the firm physical reasons: Due to a large difference in masses of electrons and nuclei, the latter have much lower velocity, and very light electrons would move much faster than nuclei. It is possible to consider the motion of electrons for some fixed (or "frozen") configuration of nuclei, which corresponds to the equilibrium positions of atoms in the considered molecule or a solid. The wave function of this "fast" electronic subsystem $\Phi(r, R)$ will depend on the nuclear coordinates as on parameters, and the nuclear wave function $\phi(R)$ will describe the states of the "slow" nuclear subsystem. To find out, which consequences will be caused by this assumption, we substitute the wave function (9.4) into Eq. (9.3):

$$-\frac{\hbar^2}{2m}\phi(R)\sum_{i=1}^{n}\Delta_i\Phi(r,R)-\frac{\hbar^2}{2}\sum_{\alpha=1}^{N}\frac{\Delta_\alpha\left(\phi(R)\Phi(r,R)\right)}{M_\alpha}+V(r,R)\phi(R)\Phi(r,R) \cdot$$
$$= E\,\phi(R)\Phi(r,R) \tag{9.5.}$$

Application of the Δ_α operator to the product $\phi(R)\Phi(r,R)$ yields:

$$\begin{aligned}\Delta_\alpha\left(\phi(R)\Phi(r,R)\right) &= \nabla_\alpha\left[\nabla_\alpha\left(\phi(R)\Phi(r,R)\right)\right]\\ &= \nabla_\alpha\left[\phi(R)\nabla_\alpha\Phi(r,R)+\Phi(r,R)\nabla_\alpha\phi(R)\right]\\ &= \Phi(r,R)\Delta_\alpha\phi(R)+\phi(R)\Delta_\alpha\Phi(r,R)+2\nabla_\alpha\Phi(r,R)\nabla_\alpha\phi(R).\end{aligned} \tag{9.6}$$

The second and the third terms in the right-hand side of Eq. (9.6) contain derivatives of the electron wave functions $\Phi(r,R)$ with respect to the nuclear coordinates. These derivatives are very small, because electrons move so fast that the motion of nuclei almost does not affect them. Therefore, these terms can be dropped. The following operator defined as

$$L_{na} = -\frac{\hbar^2}{2}\sum_{\alpha}\frac{1}{M_\alpha}\left\{\phi(R)\Delta_\alpha\Phi(r,R)+2\nabla_\alpha\Phi(r,R)\nabla_\alpha\phi(R)\right\} \tag{9.7}$$

is called the non-adiabatic operator. After neglecting these terms in Eq. (9.5) and dividing both parts by $\phi(R)\Phi(r,R)$, we have

$$-\frac{\hbar^2}{2m}\frac{1}{\Phi(r,R)}\sum_{i=1}^{n}\Delta_i\Phi(r,R)-\frac{\hbar^2}{2}\frac{1}{\phi(R)}\sum_{\alpha=1}^{N}\frac{\Delta_\alpha\phi(R)}{M_\alpha}+V(r,R)=E\,. \quad (9.8)$$

Now it is possible to collect in one side of this equation all terms depending on R only, and in another side all terms depending on both R and r:

$$-\frac{\hbar^2}{2m}\frac{1}{\Phi(r,R)}\sum_{i=1}^{n}\Delta_i\Phi(r,R)+V(r,R)=\frac{\hbar^2}{2}\frac{1}{\phi(R)}\sum_{\alpha=1}^{N}\frac{\Delta_\alpha\phi(R)}{M_\alpha}+E\,. \quad (9.9)$$

Since the last equation is an identity, it can hold true if and only if both sides are equal to the same function $W(R)$ (note that such a procedure is very similar to the separation of variables in the case of the Schrödinger equation for the hydrogen atom. However, there is a very important difference: In the present case, it is not possible to separate the variables completely, since the nuclear coordinates R enter both parts). Equating both parts of Eq. (9.9) to $W(R)$, we get two following equations:

$$\begin{aligned}&-\frac{\hbar^2}{2m}\sum_{i=1}^{n}\Delta_i\Phi(r,R)+V(r,R)\Phi(r,R)=W(R)\Phi(r,R),\\&-\frac{\hbar^2}{2}\sum_{\alpha=1}^{N}\frac{\Delta_\alpha\phi(R)}{M_\alpha}+W(R)\phi(R)=E\phi(R).\end{aligned} \quad (9.10)$$

An essential feature of these equations is that the first equation describes the motion of electrons with the parametric dependence on the nuclei coordinates, and the second equation describes the motion of nuclei in the potential, which is at the same time the eigenvalue of the electron system. Both equations are related to each other by the $W(R)$ function. Usually the first equation is solved for a fixed or "frozen" configuration of the nuclei. The $W(R)$ values obtained in this way are substituted into the second equation, and then the energy E of the nuclear sub-system is calculated.

The potential energy $W(R)$ can be decomposed into the power series with respect to R near the equilibrium coordinate R_0:

$$W(R)=W(R_0)+\left(\frac{dW}{dR}\right)_{R_0}(R-R_0)+\frac{1}{2}\left(\frac{d^2W}{dR^2}\right)_{R_0}(R-R_0)^2+\cdots \quad (9.11)$$

The first term of this decomposition is a constant, which can be set to zero by a proper choice of the system of reference. The second term is zero, since the coordinate $R = R_0$ is the equilibrium position with the minimal potential energy. If the terms with the powers of $(R - R_0)$ higher than two are neglected, then the only non-zero term is the third one, and Eq. (9.11) is reduced to the following simple form:

$$W \approx \frac{1}{2}k(R - R_0)^2, \tag{9.12}$$

with
$$k = \left(\frac{d^2W}{dR^2}\right)_{R_0}.$$

Therefore, one can assume in the harmonic approximation that the nuclei move in a parabolic potential, which is the energy of the electronic subsystem. In principle, the second equation (9.10) describes the quantum-mechanical harmonic oscillator with the equidistant spectrum shown by the horizontal lines in Fig. 9.1.

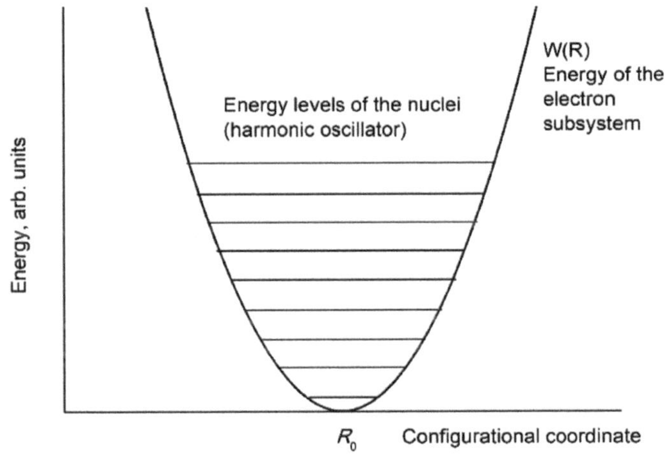

Figure 9.1 A simple illustration of the potentials involved into the adiabatic approximation. The separation between the horizontal lines representing the energy levels of the harmonic oscillator is $\hbar\omega$.

The eigenvalues of the electron subsystem $W(R)$ serve as a potential, in which the nuclei move. The nuclear motion has a confined character, which corresponds to small oscillations of nuclei around the equilibrium positions, and the energy levels

E of the nuclear subsystem can be represented in the harmonic approximation as the energy levels of the harmonic oscillator. The total energy of such a system will be just a sum of the energies of the electronic and nuclear subsystems.

The electronic states are represented by the parabolas for the ground and all other excited states. In principle, the equilibrium coordinates R_0 can be different for different electronic states—this circumstance has very important implications for the optical spectra.

Adiabatic approximation can be also applied to the description of the impurity center properties in crystals. In this case, it is also possible to divide the whole crystal into two sub-systems: the fast-moving electrons and the "frozen" nuclei. Then the Schrödinger equation can be solved for these systems, and the eigenvalues of the "electronic" Schrödinger equation form the potential, in which the slow nuclei are moving.

Using the power series expansion of the electronic energies of an impurity center (in terms of the nuclear coordinate) and considering the terms up to the second order of the coordinate only, we can represent the electronic states by the parabolas (Fig. 9.2) with the horizontal lines representing the vibrational energy levels of nuclei.

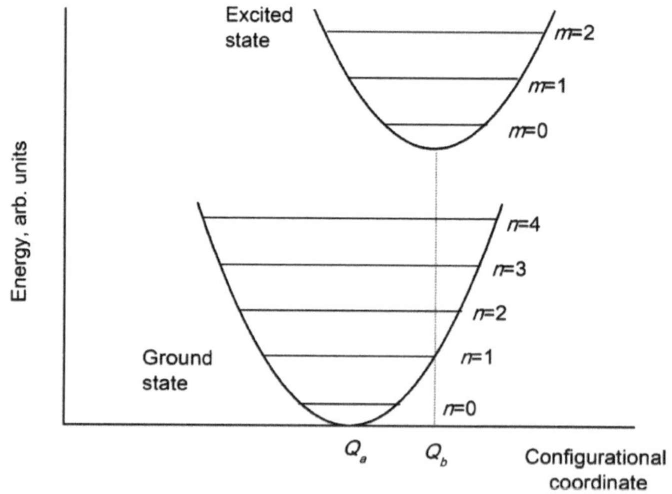

Figure 9.2 A schematic representation of two potential energy curves for the ground and excited electronic states. The vibrational levels are shown by the horizontal lines, the equilibrium coordinates for both states are denoted by Q_a and Q_b.

Such a modeling of the electronic and vibrational states is known as a single (since only one coordinate is involved) configurational coordinate model. Despite its simplicity, it gives a very good qualitative explanation of the basic features of the emission and absorption spectra.

As seen from the figure, the equilibrium positions (denoted as Q_a and Q_b) for both states are shifted with respect to each other. To explain this, let us consider an octahedral complex shown in Fig. 9.3. For the sake of simplicity, we assume that the central ion has only one electron.

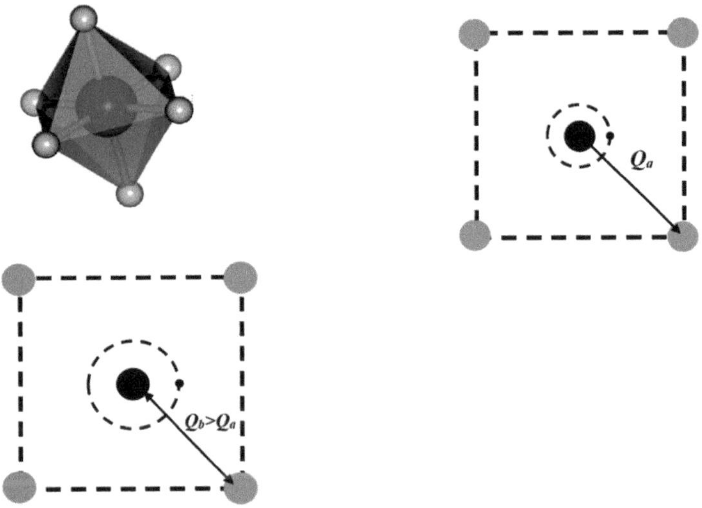

Figure 9.3 A schematic representation of an octahedral impurity center considered in the text and its view from the top in the ground and excited states. An electron orbit is shown by the dashed circle. Six ligands around an impurity ion have a negative electric charge.

The right part of Fig. 9.3 shows a view of this octahedral center from the top, and Q_a is an equilibrium distance between an impurity ion and negatively charged ligands.

After excitation an impurity ion's electron jumps to another orbit of a greater radius. Since the ligands have a negative charge, the Coulomb repulsion between an impurity ion's electron and ligands will push them a little bit apart from the impurity ion, therefore the new equilibrium Q_b distance between an impurity ion and ligands in an excited state will be greater than in the ground state (Fig. 9.3.).

This physical picture is, of course, simplified, but it clearly demonstrates the reason for a shift of the equilibrium positions of the potential energy parabolas for the ground and excited states.

Figure 9.4 shows the potential energy parabolas for the ground and excited states together with the vertical arrows representing the absorption (the upward arrow) and emission (the downward arrow) transitions. Mathematically the equations for the potential energy parabolas can be written down as follows:

$$E_a(Q) = \frac{1}{2}M\omega^2(Q - Q_a)^2 \tag{9.13}$$

for the ground state, and

$$E_b(Q) = E_{ab} - \frac{1}{2}M\omega^2(Q_b - Q_a)^2 + \frac{1}{2}M\omega^2(Q - Q_b)^2 \tag{9.14}$$

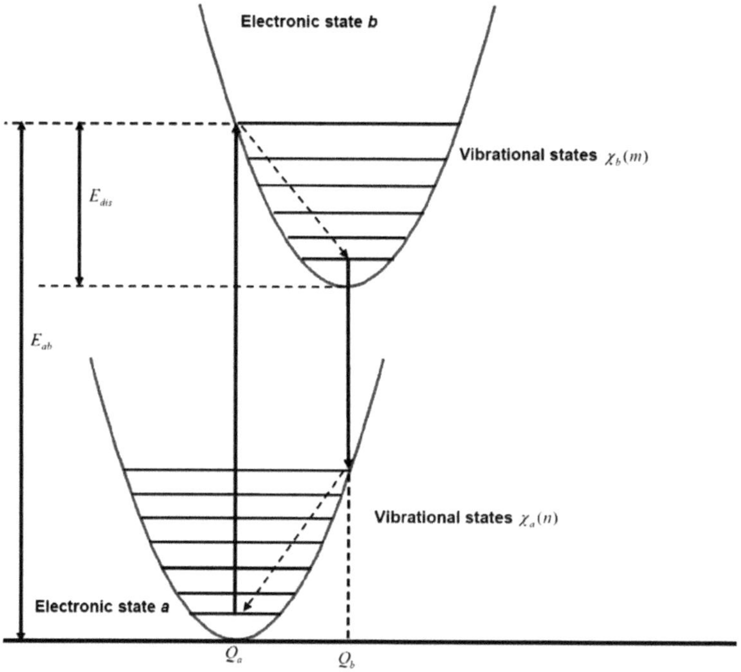

Figure 9.4 A schematic representation of the absorption and emission transitions in the configurational coordinate model. E_{ab} is the energy corresponding to the absorption from the ground state. The vibrational levels are shown by the horizontal lines; dashed arrows represent the non-radiative relaxation processes within the same electronic state.

for the excited state. Here M stands for the mass of an individual ligand, ω is the frequency of vibrations of the surrounding ligands, and E_{ab} is shown in Fig. 9.4. For simplicity, the vibrational frequencies in the ground and excited states are assumed to be equal. Generally speaking, the frequencies of vibrations in the excited state are 2–5% lower than in the ground state, but this small difference can be safely neglected in the first approximation.

One of the keystones of the adiabatic approximation is the so-called Franck-Condon principle, according to which the electronic transitions in both emission and absorption processes are so fast, that the nuclear coordinates remain practically unchanged during an electronic transition. That is why the transitions between the electronic states are depicted by the vertical arrows in the configurational coordinate model. With a simple algebra it is possible to rewrite Eq. (9.14) as

$$\begin{aligned} E_b(Q) &= E_{ab} - \frac{1}{2}M\omega^2(Q_b - Q_a)^2 + \frac{1}{2}M\omega^2(Q - Q_b)^2 \\ &= E_{ab} + \frac{1}{2}M\omega^2(Q - Q_a)^2 - M\omega^2(Q_b - Q_a)(Q - Q_a), \end{aligned} \tag{9.15}$$

or

$$E_b(Q) = E_{ab} + \frac{1}{2}M\omega^2(Q - Q_a)^2 - A\hbar\omega\sqrt{\frac{M\omega}{\hbar}}(Q - Q_a), \tag{9.16}$$

with

$$A = \sqrt{\frac{M\omega}{\hbar}}(Q_b - Q_a). \tag{9.17}$$

The A constant is a non-dimensional constant, which is proportional to the shift of the minima of the potential energy parabolas of the ground and excited electronic states. Physically it describes the difference in the electron-lattice coupling in the states a and b. It is more convenient to use the so-called Huang–Rhys parameter S:

$$S = \frac{A^2}{2} = \frac{1}{2}\frac{M\omega^2}{\hbar\omega}(Q_b - Q_a)^2 = \frac{E_{\text{dis}}}{\hbar\omega}, \tag{9.18}$$

with E_{dis} defined in Fig. 9.4. The Huang–Rhys factor is commonly used to distinguish between the systems with strong ($S > 10$), medium ($1 < S < 10$) and week ($S < 1$) electron-vibrational interaction.

9.3 Shape of the Emission and Absorption Band. Zero-Phonon Line

The probability of an electric dipole transition (μ is the electric dipole operator) from the n-th vibrational level of the a-th electronic state to the m-th vibrational level of the b-th electronic state can be found as

$$W = \left|\langle \phi_b(r,Q)\chi_b(m)|\mu|\phi_a(r,Q)\chi_a(n)\rangle\right|^2, \tag{9.19}$$

where the electronic and vibrational wave functions are denoted by $\phi_b(r, Q)$, $\phi_a(r, Q)$ and $\chi_b(m)$, $\chi_a(n)$, respectively.

By virtue of the Franck-Condon principle (assuming that the coordinates of the nuclei are fixed at Q_0) one can reduce the integral in Eq. (9.19) to the product of the integrals from the electronic and vibrational wave functions:

$$W = \left|\langle \phi_b(r,Q_0)|\mu|\phi_a(r,Q_0)\rangle\right|^2 \left|\langle \chi_b(m)|\chi_a(n)\rangle\right|^2. \tag{9.20}$$

The absorption band shape as a function of energy E will be given by the following expression:

$$I_{ab}(E) = I_0 \left\langle \sum_m \left|\langle \chi_b(m)|\chi_a(n)\rangle\right|^2 \delta(E_{b,m} - E_{a,n} - E) \right\rangle_n, \tag{9.21}$$

where the summation is to be carried over all final vibrational states m, and the averaging over all initial vibrational states n is denoted by the $\langle\rangle_n$ symbol. The electric dipole matrix element is absorbed into the I_0 pre-multiplier.

When absolute temperature $T = 0$, only the lowest vibrational level $n = 0$ is populated, and the absorption band shape can be found rather easily:

$$I_{ab}(E) = I_0 \sum_m \frac{\exp(-S)S^m}{m!} \delta(E_0 + m\hbar\omega - E). \tag{9.22}$$

Here S is the above-defined Huang–Rhys factor, and E_0 is the energy of the transition between the lowest vibrational levels of the ground and excited electronic states (Fig. 9.5).

E_0 is the zero-phonon line (ZPL) energy. It is called so since no phonons participate in this absorption transition: The vibrational state is not changed, and only the electronic state changes during this transition. The intensity of the ZPL is $I_0\exp(-S)$. If $S = 0$ (no shift of the energy parabolas), all intensity of the absorption transition is in the ZPL. If S increases, the vibrational sidebands are observed in the absorption spectrum at the energies $m\hbar\omega$ above the ZPL, where m is an integer number, those vibrational sidebands "borrow" intensity from the ZPL, decreasing its intensity.

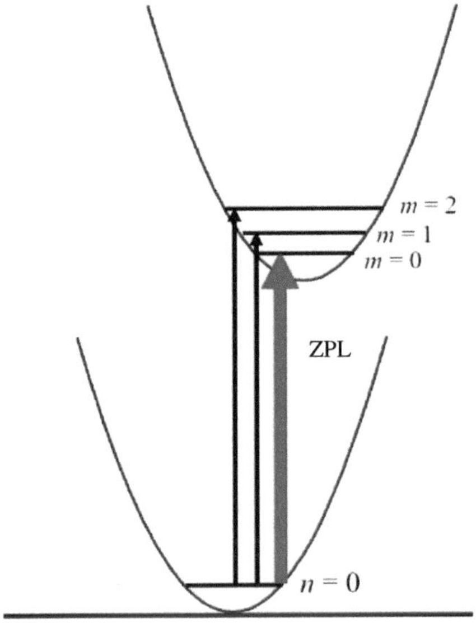

Figure 9.5 A schematic representation of the absorption transitions at $T = 0$. The absorption transition, which corresponds to the zero-phonon line (ZPL), is shown by a bold arrow.

It is instructive to analyze how the absorption band shape depends on the value of the Huang–Rhys factor S. If $S = 0$, only a narrow ZPL is observed. Increase of the Huang–Rhys factor causes appearance of an increasing number of the vibrational sidebands. The increase of the sidebands intensity occurs on account of a decrease of the ZPL intensity (Fig. 9.6.). As seen from Fig. 9.6, the maximum of absorption (or, in other words, the most probable absorption transition) corresponds to $m = S$, when the number of phonons generated in a transition is equal to the Huang–Rhys factor.

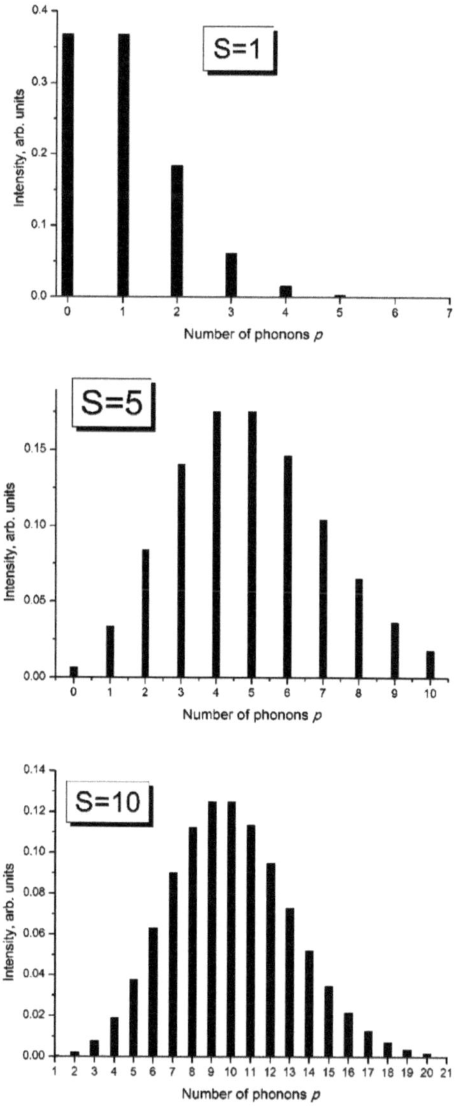

Figure 9.6 Dependence of the absorption band shape on the value of the Huang–Rhys factor (shown in the inset in each figure) at $T = 0$. The number of phonons generated in the absorption transition is shown along the horizontal axis.

When $T \neq 0$, the situation becomes more complicated, since in the ground electronic state the vibrational quantum number n can take any integer value, in principle. In this case, the absorption band shape is given by the following equation (it is also plotted in Fig. 9.7 for T = 300 K and different values of S):

$$I_{ab}(E)=I_0\sum_{p=-\infty}^{\infty}\exp(-S(1+2n))\left(\frac{1+n}{n}\right)^{p/2}I_p(2S\sqrt{n(n+1)})\delta(E_0+p\hbar\omega-E), \tag{9.23}$$

where $n=\dfrac{1}{\exp(\hbar\omega/kT)-1}$ is the mean thermal occupancy in the vibrational mode, $I_p(x)$ is the modified Bessel function of the first kind.

At $T > 0$ the vibrational sidebands with negative number of phonons appear. This means that absorption takes place at the energies lower than the ZPL energy, due to the non-zero occupation of the higher vibrational levels in the ground electronic state. The transitions from higher vibrational levels of the ground state to the lower vibrational levels of the excited state contribute to such absorption, as shown in Fig. 9.8.

The shape of the emission band at T = 0 is determined by the following equation:

$$I_{ab}(E)=I_0\sum_n\frac{\exp(-S)S^n}{n!}\delta(E_0-n\hbar\omega-E), \tag{9.24}$$

with all quantities determined above (note the change of sign before $n\hbar\omega$ compared with Eq. (9.22) for the absorption band shape). It has been established experimentally long time ago that the emission spectra are shifted to the lower energies (red-shifted) with respect to the absorption spectra.

Figure 9.9 can serve as an illustration and explanation of this experimental fact. Absorption of light from the lowest vibrational levels of the ground electronic state (which are more populated than the higher levels) is shown by a solid vertical arrow. This arrow terminates at one of the higher vibrational levels of the excited electronic state, because of the offset of the potential energy parabolas. Such absorption is followed by fast non-radiative relaxation to the lowest vibrational level of the excited electronic state (shown by a dashed arrow). The emission transition is shown by a solid downward arrow; this arrow is shorter than that one of the absorption transition. After the system goes to the ground electronic state, again a fast non-radiative relaxation to the lowest vibrational level of the ground state takes place. This is the mechanism of the red shift of the luminescence (which is also called the Stokes luminescence).

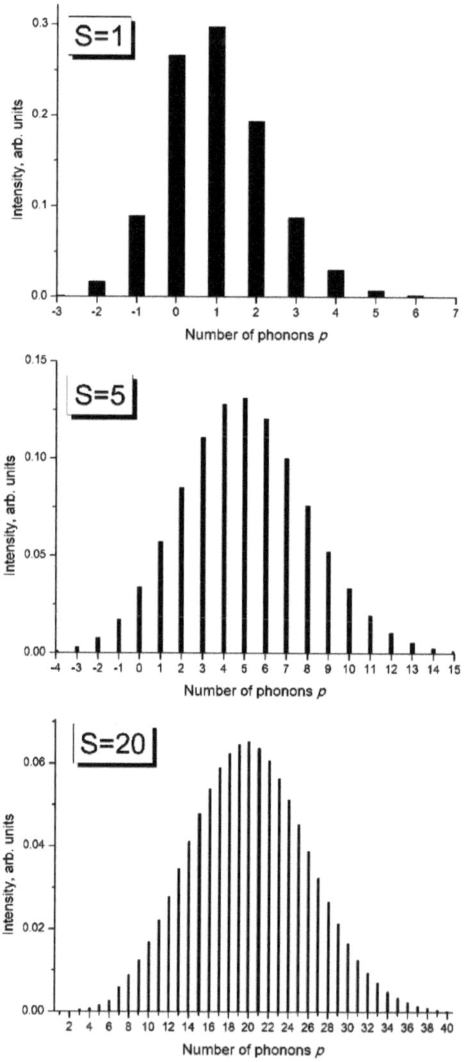

Figure 9.7 Dependence of the absorption band shape on the value of the Huang–Rhys factor (shown in the inset in each figure) at $T = 300$ K. The number of phonons generated in the absorption transition is shown along the horizontal axis.

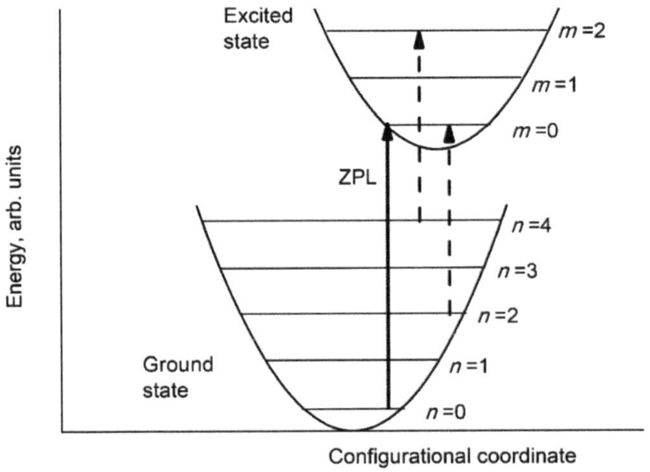

Figure 9.8 Illustration of appearance of "negative" number of phonons in the absorption spectrum at $T \neq 0$. The energies of the transitions shown by the dashed arrows are lower than the ZPL energy.

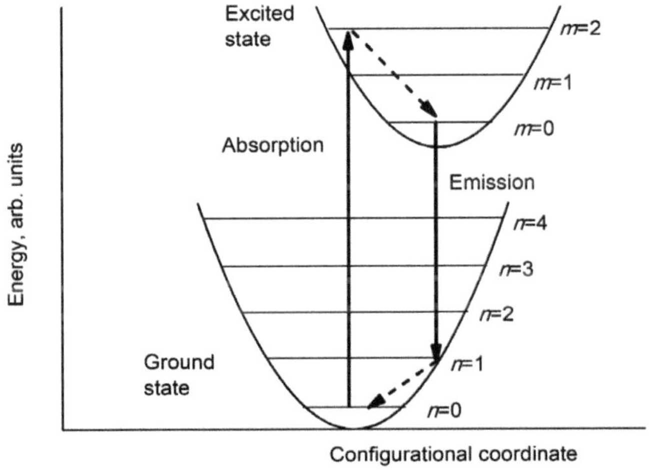

Figure 9.9 Illustration of difference in energy between the absorption (upward) and emission (downward) arrows. Non-radiative relaxation within the vibrational levels of the same electronic state is shown by the dashed arrows.

Sometimes (more rarely, however) the emitted light wavelength is shorter than that one of the absorbed light. This situation corresponds to the blue shift of the luminescence and is referred to as the anti-Stokes luminescence. It can be realized if the absorption takes place from very highly located vibrational levels of the ground electronic state. In such a case, the length of the representative arrow of the emission transition in the single configurational coordinate model will be longer than the length of the representative arrow of the absorption transition.

Figure 9.10 illustrates a correspondence between the configurational coordinate diagram and experimental absorption and emission spectra. According to the difference in length between the arrows representing the absorption and emission transitions, the maxima of the experimental absorption and emission spectra are shifted (the Stokes shift, which is shown in the figure). The maxima of the emission and absorption do not correspond to the ZPL position, as easy to see from the configurational coordinate model.

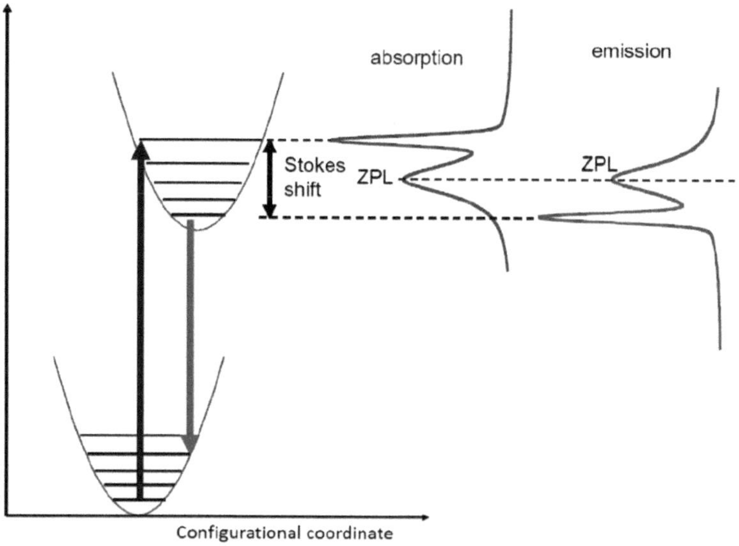

Figure 9.10 Illustration of difference in energy between the absorption (upward) and emission (downward) arrows. ZPL are shown in both emission and absorption spectra. The Stokes shift between the maxima of absorption and emission is indicated.

However, if the force constants in the ground and excited states are equal, the emission and absorption spectra are the mirror images of each other with respect to the ZPL position. This circumstance gives an opportunity to find the ZPL position as a symmetry axis, which gives a mirror symmetry between the emission and absorption spectra (Fig. 9.10).

9.4 Extraction of the Huang–Rhys Factor and Effective Phonon Energy from the Experimental Absorption and Emission Spectra

The electron-vibrational interaction between the electronic states of an impurity ion in a crystal and crystal lattice vibrations is effectively described by the following three main parameters: the Stokes shift ΔE_S (the energy difference between the absorption and emission peaks), the Huang–Rhys factor S (which is proportional to ΔE_S) and effective phonon energy $\hbar\omega$. The values of S and $\hbar\omega$ can be estimated using the following equations[1]:

$$\Delta E_S = (2S - 1)\hbar\omega, \tag{9.25}$$

$$\Gamma(T) = \sqrt{8\ln 2}\,\hbar\omega\left[S\coth\left(\frac{\hbar\omega}{2kT}\right)\right]^{1/2}, \tag{9.26}$$

where the last equation describes the full width $\Gamma(T)$ at half maximum (FWHM) of the emission band as the function of the absolute temperature T. The effective phonon energy corresponds to some averaged frequency of the crystal lattice vibrations, mostly efficiently interacting with the electronic states of an impurity center.

The values of $\Gamma(T)$ and ΔE_S can be readily extracted from the experimental emission and excitation spectra. The solution of the system of two last equations yields the values of S and $\hbar\omega$. Since Eq. (9.26) is a non-linear one, the easiest way to solve both of them simultaneously is to express $\hbar\omega$ from Eq. (9.25), substitute it into Eq. (9.26), keep the hyperbolic tangent in the right-hand side, collect all other terms in the left hand side, and plot both sides of the obtained in such a way equation as the functions of S. The point of intersection of two lines will give the value of the Huang–Rhys factor, and the estimation of $\hbar\omega$ from Eq. (9.25) is then straightforward.

It should be kept in mind that the Stokes shift is defined for the absorption/emission transitions between the same pair of energy levels, and the last equations cannot be applied, if the absorption/emission transitions occur between different pairs of energy levels.

To check out the reasonability of extracted values of S and $\hbar\omega$, the emission band shape can be modeled using the following expression[2]:

$$I = \frac{e^{-S} S^{p}}{p!}\left(1 + S^{2}\frac{e^{-\hbar\omega/kT}}{p+1}\right),\ p = \frac{E_0 - E}{\hbar\omega}, \tag{9.27}$$

where E_0 is the ZPL energy, whose position is close to the point of intersection of the absorption and emission spectra, and p is the number of the effective phonons involved into the emission transition.

As an example, we give here the results of application of Eqs. (9.25)–(9.27) to the recently reported[3] phosphor $Sr_{1-x}[(Mg_2Al_2)_{1-y}(Li_2Si_2)_yN_4]:Eu_x^{2+}$ for y = 0.3, x = 0.004. Narrow red emission of this material is due to the parity allowed $5d$-$4f$ emission of the Eu^{2+} ions. The room temperature excitation and emission spectra of this compound are shown in Fig. 9.11.

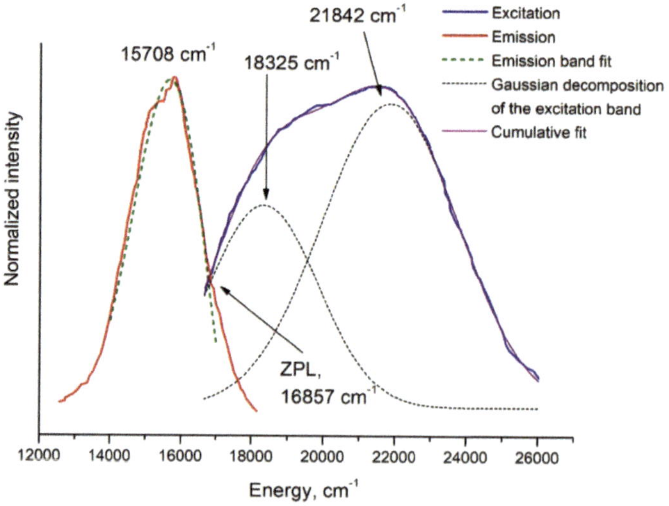

Figure 9.11 Excitation and emission spectra of $Sr_{1-x}[(Mg_2Al_2)_{1-y}(Li_2Si_2)_yN_4]:Eu_x^{2+}$ for y = 0.3, x = 0.004 phosphor (adapted from Ref. 3). See the text for more details.

The emission maximum is realized at about 15708 cm^{-1}. The excitation band is a very broad one, located in the range from about 17000 to 26000 cm^{-1} and is due to the transitions to all Eu^{2+} 5*d* states. Decomposition of the excitation band (there should be also other maxima corresponding to the Eu^{2+} 5*d* states in the higher-energy range, which are not shown here) into two Gaussians can perfectly reproduce the excitation band. Such a decomposition is needed to identify the lowest Eu^{2+} 5*d* state, which appears to be at 18325 cm^{-1}. The Stokes shift then is 2617 cm^{-1}, the emission band FWHM is 2407 cm^{-1}, and the S and $\hbar\omega$ values are 2.76 and 579 cm^{-1}. With these values, the emission band shape (Eq. (9.27)) was fitted to match the experimental emission spectrum, if the ZPL energy was taken to be 16900 cm^{-1}. As seen from Fig. 9.11, the maxima positions and FWHM of the experimental and theoretical emission bands are practically identical. The ZPL position deduced from the emission spectra fit is very close to the point of intersection of the excitation and emission spectra at about 16857 cm^{-1}. Slight deviations of the first excitation and emission band maxima from their perfect mirror symmetry can be explained by the effects of the electron-vibrational interaction, inhomogeneous broadening, anharmonic effects, etc.

The model outlined in this paragraph, even being quite simple, was proven to be successful in the description of the electron-vibrational interaction in the 3*d* states of the transition metal ions and 5*d* states of the rare-earth ions, the reader is kindly advised to look into these publications[4–11] for more details.

9.5 Non-Radiative Transitions in Impurity Centers

The non-radiative relaxation of the electron excitation energy in an impurity center can occur not only within the vibrational states of the same electronic state, but between different electronic states as well. In this case, an effective exchange of energy between the electronic and nuclear subsystems takes place, and a part of the electronic energy goes to the nuclear subsystem to be transformed into the energy of thermal vibrations of the crystal lattice ions.

These non-radiative transitions play a two-fold role: On one side, they represent a channel of the excitation energy losses and thus are not desirable. On the other side, they are essentially needed for realization of the three- and four-level schemes of laser operation.

Again, a configurational coordinate model proves out to be useful for visualization of the non-radiative transitions. Figure 9.12 shows two intersecting parabolas of two electronic states.

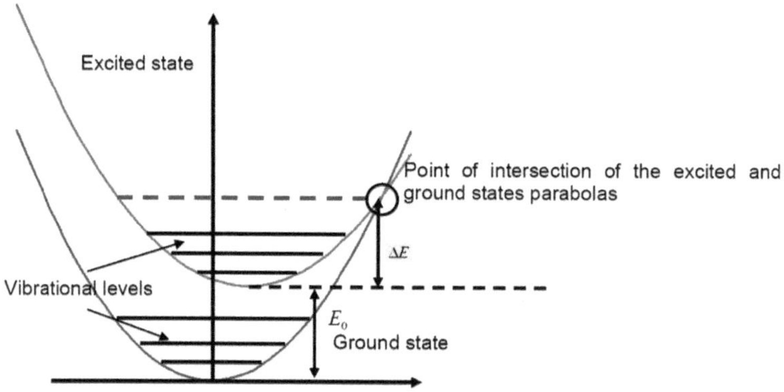

Figure 9.12 Two intersecting parabolas representing two electronic states. The energy interval between the bottom of the excited state parabola and the point of its intersection with the ground state parabola is ΔE.

At some temperature T, that vibrational level of the excited state, which coincides with the point of intersection with the ground state parabola (or is close enough to this point), can be populated. If this is the case, then the non-radiative relaxation through the vibrational levels of the ground electronic state will occur, and the energy of the electron excitation will be transformed into the energy of thermal vibrations of the crystal lattice ions.

It is intuitively clear that the probability of such a process depends on the population of that vibrational level in the excited state, which is closest to the intersection point. There is a simple equation proposed by N. F. Mott,[12] which gives the probability W_{NR} of the non-radiative transition as a function of temperature T:

$$W_{NR} = A\exp\left(-\frac{\Delta E}{kT}\right), \tag{9.28}$$

where A is an empirical constant, and ΔE is defined in Fig. 9.12.

A more precise equation for the probability W_{NR} of the non-radiative transition based on the quantum mechanical derivation is as follows:

$$W_{\mathrm{NR}} = R^2 \exp\{-S(2m+1)\}\left(\frac{1+m}{m}\right)^{\frac{p}{2}} I_p\left(2S\sqrt{m(m+1)}\right), \quad (9.29)$$

where R^2 is a square of the electronic matrix element between the states involved into a transition (usually it is a fitting parameter, since its direct calculation is a very laborious task), S is a Huang–Rhys factor, $m = \frac{1}{\exp(\hbar\omega/kT)-1}$, I_p is the modified Bessel function, and p is a mean number of phonons with energy $\hbar\omega$ needed to bridge a gap between the considered electronic levels, $p = E_0/\hbar\omega$.

Increase of temperature leads to an enhancement of the non-radiative transition probability, which can be explained by increasing population of higher vibrational states.

An example of application of Eq. (9.29) is shown in Fig. 9.13. The $^3T_2 \rightarrow {}^3A_2$ non-radiative transition probability for the Cr^{4+} ions in $Y_3Al_5O_{12}$ is calculated with the following parameters values[13]: E_0 = 8000 cm^{-1}, S = 0.85; $\hbar\omega$ = 429 cm^{-1} (such a value of the effective phonon energy results in p = 19 phonons to bridge the gap between the $^3T_2 \rightarrow {}^3A_2$ states), and $R^2 = 10^{23}$. In this case, the non-radiative transition probability is not zero even at very low temperatures, and it grows up very quickly with increasing temperature.

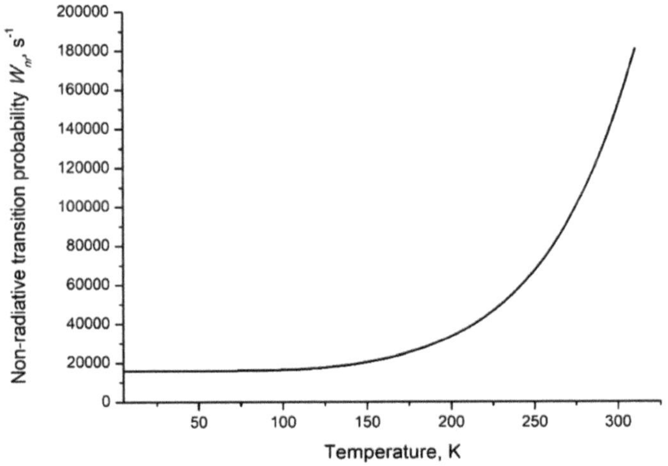

Figure 9.13 Temperature-dependent probability (Eq. 9.29) of the $^3T_2 \rightarrow {}^3A_2$ non-radiative transition of Cr^{4+} ions in $Y_3Al_5O_{12}$. See the text for the numerical values of the parameters involved.

The model of the energy levels crossing in the theory of the non-radiative transitions was developed in Refs. [14] and [15], in which the points of crossing of electronic energy levels were found from the crystal field calculations of energy levels of impurity ions for different geometrical configurations of impurity centers. The probability of the non-radiative transition was taken to be proportional to the probability of realization of such ligands configurations that are leading to the energy levels crossing. The model was applied to the case of Cr^{4+} ions in $Y_3Al_5O_{12}$ and good agreement with the experimental data was demonstrated.

9.6 Separation between the Non-Radiative and Radiative Transitions in Impurity Centers

Two kinds of transitions—radiative and non-radiative—can contribute to the depopulation of the excited state. Theoretical dependence of the excited state lifetime τ is given by the following equation:

$$\frac{1}{\tau}=\frac{1}{\tau_{r0}\exp(-\alpha T)}+s_0\exp(-\Delta E/kT), \tag{9.30}$$

where the first and the second terms describe the radiative and non-radiative decay, respectively. $\tau_{r0}\exp(-\alpha T)$ is the radiative lifetime, τ_{r0} is the radiative lifetime at 0 K, $s_0\exp(-E_{nr}/kT)$ is the probability of the non-radiative transition, and s_0 is the frequency factor. ΔE is the activation energy for the non-radiative transition, which corresponds to the energy gap between the bottom of the excited state parabola and the point of its intersection with the corresponding curve for the ground state, and T is the absolute temperature. This equation allows for a separate comparative analysis of both processes.

Table 9.1[16] contains a collection of the τ_{r0}, α, s_0, ΔE parameters (which are usually considered as fitting parameters) for the Cr^{3+} ions in several crystals.

The data for the $CeSc_3(BO_3)_4$ (CSB) crystal were used to model the temperature dependence of the $^4T_{2g}$ excited state lifetime of the Cr^{3+} ions (Fig. 9.14). As seen from the figure, agreement between the calculated and experimental data is good.

Table 9.1 Some typical values of parameters from Eq. (9.29) for the Cr^{3+} ions in different crystals

Crystal	$CeSc_3(BO_3)_4$	$ScBO_3$	$LaSc_3(BO_3)_4$	Emerald	$LiSrAlF_6$	$LiCaAlF_6$
τ_{r0}, μs	66	211	98	58	64	68
$\alpha \cdot 10^{-3}$, K^{-1}	1.3	1.3	1.9	0.4	0	0
s_0, s^{-1}	50×10^6	25×10^6	1.2×10^6	3.5×10^9	6×10^{12}	8×10^{13}
ΔE, cm^{-1}	2050	2100	650	6200	5125	8500

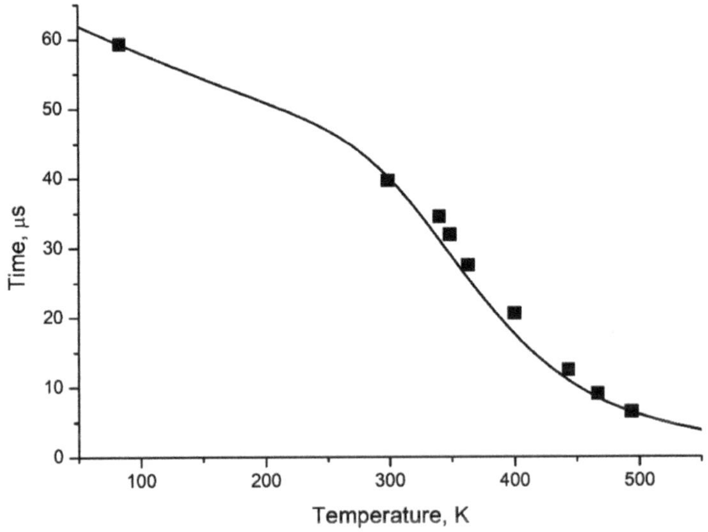

Figure 9.14 Temperature dependence of the Cr^{3+} lifetime (inverse to Eq. (9.30)) in the CSB crystal. The filled squares show experimental values, solid line is a fit with the numerical values of the required parameters from Table 9.1.

Two contributions to the total probability of the excited state deactivation (which is an inverse value to the excited state lifetime) are shown separately in Fig. 9.15. It is interesting to compare the probability of the non-radiative transitions from Fig. 9.15 with that one for Cr^{4+} in $Y_3Al_5O_{12}$ (Fig. 9.13). In the case of Cr^{3+} in CSB crystal even at about room temperature the probability of the non-radiative transitions is practically zero, in contrast with Fig. 9.13. This shows that the non-radiative energy losses are rather low in this material, which makes it a promising candidate for the solid-state laser applications.

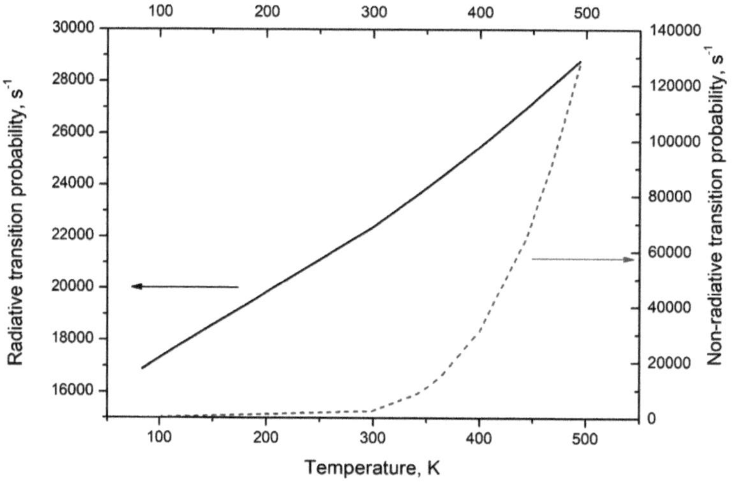

Figure 9.15 Temperature dependence of the radiative (solid line, left *y* axis) and non-radiative (dashed line, right *y* axis) probabilities of the $^4T_{2g}$–$^4A_{2g}$ transitions for Cr^{3+} ions in the CSB crystal.

9.7 Normal Vibrational Coordinates

In this paragraph, we discuss the main ideas needed to describe the vibrational motion of the atoms in a molecule or the ions in an impurity center in a crystal. This information will help to understand better the rest of the book.

Let us consider a molecule consisting of N atoms (or an impurity center in a crystal consisting of N ions). Due to the thermal vibrations, the distances between atoms and the angles between the chemical bonds are constantly changing. To have a precise description of a state of such a molecule at every moment of time, we would have to know $3N$ coordinates and their time-dependence. The total kinetic energy T of such a system is

$$T = \frac{1}{2}\sum_{i=1}^{3N} m_i \dot{x}_i^2 , \tag{9.31}$$

where the index i enumerates the atoms with masses m_i and coordinates x_i, and the dot above x_i is an accepted notation for the derivative over time t. The potential energy U of such a system can be expanded into a power series with respect to the coordinates x_i around the equilibrium position:

$$U = U_0 + \sum_{i=1}^{3N} \left(\frac{\partial U}{\partial x_i} \right)_0 x_i + \frac{1}{2} \sum_{i,j=1}^{3N} \left(\frac{\partial^2 U}{\partial x_i \partial x_j} \right)_0 x_i x_j + \cdots \tag{9.32}$$

The subscript "0" means that all derivatives are calculated at the equilibrium position. The first term U_0 in this expansion is a constant, which may be set to zero by a proper choice of the energy reference point, the second term $\sum_{i=1}^{3N} \left(\frac{\partial U}{\partial x_i} \right)_0 x_i$, which is linear in coordinates, is zero at the equilibrium position, because the potential energy should have a minimum, in other words the first derivatives over coordinates should be zero. If the harmonic approximation is used, then the third and higher powers of coordinate terms can be dropped, and the only remaining term is quadratic in the coordinates x_i.

The equations of motion for each coordinate can be obtained using the Lagrange function of the system $L = T - U$ and the Lagrange equation $\frac{d}{dt}\frac{\partial L}{\partial \dot{x}_i} - \frac{\partial L}{\partial x_i} = 0,\ i = 1,\ldots,3N$. We find then that

$$\frac{d}{dt}(m_i \dot{x}_i) + \frac{\partial U}{\partial x_i} = 0, i = 1,\ldots,3N. \tag{9.33}$$

It is a standard practice to introduce the so-called mass-weighted coordinates

$$q_i = x_i \sqrt{m_i} \tag{9.34}$$

and the coefficients

$$c_{ij} = \left(\frac{\partial^2 U}{\partial x_i \partial x_j} \right)_0 \tag{9.35}$$

In these notations the kinetic and potential energies and equations of motion will be rewritten as

$$T = \frac{1}{2} \sum_{i,j=1}^{3N} \dot{q}_i \dot{q}_j;\ U = \frac{1}{2} \sum_{i,j=1}^{3N} b_{ij} q_i q_j \tag{9.36}$$

$$\frac{d}{dt}(\dot{q}_i)+\sum_{j=1}^{3N} b_{ij}q_j=0\,,\, i=1,\ldots 3N, \tag{9.37}$$

with $b_{ij}=c_{ij}\big/\sqrt{m_i m_j}$. The solutions $q_i(t)$ of these $3N$ differential equations of the second order can be chosen as $q_i=q_{0i}\sin(t\sqrt{\lambda}+\delta)$, with q_{0i} being the amplitude of the i-th vibrations, and δ its phase. If we substitute this supposed form of solution into Eq. (9.37), we get the system of $3N$ linear algebraic equations (instead of the differential ones):

$$-\lambda q_{0i}+\sum_j b_{ij}q_{0i}=0, i=1,\ldots,3N\,, \tag{9.38}$$

or, in an extended form

$$\begin{aligned}
&(-\lambda+b_{11})q_{01}+b_{12}q_{02}+b_{13}q_{03}+\cdots b_{1,3N}q_{0,3N}=0\\
&b_{22}q_{01}+(-\lambda+b_{22})q_{02}+b_{23}q_{03}+\cdots b_{2,3N}q_{0,3N}=0\\
&\cdots\cdots\cdots\cdots\cdots\cdots\cdots\cdots\cdots\cdots\cdots\cdots\cdots\cdots\cdots\cdots\\
&b_{3N,1}q_{01}+b_{3N,2}q_{02}+\cdots+(-\lambda+b_{3N,3N})q_{0,3N}=0,
\end{aligned} \tag{9.39}$$

with the amplitudes q_{0i} being the unknown quantities. The system of linear homogeneous equations—like the system (9.39)—has a non-trivial solution, when its main determinant (also called the secular determinant) is equal to zero:

$$\begin{vmatrix} -\lambda+b_{11} & b_{12} & \cdots & b_{1,3N} \\ b_{21} & -\lambda+b_{22} & \cdots & b_{2,3N} \\ \vdots & \vdots & \vdots & \vdots \\ b_{3N,1} & b_{3N,2} & \cdots & -\lambda+b_{3N,3N} \end{vmatrix}=0\,. \tag{9.40}$$

There are $3N$ roots of the secular determinant. Six roots will be zero (five roots in the case of a linear molecule); these roots correspond to the translational and rotational motion of the molecule. The square roots of non-zero λ's give the frequencies of possible vibrations in the considered molecule. So, for a molecule consisting of N atoms we have $3N$-6 vibrational degrees of freedom (or $3N$-5 for the linear molecule, since rotation about the molecule axis in this case would not change its internal state).

It is more convenient to use the matrix formalism to calculate the vibration frequencies in a polyatomic molecule. Let us introduce the matrix **B** composed of the $b_{ij}=c_{ij}\big/\sqrt{m_i m_j}$ coefficients

$$\mathbf{B} = \begin{pmatrix} b_{11} & b_{12} & \cdots & b_{1,3N} \\ b_{21} & b_{22} & \cdots & b_{2,3N} \\ \vdots & \vdots & \vdots & \vdots \\ b_{3N,1} & b_{3N,2} & \cdots & b_{3N,3N} \end{pmatrix}, \qquad (9.41)$$

the column-vector $\mathbf{q}$ of the mass-weighted coordinates and its transposed row-vector $\mathbf{q}^T$

$$\mathbf{q} = \begin{pmatrix} q_1 \\ q_2 \\ \vdots \\ q_{3N} \end{pmatrix}, \mathbf{q}^T = \begin{pmatrix} q_1 \\ q_2 \\ \vdots \\ q_{3N} \end{pmatrix}^T = (q_1 \quad q_2 \quad \cdots \quad q_{3N}), \qquad (9.42)$$

and $3N$ by $3N$ unit matrix $\mathbf{I}$. Then Eq. (9.40) will be transformed to

$$|\mathbf{B} - \lambda\mathbf{I}| = 0. \qquad (9.43)$$

After all values of λ are found, each value λ_k has to be substituted into Eq. (9.43) to find the amplitudes of vibrations q_{0k}. Since the $3N$ equations (9.39) are linearly dependent, an additional condition should be introduced to find all q_{0k} values. The easiest and most natural way is to normalize the vector $\mathbf{q}_k$ to unity. Then we shall speak about a new vector

$$\mathbf{l}_{0k} = \begin{pmatrix} l_{01,k} \\ l_{02,k} \\ \vdots \\ l_{0,3N,k} \end{pmatrix}, \quad \sum_i l_{ik}{}^2 = 1. \qquad (9.44)$$

The set of the $\mathbf{l}_{0k}$ vectors form the $\mathbf{L}$ matrix (all individual vectors are added—the column by the column—into one matrix):

$$\mathbf{L} = \begin{pmatrix} l_{01,1} & l_{01,2} & l_{01,3} & \cdots & l_{01,3N} \\ l_{02,1} & l_{02,2} & l_{02,3} & \cdots & l_{02,3N} \\ l_{03,1} & l_{03,2} & l_{03,3} & \cdots & l_{03,3N} \\ \vdots & \vdots & \vdots & \ddots & \vdots \\ l_{0,3N,1} & l_{0,3N,2} & l_{0,3N,3} & \cdots & l_{0,3N,3N} \end{pmatrix}, \qquad (9.45)$$

where the first index after "0" denotes the component of a vector, and the second index after "0" denotes the root of Eq. (9.43). The matrix $\mathbf{L}$ is called the matrix of the eigenvectors; it and its inverse matrix transform the matrix $\mathbf{B}$ to the diagonal form:

$$\mathbf{L}^{-1}\cdot\mathbf{B}\cdot\mathbf{L}=\begin{pmatrix}\lambda_1 & 0 & 0 & \dots & 0\\ 0 & \lambda_2 & 0 & \dots & 0\\ 0 & 0 & \lambda_3 & \dots & 0\\ \vdots & \vdots & \vdots & \ddots & \vdots\\ 0 & 0 & 0 & 0 & \lambda_{3N}\end{pmatrix}=\Lambda, \tag{9.46}$$

where all diagonal elements are equal to the roots of the secular equation (9.43) and Λ stands for the diagonal matrix of the eigenvalues. One of the properties of the **L** matrix is that its inverse matrix is equal to its transposed:

$$\mathbf{L}^{-1} = \mathbf{L}^{T}. \tag{9.47}$$

It is also worthwhile to mention that the eigenvectors (9.44) form an orthogonal set:

$$\mathbf{l}_k^{\;T}\mathbf{l}_{k'}=\delta_{kk'}. \tag{9.48}$$

To illustrate the above outlined formalism, we consider a hypothetical linear molecule *ABC* (consisting of three different atoms with masses m_A, m_B, m_C; Fig. 9.16). For the sake of simplicity, we assume that the nuclei in this molecule can move only along the molecule axis. This will considerably simplify the mathematical treatment of the problem, but, at the same time, will not eliminate its physical content.

The nuclear kinetic energy T of such a system (in the Cartesian coordinates x_i and mass-weighted coordinates q_i) is

$$T=\frac{1}{2}\left(m_A\dot{x}_1^{\,2}+m_B\dot{x}_2^{\,2}+m_C\dot{x}_3^{\,2}\right)=\frac{1}{2}\left(\dot{q}_1^{\,2}+\dot{q}_2^{\,2}+\dot{q}_3^{\,2}\right), \tag{9.49}$$

and the potential energy U

$$\begin{aligned}U&=\frac{k_1(x_2-x_1)^2}{2}+\frac{k_2(x_3-x_2)^2}{2}\\ &=\frac{1}{2}\left(\frac{k_1q_1^{\,2}}{m_A}+\frac{(k_1+k_2)q_2^{\,2}}{m_B}+\frac{k_2q_3^{\,2}}{m_c}-\frac{2k_1q_1q_2}{\sqrt{m_Am_B}}-\frac{2k_2q_2q_3}{\sqrt{m_Bm_C}}\right).\end{aligned} \tag{9.50}$$

The force constants k_1, k_2 are related to the deformations of the corresponding chemical bonds. It should be pointed out that Eq. (9.50) implies that we consider interaction only between the nearest neighbors only and neglect interaction between atoms *A* and

C. It can be noted that there are the so-called "cross-terms" in the potential energy expression, containing the q_1q_2 and q_2q_3 products; this fact will be considered later from another point of view.

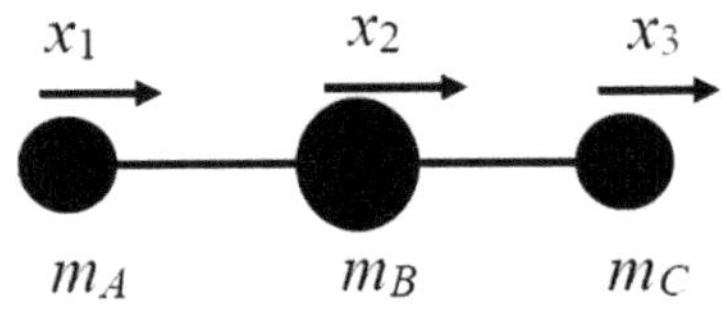

Figure 9.16 A linear molecule ABC and possible displacements of its nuclei.

The secular determinant for this molecule (we remind that its coefficients b_{ij} can be found as the second derivatives of the potential energy with respect to coordinates: $c_{ij} = \left(\frac{\partial^2 U}{\partial x_i \partial x_j}\right)_0$ $b_{ij} = c_{ij}\big/\sqrt{m_i m_j}$) is

$$\begin{vmatrix} \frac{k_1}{m_A} - \lambda & -\frac{k_1}{\sqrt{m_A m_B}} & 0 \\ -\frac{k_1}{\sqrt{m_A m_B}} & \frac{k_1 + k_2}{m_B} - \lambda & -\frac{k_2}{\sqrt{m_B m_C}} \\ 0 & -\frac{k_2}{\sqrt{m_B m_C}} & \frac{k_2}{m_C} - \lambda \end{vmatrix} = \lambda \left\{ \begin{aligned} &\lambda^2 - \lambda \left[k_1 \left(\frac{1}{m_A} + \frac{1}{m_B} \right) \right. \\ &\left. + k_2 \left(\frac{1}{m_B} + \frac{1}{m_C} \right) \right] + \\ &+ k_1 k_2 \frac{m_A + m_B + m_C}{m_A m_B m_C} \end{aligned} \right\}. = 0 \qquad (9.51)$$

There are three roots of this secular determinant, and one root (we denote it by λ_3) is zero. This root corresponds to the absence of vibrations and to the translational motion of the whole molecule along its axis. Obviously, such motion does not change the internal state of the molecule. The remaining roots λ_1 and λ_2 correspond to the stretching vibrations (which change the lengths of the A–B and B–C chemical bonds) with frequencies $\lambda_i = 4\pi^2 \nu_i^2$, $i = 1, 2$. To shed more light on the character and shape of these vibrations, at this point we introduce one additional approximation, assuming that the atoms A and C are identical (i.e., we deal with an ABA molecule. An example of such a molecule is the carbon dioxide O = C = O). Then $k_1 = k_2 = k$ and the roots of the secular determinant are

$$\lambda_1 = \frac{k}{m_A},\ \lambda_2 = \frac{k}{m_A} + \frac{2k}{m_B},\ \lambda_3 = 0\,. \tag{9.52}$$

As was mentioned previously, to find the corresponding eigenvectors we have to substitute each eigenvalue into the system of equations (9.39). In our case for $\lambda_3 = 0$ we have

$$\begin{aligned}
&\frac{k}{m_A} q_{01,3} - \frac{k}{\sqrt{m_A m_B}} q_{02,3} = 0 \\
&-\frac{k}{\sqrt{m_A m_B}} q_{01,3} + \frac{2}{m_B} q_{02,3} - \frac{k}{\sqrt{m_B m_A}} q_{03,3} = 0 \\
&-\frac{k}{\sqrt{m_A m_B}} q_{02,3} + \frac{k}{m_A} q_{03,3} = 0.
\end{aligned} \tag{9.53}$$

It is easy to see that these equations are linearly dependent. From the first and the third equations we get $\frac{q_{01,3}}{q_{02,3}} = \frac{\sqrt{m_A}}{\sqrt{m_B}};\ \frac{q_{03,3}}{q_{02,3}} = \frac{\sqrt{m_A}}{\sqrt{m_B}}$. Coming back from the mass-weighted coordinates to the Cartesian ones, we note that all Cartesian displacements are equal to each other: $x_{01,3} = x_{02,3} = x_{03,3}$, which indeed means that the root $\lambda_3 = 0$ corresponds to the translational motion of the considered molecule. The corresponding normalized eigenvector $\mathbf{l}_3$ is

$$\mathbf{l}_3 = \begin{pmatrix} \sqrt{m_A/m} \\ \sqrt{m_B/m} \\ \sqrt{m_A/m} \end{pmatrix},\ m = 2m_A + m_B\,. \tag{9.54}$$

In a similar way we find the two remaining normalized eigenvectors $\mathbf{l}_1$ and $\mathbf{l}_2$:

$$\mathbf{l}_1 = \begin{pmatrix} \sqrt{1/2} \\ 0 \\ -\sqrt{1/2} \end{pmatrix},\ \mathbf{l}_2 = \begin{pmatrix} \sqrt{m_B/2m} \\ -2\sqrt{m_A/2m} \\ \sqrt{m_B/2m} \end{pmatrix}. \tag{9.55}$$

Figure 9.17 shows all three eigenvectors (note that the eigenvectors are expressed in terms of the mass-weighted, but not Cartesian coordinates). The first eigenvector corresponds to the symmetric stretching (both *A* nuclei move in the opposite directions, whereas the middle nucleus is at rest); this vibration preserves the symmetry of the molecule. The second eigenvector corresponds to the asymmetric stretching; all three atoms participate in this vibration, but their amplitudes are different, so the symmetry of the molecule is broken. Finally, the last eigenvector, as it was mentioned previously, describes the translational motion of the molecule.

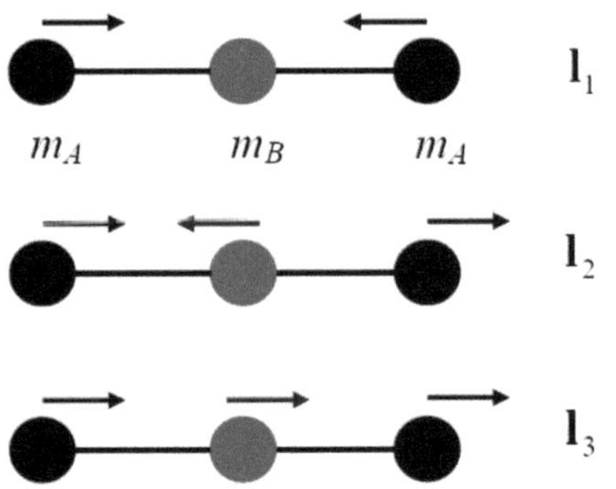

Figure 9.17 The shape of the three vibrations in the linear ABA molecule determined by Eqs. (9.54) and (9.55).

For the harmonic potentials (i.e., with no terms beyond the second power of coordinate) it is always possible to find the so-called normal coordinates Q_i, which will diagonalize the expressions for both kinetic and potential energies. If in the "old" mass-weighted coordinates q_i the kinetic and potential energies of a molecule were given by the expressions

$$T = \frac{1}{2}\sum_{i=1}^{3N} \dot{q}_i^{\,2}\,,\; U = \frac{1}{2}\sum_{i,j=1}^{3N} b_{ij} q_i q_j\,, \tag{9.56}$$

then in the "new" coordinates Q_i these equations will have the form

$$T = \frac{1}{2}\sum_{i=1}^{3N} \dot{Q}_i^{\,2}\,,\; U = \frac{1}{2}\sum_{i}^{3N} \lambda_i Q_i^{\,2}\,. \tag{9.57}$$

An essential feature of these new coordinates is that the potential energy U has no cross-terms (compare with Eq. (9.50)). The "old" coordinates $\mathbf{q}$ and "new" coordinates $\mathbf{Q}$ are connected by a linear transformation

$$\mathbf{q} = \mathbf{M}_0\mathbf{Q}. \tag{9.58}$$

Such a transformation has a simple meaning: The "old" system of coordinates (which is a 3N-dimensional one) is just rotated, and $\mathbf{M}_0$ is a matrix of such a rotation. Since the rotation of the system of coordinates does not change the total energy of a system, the eigenvalues (or vibrational frequencies) are identical in both sets of coordinates. The physical meaning of such a transformation becomes clear after we derive the equations of motion from the Lagrange equations $\frac{d}{dt}\frac{\partial L}{\partial \dot{Q}_i}-\frac{\partial L}{\partial \dot{Q}_i}=0, i=1,...,3N$ in these new coordinates. We find out immediately that we arrive at the system of 3N uncoupled homogeneous differential equations of the second order:

$$\ddot{Q}_i + \lambda_i Q_i^2 = 0, i = 1,...,3N \tag{9.59}$$

(compare with Eqs. (9.38) and (9.39)). Each normal coordinate oscillates independently; each equation of motion (which is, in fact, an equation of motion of a harmonic oscillator) can be easily solved separately. There is no energy exchange between different vibrational modes corresponding to these normal coordinates. The normal coordinates can be considered as a basis set in the 3N −6 dimensional space of nuclear displacements in a molecule. Any collective motion of atoms in the molecule can be represented as a linear combination of these normal coordinates (like any vector in the n-dimensional space is a linear combination of the n basis vectors).

The concept of the normal modes remains valid in the harmonic approximation only, and anharmonicity (the terms with the third and higher powers of coordinate in potential energy series expansion, for example, or any other anharmonic potential) will destroy this approximation resulting in impossibility to separate all vibrations into independent modes.

The question may arise how to find this transformation (in other words, how to find the $\mathbf{M}_0$ matrix). To answer this question, we express the potential energy U in these new coordinates:

$$2U = \mathbf{q}^T\mathbf{B}\mathbf{q} = \mathbf{Q}^T(\mathbf{M}_0{}^T\mathbf{B}\mathbf{M}_0)\mathbf{Q}. \tag{9.60}$$

On the other hand, $2U = \mathbf{Q}^T\Lambda\mathbf{Q}$, and $\mathbf{L}^{-1}\cdot\mathbf{B}\cdot\mathbf{L} = \Lambda$. Then it follows that

$$\mathbf{M}_0 = \mathbf{L}, \tag{9.61}$$

or, in other words, the matrix of the normalized eigenvectors is the required matrix, which would make a proper coordinates transformation.

We shall illustrate this by an example. Let us come back to the considered linear *ABA* molecule. The matrix **L** of its eigenvectors is as follows:

$$\mathbf{L} = \begin{pmatrix} \sqrt{1/2} & \sqrt{m_B/2m} & \sqrt{m_A/m} \\ 0 & -2\sqrt{m_A/2m} & \sqrt{m_B/m} \\ -\sqrt{1/2} & \sqrt{m_B/2m} & \sqrt{m_A/m} \end{pmatrix}, \tag{9.62}$$

If Q_1, Q_2, Q_3 are the normal coordinates, the "old" coordinates q_1, q_2, q_3 are

$$q = \begin{pmatrix} \frac{\sqrt{2}}{2}Q_1 + \sqrt{\frac{m_B}{2m}}Q_2 + \sqrt{\frac{m_A}{m}}Q_3 \\ -2\sqrt{\frac{m_A}{2m}}Q_2 + \sqrt{\frac{m_B}{m}}Q_3 \\ -\frac{\sqrt{2}}{2}Q_1 + \sqrt{\frac{m_B}{2m}}Q_2 + \sqrt{\frac{m_A}{m}}Q_3 \end{pmatrix}. \tag{9.63}$$

Introducing a diagonal matrix of the force constants $F = \begin{pmatrix} k & 0 & 0 \\ 0 & k & 0 \\ 0 & 0 & k \end{pmatrix}$ and remembering that the potential energy can be written as $2U = \mathbf{q}^T\mathbf{F}\mathbf{q}$, we easily get that $2U = k(Q_1^2 + Q_2^2 + Q_3^2)$, so the potential energy expression has no cross-terms in the normal coordinates.

The number and the shape of the normal modes for a particular complex are determined by the number of atoms in it and their geometrical arrangement. The normal coordinates of the octahedral complex AX_6, where A is a central ion and X denotes six ligands, are of special importance. For such a complex, there are 15 normal coordinates, which are all listed in Table 9.2 (the coordinates are determined in the system of reference shown in Fig. 8.1).

The first coordinate Q_1 is the so called fully symmetric or "breathing" mode, it has the symmetry properties of the a_{1g} irreducible representation of the O_h group (it is a common practice to denote the symmetry properties of the normal modes by the small letters, whereas the symmetry properties of the electronic states are denoted by the capital letters). It is easy to see that in this Q_1 mode the octahedral symmetry of the AX_6 complex is kept and all interatomic distances are changing equally and simultaneously. All other normal vibrations lower the symmetry of the AX_6 complex. Interaction of the impurity ions electronic states with the a_{1g}, e_g and t_{2g} modes is the strongest.

Table 9.2 Normal coordinates of the octahedral complex AX_6

Normal coordinate	Symmetry	Cartesian coordinates (Fig. 8.1)
Q_1	a_{1g}	$(X_2 - X_5 + Y_3 - Y_6 + Z_1 - Z_4)/\sqrt{6}$
Q_2	e_g	$(X_2 - X_5 - Y_3 + Y_6)/2$
Q_3	e_g	$(2Z_1 - 2Z_4 - X_2 + X_5 - Y_3 + Y_6)/2\sqrt{3}$
Q_4	t_{2g}	$(Z_3 - Z_6 + Y_1 - Y_4)/2$
Q_5	t_{2g}	$(X_1 - X_4 + Z_2 - Z_5)/2$
Q_6	t_{2g}	$(Y_2 - Y_5 + X_3 - X_6)/2$
Q_7	t_{1u}	$(X_3 + X_1 + X_6 + X_4)/2$
Q_8	t_{1u}	$(Y_2 + Y_1 + Y_5 + Y_4)/2$
Q_9	t_{1u}	$(Z_2 + Z_3 + Z_5 + Z_6)/2$
Q_{10}	t_{1u}	$(X_2 + X_5)/\sqrt{2}$
Q_{11}	t_{1u}	$(Y_3 + Y_6)/\sqrt{2}$
Q_{12}	t_{1u}	$(Z_1 + Z_4)/\sqrt{2}$
Q_{13}	t_{2u}	$(X_3 + X_6 - X_1 - X_4)/2$
Q_{14}	t_{2u}	$(Y_1 + Y_4 - Y_2 - Y_5)/2$
Q_{15}	t_{2u}	$(Z_2 + Z_5 - Z_3 - Z_6)/2$

9.8 General Picture of the Jahn–Teller Effect

The Jahn–Teller theorem states that if a molecule (or an impurity center in a crystal) is in a degenerate electronic state, the geometrical configuration of such a system is unstable. It will distort to remove this orbital degeneracy and lower the total energy of a system. The theorem was formulated by H. A. Jahn (1907–1979) and E. Teller (1908–2003) and it was proved by them using the group-theoretical concepts.[17]

The main physical content of the theorem is that due to the vibrations of atoms or ions in a molecule or in an impurity center, a new lower symmetry equilibrium configuration of atoms is formed. This symmetry lowering causes the splitting of the degenerate electronic levels. The degree of the geometrical distortion depends on two factors: decrease of the electronic energy (after the degenerated energy level is split) and increase of the vibrational energy (when the complex is deformed). Mathematically this can be written as follows:

$$U = -VQ + \frac{kQ^2}{2}, \tag{9.64}$$

where U is a potential energy of a complex, V is a constant of electron-vibrational interaction, Q is a displacement, and k is a force constant. The minimum of the potential energy (9.64) is achieved if

$$Q_{\min} = \frac{V}{k}. \tag{9.65}$$

The amount of decrease of the potential energy is called the Jahn–Teller stabilization energy E_{JT} and equals to

$$E_{JT} = -\frac{V^2}{2k}. \tag{9.66}$$

Generally, the Jahn–Teller effect can be represented as follows: A fully symmetric (which keeps the symmetry of a molecule or an impurity centers, as shown in Fig. 9.18) vibrational mode results in a shift of the potential energy surface of the excited state (Fig. 9.2), and the lower-symmetry vibrations split the degenerated electronic states.

Not all vibrations of a molecule keep its shape; most of them lower the molecule's symmetry. Due to the influence of these vibrations, the degenerated state will split, as shown in Fig. 9.19.

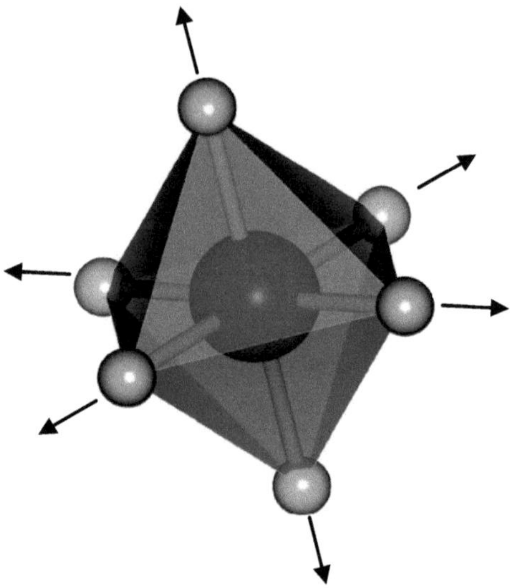

Figure 9.18 Completely symmetric mode of vibrations of an octahedral complex. All six ions at every moment of time are at the same distance from the central ion; they simultaneously move with the same amplitude either toward the center, or from the center.

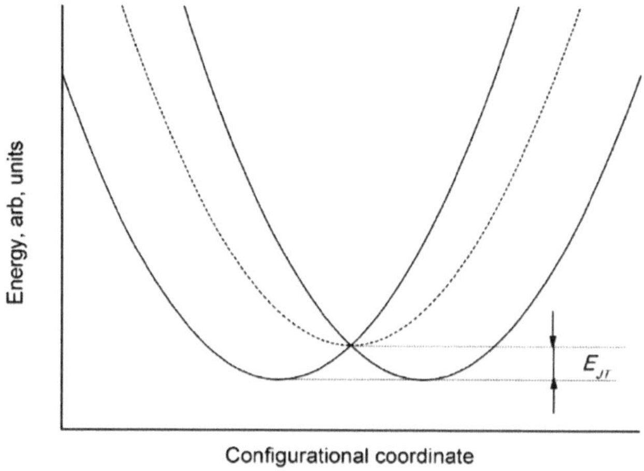

Figure 9.19 Splitting of the degenerated electronic state (dashed line) due to the non-symmetric vibrations. The Jahn–Teller stabilization energy E_{JT} is shown. The picture is a vertical cross section of the three-dimensional surfaces of potential energy.

Detailed treatment of the Jahn–Teller effect is a rather laborious problem. The numerical estimations of the Jahn–Teller stabilization energy give the results of the order of several hundred cm^{-1} for the complexes formed by the transition metal ions. For example, for the $^4T_{2g}$ state of the V^{2+} ions in $CsCaF_3$ E_{JT} = 369 cm^{-1}.[18] One example of the Jahn–Teller distortion for the $CsCaF_3$:V^{2+} system is shown in Fig. 9.20. The vanadium ions are surrounded by six fluorine ions in this host, and the ideal symmetry of the fluorine octahedron is lowered (the magnitudes of the ligands' displacements are shown in the figure). As seen from Fig. 9.20, the complex undergoes an axial stretching (along the z axis) and an equatorial expansion (in the xy plane).

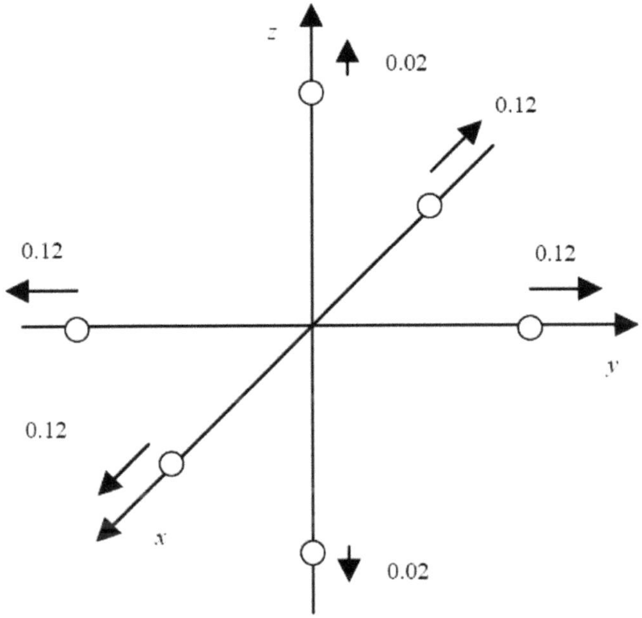

Figure 9.20 Distortion of the $[VF_6]^{4-}$ complex in the $^4T_{2g}$ excited state with respect to the ground state (directions of displacements are indicated by arrows; their magnitudes are given in Å).

Detailed description of the Jahn–Teller effect in the excited orbital triplets of the d^3 ions, which includes the second-order effective spin-Hamiltonian formalism, can be found in Refs. [19] and [20] and other papers cited there.

Bibliography

1. Henderson B, Imbush GF (2006). *Optical Spectroscopy of Inorganic Solids*. Clarendon Press.
2. Nazarov M, Tsukerblat B, Noh DY (2008). Vibronic coupling parameters and Stokes shift in thiogallate phosphors. *J Phys Chem Solids* **69**(10) 2605–2612.
3. Leaño Jr JL, Lesniewski T, Lazarowska A, et al. (2018). Thermal stabilization and energy transfer in narrow-band red-emitting $Sr[(Mg_2Al_2)_{1-y}(Li_2Si_2)_yN_4]$: Eu^{2+} phosphors, *J. Mater. Chem. C* **6**(22) 5975–5983.
4. Brik MG (2005). Crystal field analysis, electron-phonon coupling and spectral band shape modeling in MgO:Cr^{3+}, *Z Naturforsch A* **60**(10) 437–443.
5. Brik MG, Avram NM (2011). Electron-vibrational interaction in the 5*d* states of Ce^{3+} ions in halosulphate phosphors, *Mater Chem Phys* **128**(3) 326–330.
6. Bartwal KS, Ryu H, Brik MG, Sildos I (2010). Preparation and spectroscopic studies of the $Mg_xSr_{1-x}Al_2O_4$:Eu, Dy (x = 0.05–0.25) persistent phosphors, *Opt Mater* **32**(10) 1329–1332.
7. Kumar GA, Liu DX, Tian Y, Brik MG, Sardar DK (2015). 5*d*-4*f* emission of Eu^{2+} and electron-vibrational interaction in several alkaline earth sulfides doped with Eu^{2+} and Er^{3+}, *Opt Mater* **50** 199–203.
8. Bartwal KS, Ryu H, Brik MG, Sildos I (2009). Studies of electron-vibrational interaction and crystal field splitting in 5*d* states of Eu^{2+} in $CaAl_2O_4$ co-doped with Eu^{2+} and Er^{3+}, *J Phys D: Appl Phys* **42**(24) 245401.
9. Hou DJ, Ma CG, Liang HB, Brik MG (2014). Electron-Vibrational Interaction in the 5*d* States of Eu^{2+} Ions in $Sr_{6-x}Eu_xBP_5O_{20}$ (x = 0.01–0.15), *ECS J Solid State Sci Technol* **3**(4) R39–R42.
10. Ou YY, Zhou WJ, Liu CM, et al. (2018). Vacuum referred binding energy scheme, electron–vibrational interaction, and energy transfer dynamics in $BaMg_2Si_2O_7$: Ln (Ce^{3+}, Eu^{2+}) phosphors, *J Phys Chem C* **122**(5) 2959–2967.
11. de Jong M, Seijo L, Meijerink A, Rabouw FT (2015). Resolving the ambiguity in the relation between Stokes shift and Huang-Rhys parameter, *Phys. Chem. Chem. Phys.* **17**(26) 16959–16969.
12. Mott NF (1938). On the absorption of light by crystals, *Proc R Soc London Ser A* **167**(930) 384–391.

13. Brik MG, Avram NM, Avram CN, Tanaka I (2005). Effects of the spin-triplet states mixture and electron-phonon coupling in $Y_3Al_5O_{12}:Cr^{4+}$, *Eur Phys J: Appl Phys* **29**(3) 239–245.

14. Avanesov AG, Brik MG, Tumayev EN (2001). Non-radiative transitions in the oscillating field mode, *J Lumin* **92**(1–2) 133–137.

15. Brik MG, Draganescu GE, Avram NM, Avram CN (2005). Non-radiative transitions in the anharmonic oscillating field mode, *Physica B* **364**(1–4) 170–179.

16. Brik MG, Lebedev VA, Stroganova EV (2007). Spectroscopic and crystal field studies of $(Ce,Gd)Sc_3(BO_3)_4:Cr^{3+}$ crystals, *J Phys Chem Solids* **68**(9) 1796–1804.

17. Bersuker IB (2006). *The Jahn-Teller Effect.* Cambridge University Press.

18. Avram CN, Brik MG, Avram NM, Tanaka I (2005). Electron–phonon interaction in the $V^{2+}:CsCaF_3$ laser crystal: Geometry of the $[VF_6]^{4-}$ complex in the $^4T_{2g}$ excited state, *Physica B* **355**(1–4) 164–171.

19. Brik MG, Avram NM, Tanaka I (2004). Electron-phonon coupling and Jahn-Teller effect in $KMgF_3:Cr^{3+}$, *Phys Status Solidi B* **241**(13) 2982–2993.

20. Brik MG, Avram NM, Avram CN (2010). Jahn–Teler effect for 3d ions (orbital triplets in a cubic crystal field), in *The Jahn–Teller Effects: Fundamentals and Implications for Physics and Chemistry*, Köppel H, Yarkony DR, Barentzen H, eds, Springer, pp. 347–370.

Chapter 10

Combination of the First-Principles and Semi-Empirical Models for a Complete Description of the Electronic Properties of the Doped Crystals

10.1 Introduction

The inorganic solids that are used for optical applications are the dielectrics or semiconductors doped with the transition metal or rare earth ions. Pure or neat hosts, with ideal crystal structure and without any impurities, are characterized by the so-called band gaps, which show the energy difference between the empty conduction band and occupied valence band. When impurity ions are introduced into a crystal lattice of a dielectric, some of the energy levels of impurity become located within the host band gap. The position of these energy levels as well as the complicated interplay of the host's and impurity's properties determine the optical properties of a doped material.

A classic example is corundum, or aluminum oxide Al_2O_3. When it is pure, it appears to be a colorless or somewhat grayish mineral. If, however, a few percent of the Al^{3+} ions are replaced by the Cr^{3+} ion, a precious stone – ruby – is formed, with variety of red colors from light pink to deep red (and with huge range of prices as well!). But

Theoretical Spectroscopy of Transition Metal and Rare Earth Ions: From Free State to Crystal Field
Mikhail G. Brik and Chong-Geng Ma

ISBN 978-981-4800-56-3 (Hardcover), 978-0-429-27875-4 (eBook)
www.jennystanford.com

if a few percent of the Al^{3+} ions in the same corundum are replaced by the Ti^{3+} ion, the color is changed to blue, and it is then another precious stone – sapphire. So, different impurities in the same host may drastically modify the optical properties of the material and change the range of its optical absorption.

The understanding of these modifications comes from modeling the energy levels of both impurity and host material. We have seen that the energy levels splitting of an impurity ion can be calculated by using the crystal field theory. However, the crystal field theory cannot calculate the host's electronic band structure. To get the electronic properties of the solids, including the band structure etc, many computational packages based on the density functional theory (DFT) were developed. However, most of these programs are based on the one-electron approximation and cannot hold properly the multielectron effects in the impurity ions. So, a combined application of the semi-empirical and first-principles methods to the same material can make advantages of the strong features of both approaches and seems to be a promising area of research.

In this chapter, we discuss several examples of application of the semiempirical and first-principles methods (separately and jointly) to the analysis of the optical and electronic properties of the doped optical materials.

10.2 First-Principles and Semi-Empirical Modeling of Microscopic Crystal Field Effects for the 3*d* Ions

It was shown in Chapter 8 that the crystal field strength Dq depends on the distance R between the impurity ion and ligand as $1/R^5$, if the point charge model is used. However, this is a rather insufficient approximation, and actual dependence of Dq on distance for real crystals deviates from that point charge model-based prediction. It is clear, of course, that Dq should be a decreasing function of distance, and its general representation can be as follows:

$$Dq = \frac{A}{R^n}, \tag{10.1}$$

where A is a constant and n is a host- and impurity ion-specific parameter. As quantum chemical calculation show, n can take values in a wide range from 3 to 7 for various systems.[1–3] The knowledge of the n value in Eq. (10.1) is not just of an academic interest: It can help in theoretical estimations of several observable or extractable from the experimental spectra parameters, such as the Stokes shift, Huang–Rhys factor, bulk modulus and Grüneisen constant.

The Hamiltonian H_{EVI} of the electron-vibrational interaction (EVI) of the impurity ion electronic states with the a_{1g} and e_g normal modes of the octahedral complex is determined as[4]

$$H_{\mathrm{EVI}} = V_A \begin{pmatrix} Q_{a_{1g}} & 0 & 0 \\ 0 & Q_{a_{1g}} & 0 \\ 0 & 0 & Q_{a_{1g}} \end{pmatrix} + V_E \begin{pmatrix} \frac{1}{2}Q_{e_{g\theta}} - \frac{\sqrt{3}}{2}Q_{e_{g\varepsilon}} & 0 & 0 \\ 0 & \frac{1}{2}Q_{e_{g\theta}} + \frac{\sqrt{3}}{2}Q_{e_{g\varepsilon}} & 0 \\ 0 & 0 & -Q_{e_{g\theta}} \end{pmatrix}, \quad (10.2)$$

where V_A, V_E are the constants of the EVI with the a_{1g} and e_g normal modes of the octahedral complex, respectively, $Q_{e_{g\theta}} \sim 3z^2 - r^2$, $Q_{e_{g\varepsilon}} \sim x^2 - y^2$ are two normal coordinates of the e_g Jahn–Teller mode. The constants V_A, V_E can be expressed as follows:

$$V_A = -\frac{nA}{\sqrt{6}R^{n+1}}, \quad V_E = \frac{V_A}{\sqrt{2}}. \quad (10.3)$$

If the values of V_A, V_E are determined, one can proceed with the estimation of the Stokes shift produced by the i-th normal mode:

$$E_S(i) = 2S_i \hbar\omega_i = \frac{V_i^2}{M\omega_i^2}, \quad (10.4)$$

where M is a mass of a single ligand, S_i is the Huang–Rhys factor for that mode with the vibrational frequency ω_i. The total Stokes shift produced as a result of a combined action of the a_{1g} and e_g normal modes can be written as a sum of the individual Stokes shifts from each mode[5,6]:

$$E_S = E_S(a_{1g}) + E_S(e_g). \tag{10.5}$$

Moreover, if the experimental pressure dependence $\left(\frac{\partial 10Dq}{\partial P}\right)_{P_0}$ of the crystal field strength is known, estimations of the bulk modulus B from the following expression become possible:

$$\left(\frac{\partial 10Dq}{\partial R}\right)_{R=R_0} = -\left(\frac{\partial 10Dq}{\partial P}\right)_{P_0} \frac{3B}{R_0} \tag{10.6}$$

Finally, lower estimations of the Grüneisen constant $\gamma(a_{1g})$ for the a_{1g} mode can be made as[7]:

$$9\gamma(a_{1g}) - 2(n+1) > 0. \tag{10.7}$$

So, indeed, if the distance dependence of the crystal field strength is known, several spectroscopic and elastic parameters of a solid can be evaluated and compared with the experimental data.

The way to determine the value of n in Eq. (10.1) is to calculate numerically Dq values for different interionic distances and then fit the calculated results to the power law in Eq. (10.1). The calculations can be performed in different ways. The first-principles methods can be used to calculate the positions of the t_{2g} and e_g states arising from the split d-orbitals of a central ion.[1,6–14] The semi-empirical models of crystal field (and ECM, in particular) can also be used for that purpose. For the d^3 (d^7) ions in the octahedral (tetrahedral) coordination the energy interval between the $^4A_2$ and $^4T_2$ states is equal to $10Dq$. Also, for the d^8 (d^2) ions in the octahedral (tetrahedral) coordination the energy interval between the $^3A_2$ and $^3T_2$ returns the $10Dq$ value. Therefore, calculating the energy level schemes of ions with the above-mentioned electronic configurations allows to extract the $10Dq$ values and fit them to the power law from Eq. (10.1). After the value of n is determined, all Eqs. (10.2)–(10.7) can be used[15–18] (the vibrational frequencies of the normal modes can be taken from the Raman or infrared spectra and can be found in the literature).

Table 10.1 collects the values of n from Eq. (10.1) obtained for different systems. Indeed, the range of the n values is quite wide, and the difference from the value of five predicted by the point charge model is significant.

Table 10.1 Values of *n* in the distance dependence of crystal field strength *Dq* (Eq. (10.1)) for some 3*d* ions

Complex or doped crystal	n	Reference	Complex or doped crystal	n	Reference
Calculated results					
CrF_6^{3-}	4.3	19	$SrTiO_3:Cr^{3+}$	4.91	15
VF_6^{4-}	5.2	19	$SrTiO_3:Mn^{4+}$	5.80	15
MnF_6^{4-}	4.8–6.1	7	$SrTiO_3:Fe^{5+}$	6.55	15
MnF_8^{6-}	5.9	7	$K_2LiAlF_6:Cr^{3+}$	4.57	16
$MnCl_6^{4-}$	5.8	7	$MgO:Cr^{3+}$	6.40	17
$CuBr_4^{2-}$	3.4	7	$ZnS:V^{2+}$	4.42	17
$Cs_2NaYF_6:Cr^{3+}$	4.46	8	$KAl(MoO_4)_2:Cr^{3+}$	5.62	18
$Cs_2NaYCl_6:Cr^{3+}$	4.37	8	CrF_6^{2-} in Rb_2CrF_6	5.86-5.72	10
$Cs_2NaYBr_6:Cr^{3+}$	4.35	8	$Al_2O_3:Ti^{3+}$	4.67	12
$KZnF_3:Ni^{2+}$	5.40–6.02	12			
Experimental results					
$La_3Lu_2Ga_3O_{12}:Cr^{3+}$	3.45	20	$LiTaO_3:Cr^{3+}$	2.70	28
$Al_2O_3:Cr^{3+}$	3.40	21	$Y_3Al_5O_{12}:Cr^{3+}$	3.06	22
$Al_2O_3:Ti^{3+}$	2.73, 2.54	23, 24	$KNaGaF_6:Cr^{3+}$	1.50	25
$YAlO_3:Ti^{3+}$	3.18	26	$Gd_3Ga_5O_{12}:Mn^{4+}$ (site A)	3.36	27
$LiNbO_3:Cr^{3+}$	2.85	28	$Gd_3Ga_5O_{12}:Mn^{4+}$ (site B)	4.69	27

Table 10.2 Calculated EVI parameters for several crystals doped with the 3*d* ions

	R_0, Å	Normal mode frequencies, cm^{-1}		EVI constants, cm^{-1} pm^{-1}		Huang–Rhys factor		Stokes shift E_S, cm^{-1}		Ref.
		ω_{a1g}	ω_{eg}	V_A	V_E	S_A	S_E	Calc.	Exp.	
$Cs_2NaYF_6:Cr^{3+}$	1.88	501	402	−171	−121	2.05	1.99	3654	3721	8
$Cs_2NaYF_6:Cr^{3+}$	2.40	298	240	-90	-64	1.57	1.65	1663	1612	8
$Cs_2NaYF_6:Cr^{3+}$	2.44	183	144	-100	-70	3.41	3.50	2556	2218	8
$KZnF_3:Ni^{2+}$	2.07	454	387	−171	−121	0.70	0.57	1139	1400	12
$SrTiO_3:Cr^{3+}$	1.95	887	545	−166	−117	0.41	0.89	1706	—	15 *
$SrTiO_3:Mn^{4+}$	1.95	887	545	-220	−156	0.73	1.58	3020	—	15 *
$SrTiO_3:Fe^{5+}$	1.95	887	545	-276	−195	1.15	2.48	4737	—	15 *
$K_2LiAlF_6:Cr^{3+}$	1.81	560	471	−166	−117	1.39	1.16	2632	2420	16 *
$MgO:Cr^{3+}$	2.11	570	380	−194	−138	2.15	3.62	5200	4500	17 *
$KAl(MoO_4)_2:Cr^{3+}$	1.85	1003	352	−184	−130	0.35	4.11	3600	2800	18 *

Note: The calculations were based on the obtained values of *n* in the distance dependence of crystal field strength *Dq*. The calculations based on the ECM are marked with an asterisk; the remaining references are to the papers based on the first principles-calculations. All references to the relevant experimental data are in cited in those papers given in the table.

The calculated EVI parameters for several crystals containing the 3*d* ions as impurities are given in Table 10.2. The most obvious comparison of the calculated results with the experimental data is to see how the calculated Stokes shift matches the value extracted from the absorption and emission spectra. As seen from the Table, very good agreement between the experimental and estimated Stokes shift values is achieved in the most considered cases.

10.3 Determination of the Complete Energy Level Scheme of Doped Crystals (3*d* Ions)

Position of the ground and excited states of an impurity ion in the band gap of a host crystal is a crucial factor that determines the optical properties of such a doped material. It may happen that the same impurity in different hosts may or may not emit light. For example, bright 5*d* – 4*f* emission of the Ce^{3+} ions in $Y_3Al_5O_{12}$ can be detected at the ambient pressure,[29] but the same ion does not emit under the same conditions in $Y_3Ga_5O_{12}$,[30] and only application of high pressure can restore the Ce^{3+} luminescence. Apparently, this is due to the differences in positions of the Ce^{3+} 5*d* states in these garnets: In the former case, the lowest 5*d* level is in the host's band gap, whereas in the latter the lowest 5*d* level is hiding in the host's conduction band. At the increased pressure due to the increase of the band gap the Ce^{3+} lowest 5*d* "falls out" of the conduction band and becomes a luminescence level. Therefore, peculiarities of the relative positions of the impurity's energy levels and host's electronic band structure are a key-factor to understand the origin or the absence of the emission/absorption peaks in the experimental spectra.

Several empirical models have been developed by Prof. P. Dorenbos,[31–36] which allow one to identify the lowest 4*f* and 5*d* states of the di- and tri-valent lanthanides in inorganic solids, the energies of the relevant charge transfer transitions etc. Remarkable success of those models is evidenced by good agreement with the experimental data and by ever-growing number of citations to the original papers describing the models and applications.

The theoretical estimations of the location of the energy levels of the impurity ions in the host's band gap are yet very difficult. The reason is that the first-principles calculations of the electronic and optical properties of the doped compounds with realistic concentrations of impurity ions (not more that ~ several %) imply a use of large supercells and, as consequence, advanced computational facilities. More importantly, most of the modern computational use single-electron functions as the basis set, and the multielectron effects, SO interaction, effects of interaction of different electron configurations may not be considered properly. However, the multielectron configurations energy levels can be handled by the crystal field theory, although it is not capable of calculating the electronic band structure of the host.

Recently we proposed an approximate method of joint application of the crystal field theory and first-principles methods of calculations of the electronic structure to the same compound with impurity ions. The result of such a procedure is the complementary picture of the electronic and optical properties of a doped compound, which includes the crystal field-split energy levels of impurity ion superimposed onto the electronic band structure of the host. The overall algorithm of the proposed approach is as follows:

(i) Electronic and optical properties of a considered pure compound (i.e., without any impurity) is calculated. It is better to perform calculations for a larger supercell, whose size is the same as described in the next step (then comparison of the results of the first two steps is more meaningful).

(ii) A supercell is created, in which one of the host cation is substituted by a considered impurity ion. The size of the supercell depends on the size of the unit cell of a considered crystal, number of formula units in a unit cell and available computational facilities. The supercell is optimized, and the electronic properties including the band structure and density of states are calculated. The impurity states appear in the host's band gap as nearly flat almost dispersionless levels. From the density of states diagrams, the position of the lowest impurity level above the top of the valence band is found.

(iii) Conventional crystal field theory (ECM, for example) is used to calculate the crystal field splitting of all impurity ion *LS* terms.

(iv) The calculated at the previous step crystal field-split energy levels of an impurity ion are superimposed onto the band structure diagram obtained at the first step, with location of the ground impurity state at the energetical distance from the top of the valence band found at the second step.

This approach is well justified in the case of impurities with a single electron in the unfilled shell. When multielectron impurities are considered, it gives only an approximate estimate of the location of the impurity ground states.

Let us consider a couple of examples.

The cubic spinel crystal $ZnAl_2S_4$ can be efficiently doped with various transition metal ions. In particular, the V^{3+} ions can occupy the trigonally distorted Al^{3+} octahedral sites; the local site symmetry is D_{3d}. Detailed analysis of the electronic and optical properties of the pure $ZnAl_2S_4$ spinel and doped $ZnAl_2S_4$:V^{3+} was published in Ref. 37. Pure $ZnAl_2S_4$ has a wide band gap of about 3.42 eV.[38] The calculated[37] band gap was 2.5 eV, which is a usual underestimation of the band gap in the first-principles calculations. Figure 10.1a shows the calculated band structure of pure $ZnAl_2S_4$ with the corrected band gap (the conduction band was moved up, by applying the so-called scissor operator, to match the experimental value). The origin of the calculated electronic bands can be understood with the help of the density of states diagrams from Fig. 10.2a. The conduction band is mainly formed by the Al 3*s* and 3*p* states, whereas a wide valence band has two maxima: The lowest one at about −6 eV is due to the completely filled Zn 3*d* states. The upper part of the valence band is dominated by the S 3*p* states. A deeper located band and about −11 eV is composed of the S 3*s* states with minor contribution arising from the Al 3*s*, 3*p* states due to the hybridization effects.

The effects of the dopant V^{3+} ions on the electronic structure were modeled by replacing one Al^{3+} by one V^{3+} ion in a unit cell. The calculated band structure of the $ZnAl_2S_4$:V^{3+} is shown in Fig. 10.1b. An important difference from Fig. 10.1a is that there are additional states in the band gap, which are identified (with the help of the density of states diagram in Fig. 10.2b) as the 3*d* states of the vanadium ions. The lowest states arise from the t_{2g} orbitals, whereas the highest ones – from the e_g orbitals. It can be estimated from Figs. 10.1b and 10.2b that the lowest vanadium 3*d* state is located at about 1.36 eV above the top of the valence band.

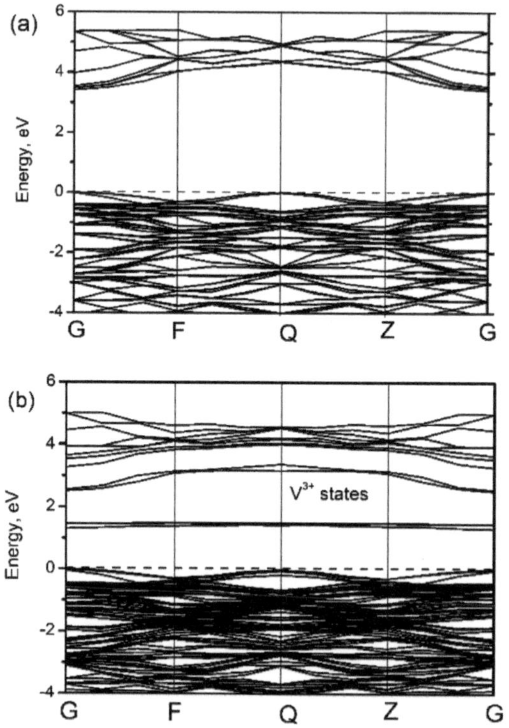

Figure 10.1 Comparison of the calculated band structures for pure $ZnAl_2S_4$ (a) and $ZnAl_2S_4$:V^{3+} (b). Coordinates of the special points of the Brillouin zone are (in units of the reciprocal lattice vectors): G(0,0,0); F(0,1/2,0); Q(0,1/2,1/2); Z(0,0,1/2). Reproduced with permission from Ref. 37.

The next step is to use the crystal field theory (the ECM model). Using the structural data on $ZnAl_2S_4$ from Ref. 39, the crystal field parameters (Table 10.3) and V^{3+} energy levels (Table 10.4) were calculated. The structure of the crystal field Hamiltonian with non-zero crystal field parameters from Table 10.3 confirms the trigonal symmetry at the vanadium position in $ZnAl_2S_4$.

The crystal field parameters from Table 10.3 were used to find the splitting of all five *LS* terms of the $3d^{\,2}$ configuration of the V^{3+} ions. All calculated energy levels are assigned in terms of the D_{3d} point group irreducible representations. Agreement with the experimentally detected V^{3+} energy levels (Table 10.4) is good.

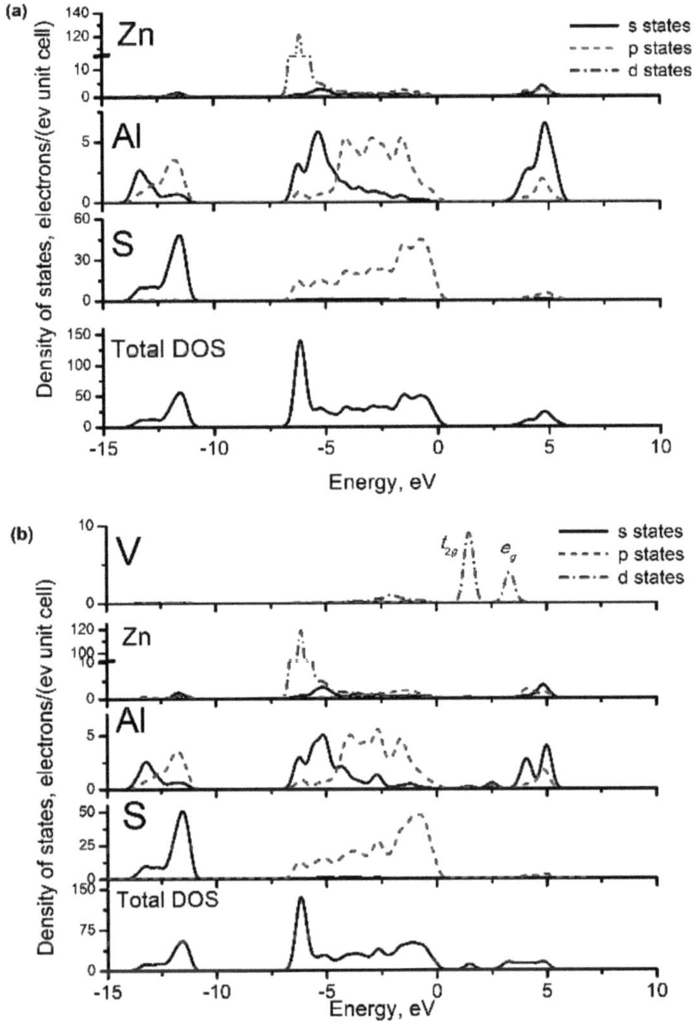

Figure 10.2 Calculated density of states diagrams for pure $ZnAl_2S_4$ (a) and $ZnAl_2S_4$:V^{3+} (b). Reproduced with permission from Ref. 37.

Finally, keeping in mind that the lowest V^{3+} level in the $ZnAl_2S_4$ band gap was identified to be at 1.36 eV above the top of the valence band and having all other V^{3+} levels from Table 10.4, it is possible to draw the combined energy level diagram in Fig. 10.3.

Table 10.3 The crystal field (in Stevens normalization) and Racah parameters (in cm^{-1}) for V^{3+} in $ZnAl_2S_4$

	$B^k_{p,q}$	$B^k_{p,S}$	B^k_p
B_2^0	-929.9	1574.5	644.6
B_4^0	-348.3	-2151.8	-2500.1
B_4^3	9991.3	53340.0.	63331.3
G (ECM parameter)	6.3		
B	440		
C	2000		

Source: Reproduced with permission from Ref. 37.
Note: G is the dimensionless ECM parameter.

Table 10.4 The calculated energy levels (in cm^{-1}) of V^{3+} in $ZnAl_2S_4$ in comparison with the experimental data

O_h group irreps. and LS parent term	D_{3d} group irreps.	Calc.	Observed[37]
$^3T_{1g}$ (^{3}F)	^3E_g, $^3A_{2g}$	0*, 409	
$^1T_{2g}$ (^{1}D)	$^1A_{1g}$, ^1E_g	6835, 6967*	
^1E_g (^{1}D)	^1E_g	7183*	
$^3T_{2g}$ (^{3}F)	^3E_g, $^3A_{1g}$	12406*, 12451 (1.5)	
$^1A_{1g}$ (^{1}G)	$^1A_{1g}$	15169 (1.88)	13873 (1.72)
$^3T_{1g}$ (^{3}P)	$^3A_{2g}$, ^3E_g	17886, (2.21) 18182* (2.25)	17906 (2.2)
$^1T_{2g}$ (^{1}G)	^1E_g, $^1A_{1g}$	20053*, 20252 (2.51)	21778 (2.7)
$^1T_{1g}$ (^{1}G)	^1E_g, $^1A_{2g}$	21681*, 21731	
$^3A_{2g}$ (^{3}F)	$^3A_{2g}$	25659 (3.18)	25649 (3.18)

Note: The orbital doublet states are denoted with an asterisk; the numbers in the parenthesis are the energies in eV. Reproduced with permission from Ref. 37.

It is clear then from Fig. 10.3 that the V^{3+} energy levels, which are in the conduction band, cannot be detected in the experimental absorption spectra. It can also be concluded then that proposed method of building up the energy level scheme of impurity ions with respect to the host's band structure is a useful tool for analysis of optical properties of doped optical materials.

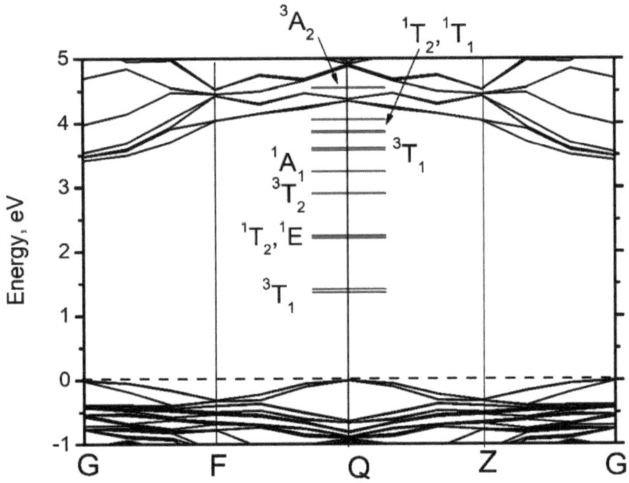

Figure 10.3 The calculated band structure of $ZnAl_2S_4$ with superimposed energy levels of the V^{3+} ions. Reproduced with permission from Ref. 37.

Another example is concerned with a single electron impurity, Ti^{3+} ions in $BeAl_2O_4$.[14] This is a well-known laser material. The initial structural data (Pnma space group, a = 9.4019 Å, b = 5.4746 Å, c = 4.4259 Å, four formula units in one unit cell) were taken from Ref. 14. The calculated direct band gap was 6.094 eV, which is close to another calculated result of 6.45 eV,[14] and which is underestimated if compared to the experimental value of about 9 eV.[40] The conduction band consists of the Be 2s, Al 3s, 3p states, whereas the oxygen 2p states produce the main contribution to the valence band. Figure 10.4 shows the calculated band structure and density of states for the pure $BeAl_2O_4$. It can be noticed that the top of the valence band is very flat, whereas the states at the bottom of the conduction band exhibit strong dispersion.

When one Al^{3+} ion in the unit cell is replaced by one Ti^{3+} ion, the electronic properties are considerably changed. Some extra states appear in the band gap (Fig. 10.5a), and the analysis of the density of states (Fig. 10.5b) unambiguously assigns these extra levels to the 3d orbitals of the Ti^{3+} ions.

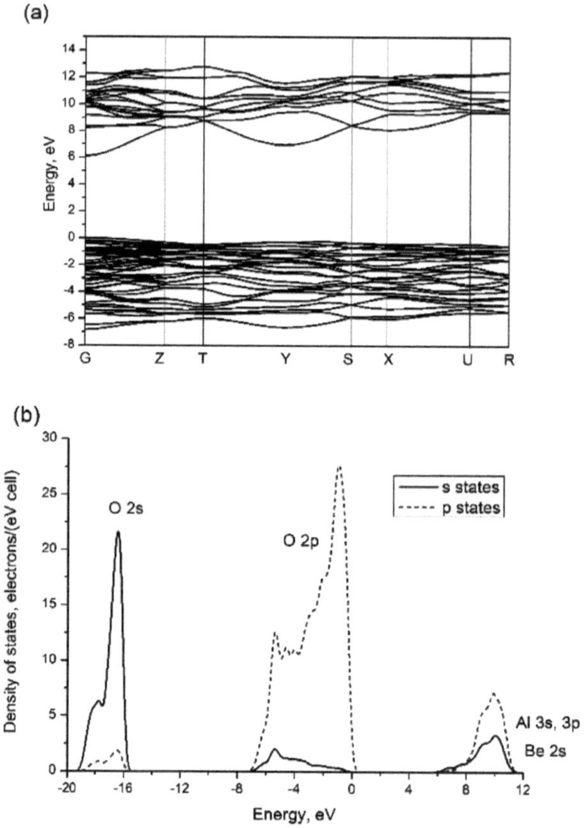

Figure 10.4 Calculated band structure (a) and density of states (b) diagram for $BeAl_2O_4$.

The conduction band in Fig. 10.5 was moved up to match the experimental value of the band gap, which is a standard way to overcome the underestimated band gaps in the DFT-based calculations. The lowest peak of the Ti^{3+} states in the band gap comes from the t_2 states and is located at about 4.4 eV above the top of the valence band. The second peak of the Ti^{3+} states in the band gap arises from the e states. The separation between them is about 2.5 eV (~20000 cm^{-1}), which is exactly $10Dq$ parameter. This result agrees well with the experimental position of the maxima of the $BeAl_2O_4$:Ti^{3+} absorption spectra at about 19980 cm^{-1}.[41]

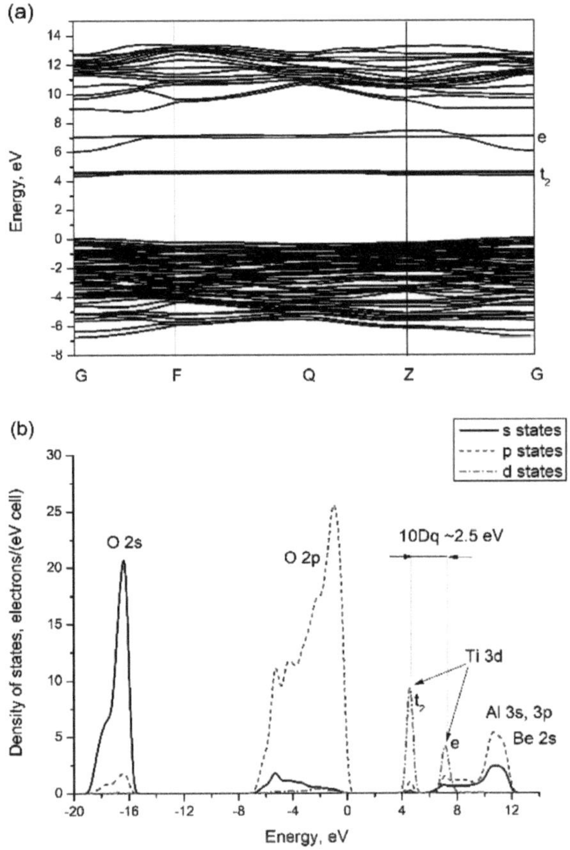

Figure 10.5 Calculated band structure (a) and density of states (b) diagram for $BeAl_2O_4:Ti^{3+}$. Reproduced with permission from Ref. 14.

The crystal field splitting of the Ti^{3+} states in $BeAl_2O_4:Ti^{3+}$ was calculated in Ref. 42. Due to a low symmetry of the Ti^{3+} position, all t_2 and e states are split completely to produce the following energy level scheme (in cm^{-1}): 0, 1450, 1684 (the t_2 states) and 19568, 22488 (the e states). Taking the lowest level to be at 4.4 eV above the valence band, it is easy then to plot the energy level scheme shown in Fig. 10.6, which combines the results of the DFT-based calculations (the electronic band structure) and crystal field theory (the $3d$ orbitals splitting).

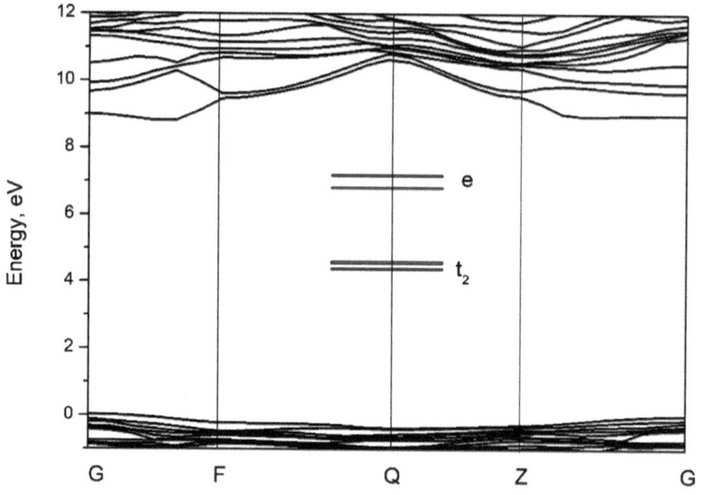

Figure 10.6 Calculated electronic band structure of $BeAl_2O_4$ and crystal field splitting of Ti^{3+} levels in $BeAl_2O_4$. Reproduced with permission from Ref. 14.

10.4 Calculations of the Oscillator Strengths of the 4*f*–4*f* Transitions: Judd–Ofelt Theory

The electronic transitions between the energy levels of the 4*f* configurations are forbidden in the electric dipole approximation, since all these states have the same parity. Nevertheless, despite this strict selection rule, these transitions are observed experimentally for the lanthanide ions in a free state, in solutions and solids. The reason is that the SO interaction produces a minor admixture of the opposite parity states from the high-lying excited 5*d*-configurations, which removes the strict prohibition for the electric dipole transitions.

The theory of the 4*f* – 4*f* transitions has been developed practically simultaneously by B. R. Judd and G. S. Ofelt[43,44] and is known nowadays as the Judd–Ofelt (JO) theory. It is quite successful in quantitative description of the experimental absorption spectra of trivalent lanthanides in crystals, glasses and aqueous solutions. Remarkable success and wide popularity of this theory can be seen from many citations of those original papers: According to the Web of Science database (as of end of July 2018), Refs. 43 and 44 were cited more than 5750 and 5220 times, respectively.

The strength of the JO theory is in its ability to calculate the oscillator strengths of the electronic transitions in the absorption and emission spectra of the $4f$ ions, branching ratios, excited states radiative lifetimes and quantum efficiency, when using only three parameters $\Omega_\lambda(\lambda = 2, 4, 6)$.

After appearance of the original papers by Judd and Ofelt in 1962, many reviews on the JO theory have been published.[45–47] We just repeat here the main ideas of the JO theory and the relevant mathematical expressions with brief explanations.

The main equation of the JO theory, which allows to calculate the oscillator strength f of a given absorption transition in the electric dipole approximation, is as follows (in the SI units):

$$f_{ED}^{abs} = \frac{8\pi^2 m\nu\chi_{ED}^{abs}}{3h(2J+1)n}\sum_{t=2,4,6}\Omega_\lambda\left|\left\langle 4f^N[\alpha SL]J\left\|U^{(\lambda)}\right\|4f^N[\alpha' S' L']J'\right\rangle\right|^2, \tag{10.8}$$

where m is the electron mass, c is the speed of light, h is Planck's constant, ν is the mean frequency for the transition, J is the total momentum quantum number of a lower state, and $\chi_{ED}^{abs} = \frac{(n^2+2)^2}{9}$ with n being the refractive index of the medium at the transition frequency. $\left\langle 4f^N[\alpha SL]J\left\|U^{(\lambda)}\right\|4f^N\left|[\alpha' S' L']J'\right.\right\rangle$ are the matrix elements of the unit tensor operators that are tabulated in Ref. 48, and $\Omega_\lambda(\lambda = 2, 4, 6)$ are the phenomenological parameters of the model, which are determined by the least squares fitting of the calculated oscillator strengths to those obtained from the analysis of the experimental spectra. Thus, only three parameters are needed to describe the whole absorption spectrum, which provides a very convenient and physically well-based way of a description of electric dipole transitions of rare earth ions in various environments.

If the magnetic dipole transitions are considered, their oscillator strength is calculated by the following equation:

$$f_{MD}^{abs} = \frac{h\nu n}{6mc^2(2J+1)}\left|\left\langle 4f^N[\alpha SL]J\left\|L+gS\right\|4f^N[\alpha' S' L']J'\right\rangle\right|^2, \tag{10.9}$$

where $g = 2.002319$ is the electron's g-factor (it can be assumed to be equal to two). The matrix elements of the magnetic dipole operator $L + gS$ are calculated as follows[49]:

$$\left\langle 4f^N[\alpha SL]J\|L+gS\|4f^N[\alpha' S' L']J'\right\rangle = g\sqrt{J(J+1)(2J+1)} \text{ for } J = J', \tag{10.10}$$

$$\left\langle 4f^N[\alpha SL]J\|L+gS\|4f^N[\alpha' S' L']J'\right\rangle = (g-1)\sqrt{\frac{(S+L+J+1)(J+L-S)(J+S-L)(S+L-J+1)}{4J}}, \text{ for } J' = J-1. \tag{10.11}$$

Very often in the calculations, the dispersion of the refractive index is neglected, and some constant value of n is taken. This approach, of course, simplifies the whole calculations; however, it is better to consider how the refractive index depends on the wavelength, e.g., in the form of the Sellmeier equation with infrared correction[50]

$$n = A + \frac{B}{1-\left(\frac{C}{\lambda}\right)^2} - D\lambda^2, \tag{10.12}$$

where A, B, C, D are the constants, which can be obtained by fitting Eq. (10.10) to a number of the experimentally obtained refractive index values at different wavelengths.

The experimental oscillator strengths are calculated from the experimental absorption spectra using the following equation:

$$f_{\text{exp}} = 4.319\times10^{-9}\frac{\text{mol}\cdot\text{cm}^2}{\text{l}}\int\varepsilon(\bar{\nu})d\bar{\nu}, \tag{10.13}$$

where $\varepsilon(\bar{\nu})$ is the molar extinction coefficient in l/(mol cm) (in fact, it is equal to the concentration of the absorbing rare earth ions in liter/mole divided by the sample thickness in cm), and $\bar{\nu}$ is the energy in cm^{-1}. For the practical applications of Eq. (10.13) the peaks in the absorption spectra of a considered lanthanide ion (obviously, the JO theory cannot be applied to the spectra of the Ce^{3+} and Yb^{3+} ions, because the number of the transitions in the spectra of these ions in insufficient for determination of the parameters Ω_λ) should be assigned according to Dieke's diagram. The background in the absorption spectra plotted in the wavenumber cm^{-1} scale should be subtracted, and each absorption peak should be integrated numerically, and Origin, KaleidaGraph or any other software than can find areas under the peaks can be used to do that.

After the experimental oscillator strengths from Eq. (10.13) are calculated, the fit of the calculated oscillator strengths from Eq. (10.12) (sometimes Eq. (10.9) should also be added, if the magnetic dipole transitions are strong ones) should be performed to minimize either absolute or relative root-mean-square (RMS) deviation between both sets of data:

$$\mathrm{RMS}_{abs} = \sqrt{\frac{1}{N-p}\sum_{k=1}^{N}\left(f_k^{\mathrm{exp}} - f_k^{\mathrm{calc}}\right)^2}, \tag{10.14}$$

$$\mathrm{RMS}_{\mathrm{rel}} = \sqrt{\frac{1}{N-p}\sum_{k=1}^{N}\left(\frac{f_k^{\mathrm{exp}} - f_k^{\mathrm{calc}}}{f_k^{\mathrm{exp}}}\right)^2}. \tag{10.15}$$

The intensity parameters $\Omega_\lambda(\lambda = 2, 4, 6)$ extracted in this way have dimension of area (usually they are expressed in cm^2) and can be used further to estimate the $SLJ \rightarrow S'L'J'$ radiative transition rates (in the SI units):

$$A_{SLJ\rightarrow S'L'J'}^{ED} = \frac{16\pi^3 e^2 \nu^3}{3\varepsilon_0 hc^3(2J+1)}\chi_{ED}^{em} \sum_{t=2,4,6}\Omega_\lambda\left|\left\langle 4f^N[\alpha SL]J\left\|U^{(\lambda)}\right\|4f^N[\alpha' S'L']J'\right\rangle\right|^2, \tag{10.16}$$

$$A_{SLJ\rightarrow S'L'J'}^{MD} = \frac{\pi h e^2 \nu^3}{3\varepsilon_0 c^5 m^2(2J+1)}n^3 \sum_{t=2,4,6}\Omega_\lambda\left|\left\langle 4f^N[\alpha SL]J\left\|U^{(\lambda)}\right\|4f^N[\alpha' S'L']J'\right\rangle\right|^2, \tag{10.17}$$

with $\chi_{ED}^{em} = n^2\chi_{ED}^{abs}$.

Summing up all radiative transition rates and taking an inverse of that value gives the calculated radiative lifetime $\tau_{\mathrm{rad,calc}}$ of an excited level SLJ:

$$\tau_{\mathrm{rad,calc}} = \frac{1}{\sum_{S'L'J'}\left(A_{SLJ\rightarrow S'L'J'}^{ED} + A_{SLJ\rightarrow S'L'J'}^{MD}\right)}. \tag{10.18}$$

It should be mentioned that $\tau_{\mathrm{rad,calc}}$ does not include the non-radiative transitions etc., so the experimental lifetimes τ_{exp} are shorter than the calculated ones. Their ratio $\tau_{exp}/\tau_{rad,calc}$ gives the quantum efficiency for the corresponding transition.

Since there can be many radiative transitions from an excited state to the lower energy levels (all the way down to the ground state), their probabilities are generally different and the so-called branching ratios

$$\beta_{SLJ \to S'L'J'} = \frac{A^{ED}_{SLJ \to S'L'J'} + A^{MD}_{SLJ \to S'L'J'}}{\sum\limits_{S'L'J'} \left(A^{ED}_{SLJ \to S'L'J'} + A^{MD}_{SLJ \to S'L'J'} \right)} \quad (10.19)$$

can be calculated. They show the partial contribution of the $SLJ \to S'L'J'$ radiative transition to the total radiative decay rate and can help in determination of the dominant radiative transition in the radiative deactivation of the excited SLJ state. The values of $\beta_{SLJ \to S'L'J'}$ can be directly compared with the experimental data on the ratios of intensities of various emission transitions from this SLJ excited state in the emission spectrum. The transitions with the values of $\beta_{SLJ \to S'L'J'} > 0.5$ are of a particular interest since they are considered as good potential candidates for laser transitions.[51,52]

Sometimes application of the JO theory to the case of the Pr^{3+} ion results in the negative value of the Ω_2 parameter, which is not physical and can even lead to the negative values of the calculated radiative life times. To overcome this difficulty, several modifications to the standard JO theory were suggested to include the influence of the 5*d* states[53,54] or add the odd intensity parameters $\Omega_\lambda (\lambda = 1, 3, 5)$.[55,56]

Numerous examples of application of the JO theory can be found in the literature[57–64]; the reader is kindly advised to look into these and other relevant publications for more details of applications of the JO theory to the calculations of the 4*f*–4*f* transition intensities, excited states radiative lifetimes, branching ratios in different materials.

10.5 Determination of the Complete Energy Level Scheme of Doped Crystals (4*f* Ions)

In this paragraph we consider the yttrium aluminum perovskite (YAP) $YAlO_3$ crystal doped with the Ce^{3+} ions. $YAlO_3$ is an important material that finds its application in scintillators,[65] solid-state lasers,[66] optical data storage,[67] etc. In addition, YAP has a wide band gap equal to about 8.8 eV,[68,69] which makes it suitable for doping with the transition metal and rare earth ions. The crystal field effects for the YAP with various dopants were studied by us earlier, e.g., Mn^{4+} and Mn^{5+},[70] Nd^{3+},[71] Ce^{3+}.[72,73] Here we focus on the identification of the cerium 4*f* and 5*d* energy levels with respect to the YAP electronic bands.

All pertinent details of calculations of the YAP:Ce^{3+} electronic structure can be found in Ref. 72. The 2×1×1 YAP supercell with one yttrium ion replaced by the cerium ion was optimized, and the electronic band structure scheme shown in Fig. 10.7. Since the density functional theory–based methods underestimate the band gap, in the present case the so-called scissor operator of 3.5 eV was applied to bring the separation between the valence and conduction bands in correspondence with the experimental band gap value.

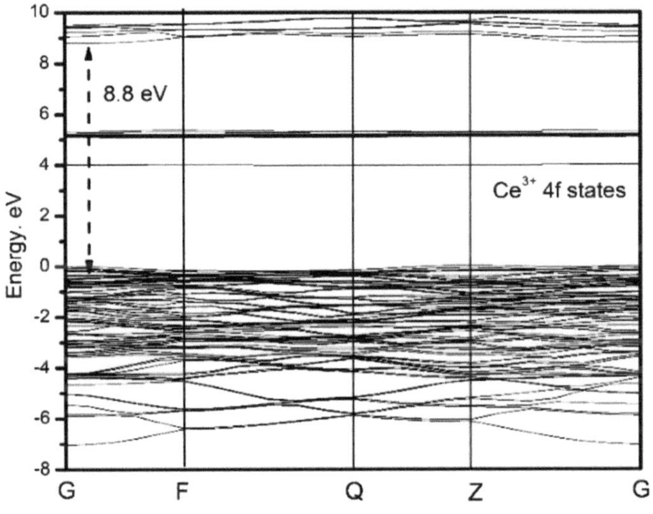

Figure 10.7 Calculated electronic band structure of YAP:Ce^{3+}.

Analysis of the density of electronic states in YAP:Ce^{3+} (Fig. 10.8) indicates that the Ce 4*f* states appear in the band gap, at about 4 eV above the top of the valence band. Such positioning of the Ce^{3+} 4*f* ground state in the YAP band gap agrees well with the value of 3.5 eV obtained by P. Dorenbos.[74]

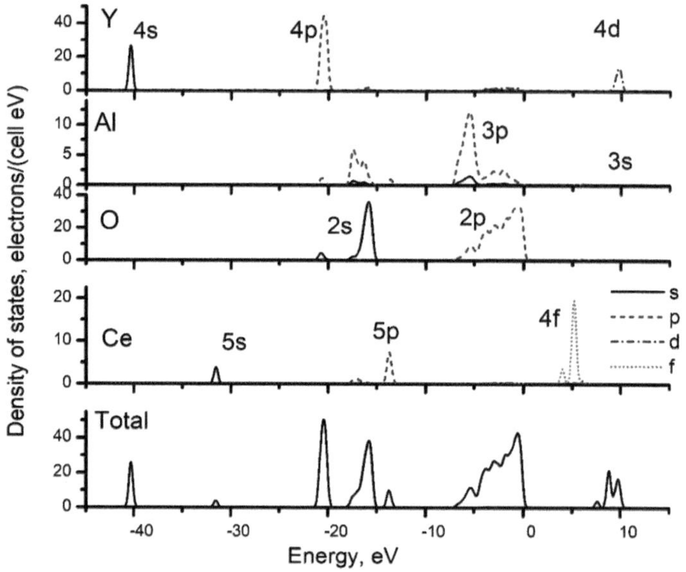

Figure 10.8 Calculated density of states diagrams for YAP:Ce^{3+}.

Now, after the position of the Ce^{3+} $4f$ ground state in the band gap is fixed, it is possible to plot the combined energy level diagram, which includes the YAM band structure and location of all Ce^{3+} energy levels in the band gap, placing the Ce^{3+} ground state at 4 eV above the valence band top. According to Ref. 73, the calculated (in the framework of the crystal field theory) Ce^{3+} energy levels in YAM are (in cm^{-1}): 0, 334, 722 ($^2F_{5/2}$ state of the $4f^1$ configuration); 2107, 2432, 2772, 3209 ($^2F_{7/2}$ state of the $4f^1$ configuration), 32371, 33965, 36583, 43986, 45415 (^2D state of the $5d^1$ configuration). Plotting these energy levels together with the YAM band structure yields the following combined energy level diagram of YAM and Ce^{3+} shown in Fig. 10.9.

Three lowest $5d$ states of the Ce^{3+} ions are below the conduction band, which indicates the possibility of getting broadband emission arising from the parity and spin-allowed $5d$ – $4f$ Ce^{3+} transitions. Two highest $5d$ Ce^{3+} states are located in the conduction band (Fig. 10.9); therefore, absorption to these states from the ground $4f$ state may be difficult to be observed experimentally.

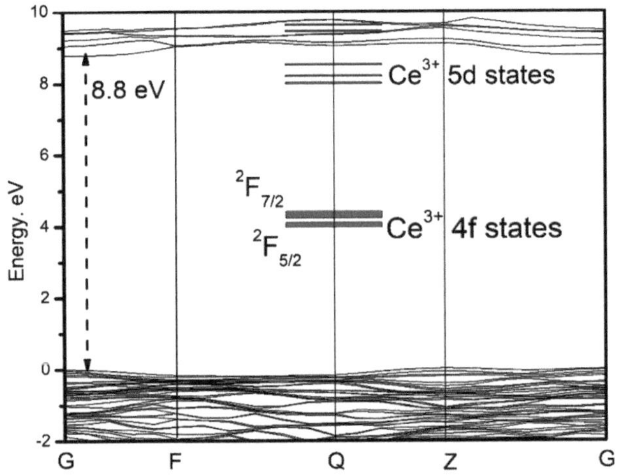

Figure 10.9 The calculated band structure of YAM with superimposed energy levels of the Ce^{3+} ions.

In a similar way the location of the Tm^{2+} 4*f* states in the band gap of SrF_2 was analyzed in Ref. 75, where it was shown that the ground 4*f* Tm^{2+} states are at about 0.742 eV above the valence band top.

10.6 4*f*–5*d* Transitions of Trivalent Lanthanide Ions in Crystals

Pioneering works of G. Dieke[76] allowed to find experimentally the energy levels of all trivalent lanthanides ions up to about 40000 cm^{-1}; the energy level diagram obtained in that way is known as Dieke's diagram. However, with the development of the synchrotron radiation sources it became possible to test the high-lying energy levels of rare-earth ions, including those ones of the opposite parity, e.g., the states of the $4f^{n-1}5d$ electron configurations. Apart from purely scientific curiosity to study those states, there is also a practical interest that drives experimental and theoretical research in this area: The electronic transitions between the states of the ground $4f^n$ and excited $4f^{n-1}5d$ electron configurations are parity allowed and are much more intensive that the parity forbidden intraconfigurational $4f^n$–$4f^n$ transitions. Besides, since the 5*d* states are the outer ones and strongly interact with the crystalline environment, the corresponding emission bands are very broad, which opens a principal opportunity either to cover a broad emission range (in the case of the Ce^{3+} or Eu^{2+} ions, whose emission is in the visible part of the spectrum) or realize the tunable UV or VUV laser, in the case of other trivalent lanthanides.

First systematic experimental and crystal field studies of the $4f^{n-1}5d$ electron configurations for all trivalent lanthanides were reported in Refs. 77–80, where the crystal field Hamiltonian parameters for the mixed electron configurations were given and the trends across the rare-earth elements group were identified. The ab initio calculations of the $4f^N \rightarrow 4f^{N-1}5d$ spectra of the trivalent rare earth ions based on the DV-X_α method were performed in Refs. 81–84. As an additional result of these research activities, Dieke's diagram was extended to include all energetic states of the $4f^n$ electron configurations (up to about 160000 cm^{-1})[85–87] and $4f^{N-1}5d$ electron configurations (up to about 350000 cm^{-1}).[88] Recently complete energy level diagrams of the f^n electron configurations for all di-, tri-, and tetra-valent 4f and 5f ions were calculated and published in Ref. 89. Chapter 6 in this book dealt with the trends of the free ion Hamiltonian parameters for the $f^{N-1}d$ mixed configurations in a free state. In this section some examples of calculations of the $4f^N \rightarrow 4f^{N-1}5d$ spectra of the trivalent lanthanides in crystals are discussed.

The free ion Hamiltonian for the $f^{N-1}d$ mixed configurations is given by Eqs. (6.10) and (6.11) in this book. The structure of the crystal field Hamiltonian depends on the point symmetry of the crystal lattice site and non-zero CFPs are listed in Table 8.3. We start with the simplest example of the Ce^{3+} ion in the YAP crystal.[73] The crystal field Hamiltonian for its single $5d$ electron in the excited electron configuration is as follows:

$$\begin{aligned}\mathbf{H}(5d^1) = {} & \Delta_E(fd) + \zeta_{5d}\mathbf{s}_d \cdot \mathbf{l}_d + B_0^2(\mathrm{d})\mathbf{C}_0^{(2)} + B_1^2(\mathrm{d})(\mathbf{C}_1^{(2)} + \mathbf{C}_{-1}^{(2)}) \\ & + B_2^2(\mathrm{d})(\mathbf{C}_2^{(2)} + \mathbf{C}_{-2}^{(2)}) + B_0^4(\mathrm{d})\mathbf{C}_0^{(4)} + B_1^4(\mathrm{d})(\mathbf{C}_1^{(4)} + \mathbf{C}_{-1}^{(4)}) \\ & + B_2^4(\mathrm{d})(\mathbf{C}_2^{(4)} + \mathbf{C}_{-2}^{(4)}) + B_3^4(\mathrm{d})(\mathbf{C}_3^{(4)} + \mathbf{C}_{-3}^{(4)}) + B_4^4(\mathrm{d})(\mathbf{C}_4^{(4)} + \mathbf{C}_{-4}^{(4)}),\end{aligned} \tag{10.20}$$

with $\Delta_E(fd)$ being the energy difference between the f^1 and d^1 configurations of the Ce^{3+} ions; ζ_{5d} stands for the constant of SO interaction in the $5d$ states, and all the remaining terms describe the crystal field interactions. The ECM was used to calculate the Ce^{3+} $5d$

states splitting in YAP. The overlap integrals for the Ce^{3+} 5d functions and O^{2-} 2s, 2p functions were calculated numerically for different interionic separations R and then they were approximated by the following functions (R is measured in the atomic units and is in the range from 0 to 12 a.u.)[90,91]:

$$S_s = \langle d0|s0\rangle = 0.41410(R+0.89658)\exp(-1.44528R)R^{2.640\,21},$$
$$S_\sigma = \langle d0|p0\rangle = -0.99616(R-2.49277)\exp(-1.23854R)R^{2.06605},$$
$$S_\pi = \langle d\pm1|p\pm1\rangle = 1.11348(R+0.28061)\exp(-1.23219R)R^{1.20503}. \quad (10.21)$$

Figure 10.10 visualizes the calculated dependences from Eq. (10.21).

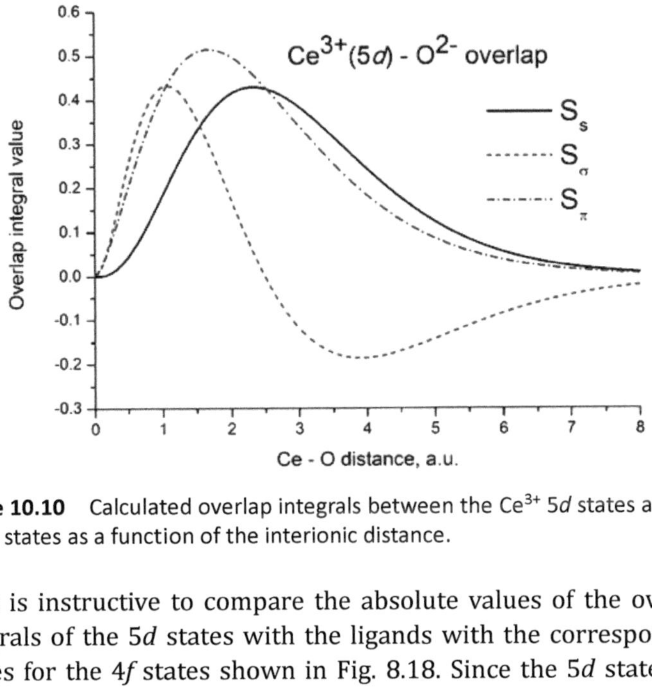

Figure 10.10 Calculated overlap integrals between the Ce^{3+} 5d states and O^{2-} 2s, 2p states as a function of the interionic distance.

It is instructive to compare the absolute values of the overlap integrals of the 5d states with the ligands with the corresponding values for the 4f states shown in Fig. 8.18. Since the 5d states are the outer ones and extend considerably further from the nucleus than the 4f states, their overlap with the ligand's wave functions is noticeably enhanced.

Using the YAP structural data, the following CFPs values were calculated (in cm^{-1}, Wybourne notation): $B_0^2 = 1630$, $B_1^2 = -3448$, $B_2^2 = 908$, $B_0^4 = 2830$, $B_1^4 = 18123$, $B_2^4 = -2321$, $B_3^4 = 5710$, $B_4^4 = -4082$.

Besides, $\Delta_E(fd)$ = 36982 cm^{-1}, ζ_{5d} = 1082 cm^{-1}. Diagonalization of the crystal field Hamiltonian with these parameters in the basis set spanned by the five wave functions of the ^{2}D state led to the following calculated energy level scheme (Table 10.5), which agrees well with the corresponding experimental data.

Table 10.5 Calculated and experimental 5*d* energy levels of the Ce^{3+} ions in the YAM crystal[73]

Level No.	^2L	E_{calc}	E_{exp}
1	^{2}D	32371	33360
2		33965	34470
3		36583	35700
4		43986	42590
5		45415	46200

More details related to similar calculations of the Ce^{3+} [72,73,90–92] and Eu^{2+} [93,94] 5*d* levels splittings and in different crystals are described in the cited references. Reference[95] can be mentioned as an example of consistent calculations of the $4f^N \rightarrow 4f^{N-1}5d$ transitions (including energy levels and transition intensities) of all lanthanide ions in one host of K_3YF_6. The systematic variations of the Hamiltonian parameters across the considered series were described in the framework of the unified approach. It should be emphasized that due to considerably increased number of the electronic states in the mixed $4f^{N-1}5d$ electron configurations, such calculations are more complicated and time-consuming compared with the analysis of the energy levels and transitions within the $4f^N$ electron configuration.

Bibliography

1. Moreno M, Barriuso MT, Aramburu JA (1994). The dependence of 10*Dq* upon the metal–ligand distance, *R*, for transition-metal complexes, What is its microscopic origin? *Int J Quantum Chem* **52**(4) 829–835.
2. Wissing K, Aramburu JA, Barriuso MT, Moreno M (1998). Optical properties due to Cr^{4+} in oxides: Density functional study, *Solid State Commun* **108**(12) 1001–1005.

3. Garcia-Lastra JM, Wesolowski T, Barriuso MT, Aramburu JA, Moreno M (2006). Optical and vibrational properties of MnF_6^{4-} complexes in cubic fluoroperovskites: Insight through embedding calculations using Kohn–Sham equations with constrained electron density, *J Phys: Condens Matter* **18**(5) 1519–1534.
4. Sturge MD (1968). The Jahn-Teller effect in solids, in *Solid State Physics*, Seitz F, Turnbull D, Ehrenreich H, eds, 20 pp. 91–211.
5. Güdel HU, Snellgrove TR (1978). Jahn-Teller effect in the $^4T_{2g}$ state of chromium(III) in dicesium sodium indium(III) hexachloride, *Inorg Chem* **17**(6) 1617–1620.
6. Barriuso MT, Aramburu JA, Moreno M (1996). Coupling with the Jahn-Teller mode for triplet states of MF_6 ($M = Mn^{2+}$, Cr^{3+}) complexes: Dependence on the M-F distance and influence on the Stokes shift, *Phys Stat Solidi B* **196**(1) 193–208.
7. Moreno M, Barriuso MT, Aramburu JA (1992). The Huang-Rhys factor $S(a_{1g})$ for transition-metal impurities: A microscopic insight, *J Phys: Condens Matter* **4**(47) 9481–9488.
8. Brik MG, Ogasawara K (2006). Microscopic analysis of the crystal field strength and lowest charge transfer energies in the elpasolite crystals Cs_2NaYX_6 (X = F, Cl, Br) doped with Cr^{3+}, *Phys Rev B* **74**(4) 045105.
9. Brik MG (2007). Influence of chemical bond length changes on the crystal field strength and "ligand–metal" charge transfer transitions in Cs_2GeF_6 doped with Mn^{4+} and Os^{4+} ions, *J Phys Chem Solids* **68**(7) 13411–1347.
10. Brik MG, Avram NM, Ma CG (2011). First-principles calculations of structural, electronic, optical, elastic properties and microscopic crystal field effects in Rb_2CrF_6, *Comput Mater Sci* **50**(8) 2482–2487.
11. Brik MG, Avram NM (2007). First-principles calculations of parameters of electron-vibrational interaction and estimations of Jahn-Teller stabilization energy for Cr^{3+} ion in elpasolites, *J Molec Struct* **838**(1–3) 193–197.
12. Brik MG, Kumar GA, Sardar DK (2012). Ab initio, crystal field and experimental spectroscopic studies of pure and Ni^{2+}-doped $KZnF_3$ crystals, *Mater Chem Phys* **136**(1) 90–102.
13. Brik MG (2018). Ab-initio studies of the electronic and optical properties of Al_2O_3:Ti^{3+} laser crystals, *Physica B* **532** 178–183.
14. Su P, Ma CG, Brik MG, Srivastava AM (2018). A short review of theoretical and empirical models for characterization of optical materials doped with the transition metal and rare earth ions, *Opt Mater* **79** 129–136.

15. Brik MG, Avram NM (2009). Microscopic analysis of the crystal field strength and electron-vibrational interaction in cubic $SrTiO_3$ doped with Cr^{3+}, Mn^{4+} and Fe^{5+} ions, *J Phys: Condens Matter* **21**(15) 15502.
16. Reisz AM, Brik MG, Avram CN, Avram NM (2010). Crystal field effects and electron–phonon interaction in K_2LiAlF_6:Cr^{3+}, *Physica B* **405**(5) 1244–1247.
17. Avram NM, Brik MG, Kityk IV (2010). Studies of variation of interionic distances and crystal field effects in ZnS:V^{2+} and MgO:Cr^{3+}, *Opt Mater* **32**(12) 1668–1670.
18. Brik MG, Avram CN (2011). Exchange charge model and analysis of the microscopic crystal field effects in $KAl(MoO_4)_2$:Cr^{3+}, *J Lumin* **131**(12) 2642–2645.
29. Naik SR, Shripathi T, Salker AV (2015). Preparation, characterization and photoluminescent studies of Cr and Nd co-doped Ce:YAG compounds, *J Lumin* **161** 335–342.
30. Wittlin A, Przybylińska H, Berkowski M, et al. (2015). Ambient and high pressure spectroscopy of Ce^{3+} doped yttrium gallium garnet, *Opt Mater Express* **5**(8) 1868–1880.
31. Dorenbos P (2000). The $4f^n \leftrightarrow 4f^{n-1}5d$ transitions of the trivalent lanthanides in halogenides and chalcogenides, *J Lumin* **91**(1–2) 91–106.
32. Dorenbos P (2000). The 5d level positions of the trivalent lanthanides in inorganic compounds, *J Lumin* **91**(3–4) 155–176.
33. Dorenbos P (2000). Energy of the first $4f^7 \rightarrow 4f^6 5d$ transition of Eu^{2+} in inorganic compounds, *J Lumin* **104**(4) 239–260.
34. Dorenbos P (2013). A review on how lanthanide impurity levels change with chemistry and structure of inorganic compounds, *ECS J Solid State Sci Technol* **2**(2) R3001–R3011.
35. Dorenbos P (2013). Ce^{3+} 5*d*-centroid shift and vacuum referred 4f-electron binding energies of all lanthanide impurities in 150 different compounds, *J Lumin* **135** 93–104.
36. Dorenbos P (2017). Charge transfer bands in optical materials and related defect level location, *Opt Mater* **69** 8–22.
37. Brik MG, Nazarov M, Ahmad-Fauzi MN, et al. (2012). Comparative first-principle analysis of un-doped and V^{3+}-doped α-$ZnAl_2S_4$ spinel, *J Lumin* **132**(9) 2489–2494.
38. Güner S, Yildiz F, Rameev B, Aktaş B (2005). Electron paramagnetic resonance study of $ZnAl_2S_4$ spinel, *J Phys: Condens Matter* **17**(25) 3943–3952.

39. Berthold HJ, Köhler K, Wartchow R (1983). Kristallstrukturverfeinerung des Zinkaluminiumsulfids $ZnAl_2S_4$ (normale Spinellstruktur) mit Röntgen-Einkristalldaten, *Zeitschrift für Anorganische und Allgemeine Chemie* **496**(1) 7–20.

40. Ivanov VYu, Pustovarov VA, Shlygin ES, Korotaev AV, Kruzhalov AV (2005). Electronic excitations in $BeAl_2O_4$, Be_2SiO_4, and $Be_3Al_2Si_6O_{18}$ crystals, *Phys Solid State* **47**(3) 466–473.

41. Sugimoto A, Segawa Y, Kim PH, Namba S, Yamagishi K, Anzai Y, Yamaguchi Y (1989). Spectroscopic properties of Ti^{3+}-doped $BeAl_2O_4$, *J Opt Soc Am B* **6**(12) 2334–2337.

42. Brik MG, Avram NM, Avram CN (2013). In *Optical Properties of 3d-Ions in Crystals: Spectroscopy and Crystal Field Analysis.* Avram NM, Brik MG, eds, Tsinghua University Press, Beijing and Springer-Verlag Berlin Heidelberg, Chapter 2, p. 38.

43. Judd BR (1962). Optical absorption intensities of rare-earth ions, *Phys Rev* **127**(3) 750–761.

44. Ofelt GS (1962). Intensities of crystal spectra of rare-earth ions, *J Chem Phys* **37**(3) 511–520.

45. Görller-Walrand C, Binnemans K (1998). Spectral intensities of *f–f* transitions, In *Handbook on the Physics and Chemistry of Rare Earths.* Gschneiner Jr KA, Eyring L, eds, **25**, North-Holland, Amsterdam, p. 101.

46. Reid MF (2000). *Transition Intensities, Crystal Field Handbook.* Cambridge University Press, Cambridge, 190 p.

47. Hehlen MP, Brik MG, Krämer KW (2013). 50th anniversary of the Judd–Ofelt theory: An experimentalist's view of the formalism and its application, *J Lumin* **136** 221–239.

48. Nielsen CW, Koster GF (1963). *Spectroscopic Coefficients for the p^n, d^n, and f^n Configurations.* MIT Press, Cambridge, Massachusetts.

49. Wybourne BG (1965). *Spectroscopic Properties of Rare Earths.* Wiley, New York.

50. Pujol MC, Rico M, Zaldo C, et al. (1999). Crystalline structure and optical spectroscopy of Er^{3+}-doped $KGd(WO_4)_2$ single crystals, *Appl Phys B* **68**(2) 187–197.

51. Adam JL, Sibley WA (1985). Optical transitions of Pr^{3+} ions in fluorozirconate glass, *J Non-Cryst Solids* **76**(2–3) 267–279.

52. Kumar VR, Veeraiah N, Rao BA, Buddhudu S (1998). Optical absorption and photoluminescence properties of Eu^{3+}-doped ZnF_2–PbO–TeO_2 glasses, *J Mater Sci* **33**(10) 2659–2662.

53. Kornienko AA, Kaminskii AA, Dunina EB (1990). Dependence of the line strength of f–f transitions on the manifold energy. I. Projector on the basis of nonorthogonal functions, *Phys Stat Solidi B* **157**(1) 261–266.

54. Kornienko AA, Kaminskii AA, Dunina EB (1990). Dependence of the line strength of *f–f* transitions on the manifold energy. II. Analysis of Pr^{3+} in $KPrP_4O_{12}$, *Phys Stat Solidi B* **157**(1) 267–273.

55. Auzel F, Hubert S, Delamoye P (1982). Absolute oscillator strengths of 5f–5f transitions of U^{4+} in $ThBr_4$ and in hydrobromic acid solutions, *J Lumin* **26**(3) 251–262.

56. Florez A, Brik MG, Plata O (2009). Effect of the odd intensity parameters and optical properties of Tb^{3+} doped fluoroindate glasses, *J Mater Sci Mater Electron* **20** (2009) S230–S234.

57. Brik MG, Ishii T, Tkachuk AM, Ivanova SE, Razumova IK (2004). Calculations of the transitions intensities in the optical spectra of Dy^{3+}:$LiYF_4$, *J Alloys Compds* **374**(1–2) 63–68.

58. Lakshminarayana G, Qiu JR, Brik MG, Kumar GA, Kityk IV (2008). Spectral analysis of Er^{3+}-, Er^{3+}/Yb^{3+}- and $Er^{3+}/Tm^{3+}/Yb^{3+}$-doped TeO_2–ZnO–WO_3–TiO_2–Na_2O glasses, *J Phys: Condens Matter* **20**(37) 375101.

59. El Okr M, Farouk M, El-Sherbiny M, El-Fayoumi MAK, Brik MG (2010). Spectroscopic studies of the Pr^{3+}-doped borovanadate glass, *J Alloys Compds* **490**(1–2) 184–189.

60. Kasprowicz D, Brik MG, Majchrowski A, Michalski E, Mykowska E (2010). Spectroscopic studies of Nd^{3+} and Er^{3+} in $KGd(WO_4)_2$ single crystals, *J Lumin* **130**(4) 632–630.

61. Bettinelli M, Speghini A, Brik MG (2010). Spectroscopic studies of emission and absorption properties of $38PbO$-$62SiO_2$:Nd^{3+} glass, *Opt Mater* **32**(12) 1592–1596.

62. Rao PR, Krishna GM, Brik MG, Gandhi Y, Veeraiah N (2011). Fluorescence features of Sm^{3+} ions in Na_2SO_4-MO-P_2O_5 glass system-influence of modifier oxide, *J Lumin* **131**(2) 212–217.

63. Brik MG, Antić **ŽM,** Vuković K, Dramićanin MD (2015). Judd–Ofelt analysis of Eu^{3+} emission in TiO_2 anatase nanoparticles, *Mater Trans* **56**(2015) 1416–1418.

64. Kitagawa Y, Ueda J, Brik MG, Tanabe, S (2018). Intense hypersensitive luminescence of Eu^{3+}-doped $YSiO_2N$ oxynitride with near-UV excitation, *Opt Mater* **83** 111–117.

65. Nikl M (2000). Wide band gap scintillation materials: Progress in the technology and material understanding, *Phys Stat Sol A* **178**(2) 595–620.

66. Zhydachevskii Ya, Galanciak D, Kobyakov S, Berkowski M, Kamińska A, Suchocki A, Zakharko Ya, Durygin A (2006). Photoluminescence studies of Mn^{4+} ions in $YAlO_3$ crystals at ambient and high pressure, *J Phys: Condens Matter* **18**(49) 11385.

67. Loutts GB, Warren M, Taylor R, et al. (1998). Manganese-doped yttrium orthoaluminate: A potential material for holographic recording and data storage, *Phys Rev B* **57**(7) 3706.

68. Lushchik C, Feldbach E, Frorip A, Kirm M, Lushchik A, Maaros A, Martinson I (1994). Multiplication of electronic excitations in CaO and $YAlO_3$ crystals with free and self-trapped excitons, *J Phys: Condens Matter* **6**(50) 11177.

69. Basun SA, Danger T, Kaplyanskii AA, McClure DS, Petermann K, Wong WC (1996). Optical and photoelectrical studies of charge-transfer processes in $YAlO_3$:Ti crystals, *Phys Rev B* **54**(9) 6141–6149.

70. Brik MG, Sildos I, Berkowski M, Suchocki A (2009). Spectroscopic and crystal field studies of $YAlO_3$ single crystals doped with Mn ions, *J Phys: Condens Matter* **21**(2) 025404.

71. Brik MG (2009). Modeling of optical properties of 3d and 4f ions, *ECS Trans* **25**(9) 25–37.

72. Brik MG, Sildos I, Kiisk V (2011). Calculations of physical properties of pure and doped crystals: Ab initio and semi-empirical methods in application to $YAlO_3$:Ce^{3+} and TiO_2, *J Lumin* **131**(3) 396–403.

73. Wang YJ, Włodarczyk D, Li LA, et al. (2017). Electronic structure of Ce^{3+} in yttrium and lutetium orthoaluminate crystals and single crystal layers, *J Alloys Compds* **723** 157–163.

74. Dorenbos P (2013). The electronic level structure of lanthanide impurities in $REPO_4$, $REBO_3$, $REAlO_3$, and RE_2O_3 (RE = La, Gd, Y, Lu, Sc) compounds, *J Phys: Condens Matter* **25**(22) 225501.

75. Ma CG, Brik MG (2013). First-principles calculations of structural and electronic properties of pure and Tm^{2+}-doped $SrCl_2$, *Phys Stat Solidi A* **250**(4) 858–863.

76. Dieke GH, Crosswhite HM (1963). The spectra of the doubly and triply ionized rare earths, *Appl Opt* **2**(7) 675–686.

77. Reid MF, van Pieterson L, Wegh RT, Meijerink A (2000). Spectroscopy and calculations for $4f^N$-> $4f^{N-1}5d$ transitions of lanthanide ions in $LiYF_4$, *Phys Rev B* **62**(22) 14744–14749.

78. van Pieterson L, Reid MF, Wegh RT, Soverna S, Meijerink A (2002). $4f^N$-> $4f^{N-1}5d$ transitions of the light lanthanides: Experiment and theory, *Phys Rev B* **65**(4) 045113.

79. van Pieterson L, Reid MF, Burdick GW, Meijerink A (2002). $4f^N$-> $4f^{N-1}5d$ transitions of the heavy lanthanides: Experiment and theory, *Phys Rev B* **65**(4) 045114.

80. Reid MF, van Pieterson L, Meijerink A (2002). Trends in parameters for the $4f^N$-> $4f^{N-1}5d$ spectra of lanthanide ions in crystals, *J Alloys Compds* **344**(1–2) 240–245.

81. Ishii T, Fujimura K, Sato K, Brik MG, Ogasawara K (2004). First-principles analysis method for the *f-d* transitions of heavy metal ions, *J Alloys Compds* **374**(1–2) 18–21.

82. Ishii T, Brik MG, Ogasawara K (2004). First-principles analysis method for the multiplet structures of rare-earth ions in solids, *J Alloys Compds* **380**(1–2) 136–140.

83. Ogasawara K, Watanabe S, Toyoshima H, Ishii T, Brik MG, Ikeno H, Tanaka I (2005). Optical spectra of trivalent lanthanides in $LiYF_4$ crystal, *J Solid State Chem* **178**(2) 412–418.

84. Ogasawara K, Watanabe S, Toyoshima H, Brik MG (2007). First-principles calculations of $4f^N$-> $4f^{N-1}5d$ transition spectra, in *Handbook on the Physics and Chemistry of Rare Earths*. vol. **37**, Gschneindner KA, Bünzli JCG, Pecharsky VK, eds., pp. 1–59.

85. Wegh RT, Meijerink A, Lamminmaki RJ, Holsa J (2000). Extending Dieke's diagram, *J Lumin* **87–89** 1002–1004.

86. Peijzel PS, Meijerink A, Wegh RT, Reid MF, Burdick GW (2005). A complete $4f^n$ energy level diagram for all trivalent lanthanide ions, *J Sol State Chem* **178**(2) 448–453.

87. Ogasawara K, Watanabe S, Ishii T, Brik MG (2005). Relativistic calculations of complete $4f^n$ energy level schemes of free trivalent rare-earth ions, *Jpn J Appl Phys* **44**(10) 7488–7490.

88. Ogasawara K, Watanabe S, Sakai Y, Toyoshima H, Ishii T, Brik MG, Tanaka I (2004). Calculations of complete $4f^n$ and $4f^{n-1}5d$ energy level schemes of free trivalent rare-earth ions, *Jpn J Appl Phys* **43**(5A) L611–L613.

89. Ma CG, Brik MG, Liu DX, Feng B, Tian Y, Suchocki A (2016). Energy level schemes of f^N electronic configurations for the di-, tri-, and tetravalent lanthanides and actinides in a free state, *J Lumin* **170** 369–374.

90. Ma CG, Trevisani M, Piccinelli F, Ivanovskikh KV, Bettinelli M, Brik MG (2013). Analysis of vacuum ultraviolet electronic spectra of Ce^{3+} and

Pr^{3+} ions in $Ca_9Lu(PO_4)_7$: Crystal-field calculations and simulation of optical spectra, *J Phys: Condens Matter* **25**(16) 165503.

91. Przybylińska H, Ma CG, Brik MG, et al. (2013). Evidence of multicenter structure of cerium ions in gadolinium gallium garnet crystals studied by infrared absorption spectroscopy, *Phys Rev B* **87** 045114.

92. Omelkov SI, Brik MG, Kirm M, et al. (2011). A luminescence spectroscopy and theoretical study of 4*f*-5*d* transitions of Ce^{3+} ions in $SrAlF_5$ crystals, *J Phys Condens Matter* **23**(10) 105501.

93. Nazarov M, Brik MG, Spassky D, Tsukerblat B, Palii A, Nazida AN, Ahmad-Fauzi MN (2013). Phonon-assisted optical bands of nanosized powdery $SrAl_2O_4:Eu^{2+}$ crystals: Evidence of a multimode Pekarian, *Phys Lett A* **377**(43) 3710–3178.

94. Nazarov M, Brik MG, Spassky D, Tsukerblat B (2017). Crystal field splitting of 5*d* states and luminescence mechanism in $SrAl_2O_4:Eu^{2+}$ phosphor, *J Lumin* **182** 79–86.

95. Ma CG, Brik MG, Ryba-Romanowski W, Swart HC, Gusowski MA (2012). Spectroscopy and calculations for $4f^N \rightarrow 4f^{N-1}5d$ transitions of lanthanide ions in K_3YF_6, *J Phys Chem A* **116**(36) 9158–9180.

Concluding Remarks

Spectroscopic properties of the transition metal and rare earth ions in a free state and in solids were theoretically considered in this book. The emphasis was placed on the formation of the clear picture of how the electronic energy levels are formed, how they can be described and labeled in a unique way, and how their positions can be calculated. No experimental aspects of the transition metal and lanthanide spectroscopy were discussed. The theoretical foundations were, however, presented in a simple way—wherever it was possible—to make the book understandable for the undergraduate and graduate students, researchers, and experimental scientists, whose major is not in the theoretical field. We hope that at least to some extent the book will be useful for this category of the readers. We also would like to stress that there is no ideal single book on any topic, and this book is not an exception. At the same time, we believe that this book can serve as a good starting point for the beginners and/or for those who want just to refresh some forgotten details. If the readers will read or at least look into other books recommended for the additional reading, they will gain a deeper knowledge of the subject. The original research papers cited in the book also contain more details, which simply could not be accommodated within the book chapters. Reading those papers is also strongly recommended.

We shall be grateful to receive the readers' feedback. Therefore, those who are interested in sharing their opinions with us, are encouraged to contact us via e-mail: Prof. Mikhail G. Brik: mikhail.brik@ut.ee, mikhail_brik@yahoo.com; Prof. Chong-Geng Ma: cgma.ustc@gmail.com

Appendix I

The Angular Parts of Wave Functions of the d^2 Electron Configuration

The $|LM\rangle$ notation is used, where L and M stand for the orbital and magnetic quantum numbers of the corresponding term, respectively, the Y_{lm} are the spherical functions (angular parts) of the one-electron d wave functions (l = 2, -2 ≤ m ≤ 2), and the arguments of these spherical functions denote the spherical coordinates θ, φ of the electrons with the numbers 1 and 2.

The ^{3}F term

$$|3-3\rangle = \frac{1}{\sqrt{2}}\left(Y_{2-1}(1)Y_{2-2}(2) - Y_{2-2}(1)Y_{2-1}(2)\right)$$

$$|3-2\rangle = \frac{1}{\sqrt{2}}\left(Y_{20}(1)Y_{2-2}(2) - Y_{2-2}(1)Y_{20}(2)\right)$$

$$|3-1\rangle = -\frac{\sqrt{30}}{10}Y_{2-2}(1)Y_{21}(2) - \frac{\sqrt{5}}{5}Y_{2-1}(1)Y_{20}(2) + \frac{\sqrt{5}}{5}Y_{20}(1)Y_{2-1}(2) + \frac{\sqrt{30}}{10}Y_{21}(1)Y_{2-2}(2)$$

$$|30\rangle = -\frac{\sqrt{10}}{10}Y_{2-2}(1)Y_{22}(2) - \frac{\sqrt{10}}{5}Y_{2-1}(1)Y_{21}(2) + \frac{\sqrt{10}}{5}Y_{21}(1)Y_{2-1}(2) + \frac{\sqrt{10}}{10}Y_{22}(1)Y_{2-2}(2)$$

$$|31\rangle = -\frac{\sqrt{30}}{10}Y_{2-1}(1)Y_{22}(2) - \frac{\sqrt{5}}{5}Y_{20}(1)Y_{21}(2) + \frac{\sqrt{5}}{5}Y_{21}(1)Y_{20}(2) + \frac{\sqrt{30}}{10}Y_{22}(1)Y_{2-1}(2)$$

$$|32\rangle = \frac{1}{\sqrt{2}}\left(Y_{22}(1)Y_{20}(2) - Y_{20}(1)Y_{22}(2)\right)$$

$$|33\rangle = \frac{1}{\sqrt{2}}\left(Y_{22}(1)Y_{21}(2) - Y_{21}(1)Y_{22}(2)\right)$$

The ^{3}P term

$$|1-1\rangle = -\frac{\sqrt{5}}{5}Y_{2-2}(1)Y_{21}(2) + \frac{\sqrt{30}}{10}Y_{2-1}(1)Y_{20}(2) - \frac{\sqrt{30}}{10}Y_{20}(1)Y_{2-1}(2) + \frac{\sqrt{5}}{5}Y_{21}(1)Y_{2-2}(2)$$

$$|10\rangle = -\frac{\sqrt{10}}{5}Y_{2-2}(1)Y_{22}(2) + \frac{\sqrt{10}}{10}Y_{2-1}(1)Y_{21}(2) - \frac{\sqrt{10}}{10}Y_{21}(1)Y_{2-1}(2) + \frac{\sqrt{10}}{5}Y_{22}(1)Y_{2-2}(2)$$

$$|11\rangle = -\frac{\sqrt{5}}{5}Y_{2-1}(1)Y_{22}(2) + \frac{\sqrt{30}}{10}Y_{20}(1)Y_{21}(2) - \frac{\sqrt{30}}{10}Y_{21}(1)Y_{20}(2) + \frac{\sqrt{5}}{5}Y_{22}(1)Y_{2-1}(2)$$

The ^{1}G term

$$|4-4\rangle = Y_{2-2}(1)Y_{2-2}(2)$$

$$|4-3\rangle = \frac{1}{\sqrt{2}}\left(Y_{2-1}(1)Y_{2-2}(2) + Y_{2-2}(1)Y_{2-1}(2)\right)$$

$$|4-2\rangle = \frac{\sqrt{42}}{14}Y_{20}(1)Y_{2-2}(2) + \frac{2\sqrt{7}}{7}Y_{2-1}(1)Y_{2-1}(2) + \frac{\sqrt{42}}{14}Y_{2-1}(1)Y_{20}(2)$$

$$|4-1\rangle = \frac{\sqrt{14}}{14}Y_{21}(1)Y_{2-2}(2) + \frac{\sqrt{21}}{7}Y_{20}(1)Y_{2-1}(2) + \frac{\sqrt{21}}{7}Y_{2-1}(1)Y_{20}(2) + \frac{\sqrt{14}}{14}Y_{2-2}(1)Y_{21}(2)$$

$$|40\rangle = \frac{\sqrt{70}}{70}Y_{22}(1)Y_{2-2}(2) + \frac{2\sqrt{70}}{35}Y_{21}(1)Y_{2-1}(2) + \frac{3\sqrt{70}}{35}Y_{20}(1)Y_{20}(2) + \frac{2\sqrt{70}}{35}Y_{2-1}(1)Y_{21}(2) + \frac{\sqrt{70}}{70}Y_{2-2}(1)Y_{22}(2)$$

$$|41\rangle = \frac{\sqrt{14}}{14}Y_{22}(1)Y_{2-1}(2) + \frac{\sqrt{21}}{7}Y_{21}(1)Y_{20}(2)$$
$$+\frac{\sqrt{21}}{7}Y_{20}(1)Y_{21}(2) + \frac{\sqrt{14}}{14}Y_{2-1}(1)Y_{22}(2)$$

$$|42\rangle = \frac{\sqrt{42}}{14}Y_{22}(1)Y_{20}(2) + \frac{2\sqrt{7}}{7}Y_{21}(1)Y_{21}(2) + \frac{\sqrt{42}}{14}Y_{20}(1)Y_{22}(2)$$

$$|43\rangle = \frac{1}{\sqrt{2}}\left(Y_{22}(1)Y_{21}(2) + Y_{21}(1)Y_{22}(2)\right)$$

$$|44\rangle = Y_{22}(1)Y_{22}(2)$$

The ^{1}S term

$$|00\rangle = \frac{\sqrt{5}}{5}Y_{22}(1)Y_{2-2}(2) - \frac{\sqrt{5}}{5}Y_{21}(1)Y_{2-1}(2) + \frac{\sqrt{5}}{5}Y_{20}(1)Y_{20}(2)$$
$$-\frac{\sqrt{5}}{5}Y_{2-1}(1)Y_{21}(2) + \frac{\sqrt{5}}{5}Y_{2-2}(1)Y_{22}(2)$$

The ^{1}D term

$$|2-2\rangle = \frac{\sqrt{14}}{7}Y_{20}(1)Y_{2-2}(2) - \frac{\sqrt{21}}{7}Y_{2-1}(1)Y_{2-1}(2) + \frac{\sqrt{14}}{7}Y_{2-2}(1)Y_{20}(2)$$

$$|2-1\rangle = \frac{\sqrt{21}}{7}Y_{21}(1)Y_{2-2}(2) - \frac{\sqrt{14}}{14}Y_{20}(1)Y_{2-1}(2)$$
$$-\frac{\sqrt{14}}{14}Y_{2-1}(1)Y_{20}(2) + \frac{\sqrt{21}}{7}Y_{2-2}(1)Y_{21}(2)$$

$$|20\rangle = \frac{\sqrt{14}}{7}Y_{22}(1)Y_{2-2}(2) + \frac{\sqrt{14}}{14}Y_{21}(1)Y_{2-1}(2) - \frac{\sqrt{14}}{7}Y_{20}(1)Y_{20}(2)$$
$$+\frac{\sqrt{14}}{14}Y_{2-1}(1)Y_{21}(2) + \frac{\sqrt{14}}{7}Y_{2-2}(1)Y_{22}(2)$$

$$|21\rangle = \frac{\sqrt{21}}{7}Y_{22}(1)Y_{2-1}(2) - \frac{\sqrt{14}}{14}Y_{21}(1)Y_{20}(2) - \frac{\sqrt{14}}{14}Y_{20}(1)Y_{21}(2)$$
$$+\frac{\sqrt{21}}{7}Y_{2-1}(1)Y_{22}(2)$$

$$|2-2\rangle = \frac{\sqrt{14}}{7}Y_{22}(1)Y_{20}(2) - \frac{\sqrt{21}}{7}Y_{21}(1)Y_{21}(2) + \frac{\sqrt{14}}{7}Y_{20}(1)Y_{22}(2)$$

Appendix II

The Angular Parts of Wave Functions of the f^2 Electron Configuration

The $|LM\rangle$ notation is used, where L and M stand for the orbital and magnetic quantum numbers of the corresponding term, respectively, the Y_{lm} are the spherical functions (angular parts) of the one-electron f wave functions ($l = 3$, $-3 \le m \le 3$), and the arguments of these spherical functions denote the spherical coordinates θ, φ of the electrons with the numbers 1 and 2.

The ^{3}H term

$$|5-5\rangle = \frac{1}{\sqrt{2}}\left(Y_{3-2}(1)Y_{3-3}(2) - Y_{3-3}(1)Y_{3-2}(2)\right)$$

$$|5-4\rangle = \frac{1}{\sqrt{2}}\left(Y_{3-1}(1)Y_{3-3}(2) - Y_{3-3}(1)Y_{3-1}(2)\right)$$

$$|5-3\rangle = -\frac{\sqrt{3}}{3}Y_{3-3}(1)Y_{30}(2) - \frac{\sqrt{6}}{6}Y_{3-2}(1)Y_{3-1}(2) + \frac{\sqrt{6}}{6}Y_{3-1}(1)Y_{3-2}(2) + \frac{\sqrt{3}}{3}Y_{30}(1)Y_{3-3}(2)$$

$$|5-2\rangle = -\frac{\sqrt{6}}{6}Y_{3-3}(1)Y_{31}(2) - \frac{\sqrt{3}}{3}Y_{3-2}(1)Y_{30}(2) + \frac{\sqrt{3}}{3}Y_{30}(1)Y_{3-2}(2) + \frac{\sqrt{6}}{6}Y_{31}(1)Y_{3-3}(2)$$

$$|5-1\rangle = -\frac{\sqrt{105}}{42}Y_{3-3}(1)Y_{32}(2) - \frac{3\sqrt{7}}{14}Y_{3-2}(1)Y_{31}(2) - \frac{\sqrt{210}}{42}Y_{3-1}(1)Y_{30}(2) + \frac{\sqrt{210}}{42}Y_{30}(1)Y_{3-1}(2) + \frac{3\sqrt{7}}{14}Y_{31}(1)Y_{3-2}(2) + \frac{\sqrt{105}}{42}Y_{32}(1)Y_{3-3}(2)$$

$$|50\rangle = -\frac{\sqrt{21}}{42}Y_{3-3}(1)Y_{33}(2) - \frac{2\sqrt{21}}{21}Y_{3-2}(1)Y_{32}(2) - \frac{5\sqrt{21}}{42}Y_{3-1}(1)Y_{31}(2) + \frac{5\sqrt{21}}{42}Y_{31}(1)Y_{3-1}(2) + \frac{2\sqrt{21}}{21}Y_{32}(1)Y_{3-2}(2) + \frac{\sqrt{21}}{42}Y_{33}(1)Y_{3-3}(2)$$

$$|51\rangle = -\frac{\sqrt{105}}{42}Y_{3-2}(1)Y_{33}(2) - \frac{3\sqrt{7}}{14}Y_{3-1}(1)Y_{32}(2) - \frac{\sqrt{210}}{42}Y_{30}(1)Y_{31}(2) + \frac{\sqrt{210}}{42}Y_{31}(1)Y_{30}(2) + \frac{3\sqrt{7}}{14}Y_{32}(1)Y_{3-1}(2) + \frac{\sqrt{105}}{42}Y_{33}(1)Y_{3-2}(2)$$

$$|52\rangle = -\frac{\sqrt{6}}{6}Y_{3-1}(1)Y_{33}(2) - \frac{\sqrt{3}}{3}Y_{30}(1)Y_{32}(2) + \frac{\sqrt{3}}{3}Y_{32}(1)Y_{30}(2) + \frac{\sqrt{6}}{6}Y_{33}(1)Y_{3-1}(2)$$

$$|53\rangle = -\frac{\sqrt{3}}{3}Y_{30}(1)Y_{33}(2) - \frac{\sqrt{6}}{6}Y_{31}(1)Y_{32}(2) + \frac{\sqrt{6}}{6}Y_{32}(1)Y_{31}(2) + \frac{\sqrt{3}}{3}Y_{33}(1)Y_{30}(2)$$

$$|54\rangle = \frac{1}{\sqrt{2}}\left(Y_{33}(1)Y_{31}(2) - Y_{31}(1)Y_{33}(2)\right)$$

$$|55\rangle = \frac{1}{\sqrt{2}}\left(Y_{33}(1)Y_{32}(2) - Y_{32}(1)Y_{33}(2)\right)$$

The ^{3}F term

$$|3-3\rangle = -\frac{\sqrt{6}}{6}Y_{3-3}(1)Y_{30}(2) + \frac{\sqrt{3}}{3}Y_{3-2}(1)Y_{3-1}(2) - \frac{\sqrt{3}}{3}Y_{3-1}(1)Y_{3-2}(2) + \frac{\sqrt{6}}{6}Y_{30}(1)Y_{3-3}(2)$$

$$|3-2\rangle = -\frac{\sqrt{3}}{3}Y_{3-3}(1)Y_{31}(2) + \frac{\sqrt{6}}{6}Y_{3-2}(1)Y_{30}(2) - \frac{\sqrt{6}}{6}Y_{30}(1)Y_{3-2}(2) + \frac{\sqrt{3}}{3}Y_{31}(1)Y_{3-3}(2)$$

$$|3-1\rangle = -\frac{\sqrt{3}}{3}Y_{3-3}(1)Y_{32}(2) + \frac{\sqrt{6}}{6}Y_{3-1}(1)Y_{30}(2) - \frac{\sqrt{6}}{6}Y_{30}(1)Y_{3-1}(2) + \frac{\sqrt{3}}{3}Y_{32}(1)Y_{3-3}(2)$$

$$|30\rangle = \frac{\sqrt{6}}{6}\begin{pmatrix} -Y_{3-3}(1)Y_{33}(2) - Y_{3-2}(1)Y_{32}(2) + Y_{3-1}(1)Y_{31}(2) \\ -Y_{31}(1)Y_{3-1}(2) + Y_{32}(1)Y_{3-2}(2) + Y_{33}(1)Y_{3-3}(2) \end{pmatrix}$$

$$|31\rangle = -\frac{\sqrt{3}}{3}Y_{3-2}(1)Y_{33}(2) + \frac{\sqrt{6}}{6}Y_{30}(1)Y_{31}(2) - \frac{\sqrt{6}}{6}Y_{31}(1)Y_{30}(2) + \frac{\sqrt{3}}{3}Y_{33}(1)Y_{3-2}(2)$$

$$|32\rangle = -\frac{\sqrt{3}}{3}Y_{3-1}(1)Y_{33}(2) + \frac{\sqrt{6}}{6}Y_{30}(1)Y_{32}(2) - \frac{\sqrt{6}}{6}Y_{32}(1)Y_{30}(2) + \frac{\sqrt{3}}{3}Y_{33}(1)Y_{3-1}(2)$$

$$|33\rangle = -\frac{\sqrt{6}}{6}Y_{30}(1)Y_{33}(2) + \frac{\sqrt{3}}{3}Y_{31}(1)Y_{32}(2) - \frac{\sqrt{3}}{3}Y_{32}(1)Y_{31}(2) + \frac{\sqrt{6}}{6}Y_{33}(1)Y_{30}(2)$$

The ^{3}P term

$$|1-1\rangle = \frac{\sqrt{21}}{14}Y_{32}(1)Y_{3-3}(2) - \frac{\sqrt{35}}{14}Y_{31}(1)Y_{3-2}(2) + \frac{\sqrt{42}}{14}Y_{30}(1)Y_{3-1}(2) - \frac{\sqrt{42}}{14}Y_{3-1}(1)Y_{30}(2) + \frac{\sqrt{35}}{14}Y_{3-2}(1)Y_{31}(2) - \frac{\sqrt{21}}{14}Y_{3-3}(1)Y_{32}(2)$$

$$|10\rangle = \frac{3\sqrt{7}}{14}Y_{33}(1)Y_{3-3}(2) - \frac{\sqrt{7}}{7}Y_{32}(1)Y_{3-2}(2) + \frac{\sqrt{7}}{14}Y_{31}(1)Y_{3-1}(2) - \frac{\sqrt{7}}{14}Y_{3-1}(1)Y_{31}(2) + \frac{\sqrt{7}}{7}Y_{3-2}(1)Y_{32}(2) - \frac{3\sqrt{7}}{14}Y_{3-3}(1)Y_{33}(2)$$

$$|11\rangle = \frac{\sqrt{21}}{14}Y_{33}(1)Y_{3-2}(2) - \frac{\sqrt{35}}{14}Y_{32}(1)Y_{3-1}(2) + \frac{\sqrt{42}}{14}Y_{31}(1)Y_{30}(2) - \frac{\sqrt{42}}{14}Y_{30}(1)Y_{31}(2) + \frac{\sqrt{35}}{14}Y_{3-1}(1)Y_{32}(2) - \frac{\sqrt{21}}{14}Y_{3-2}(1)Y_{33}(2)$$

The ^{1}S term

$$|00\rangle = \frac{\sqrt{7}}{7}\begin{pmatrix} Y_{33}(1)Y_{3-3}(2) - Y_{32}(1)Y_{3-2}(2) + Y_{31}(1)Y_{3-1}(2) \\ -Y_{30}(1)Y_{30}(2) + Y_{3-1}(1)Y_{31}(2) - Y_{3-2}(1)Y_{32}(2) \\ +Y_{3-3}(1)Y_{33}(2) \end{pmatrix}$$

The ^{1}D term

$$|2-2\rangle = \frac{\sqrt{210}}{42}Y_{31}(1)Y_{3-3}(2) - \frac{\sqrt{105}}{21}Y_{30}(1)Y_{3-2}(2) + \frac{\sqrt{14}}{7}Y_{3-1}(1)Y_{3-1}(2) - \frac{\sqrt{105}}{21}Y_{3-2}(1)Y_{30}(2) + \frac{\sqrt{210}}{42}Y_{3-3}(1)Y_{31}(2)$$

$$|2-1\rangle = \frac{5\sqrt{21}}{42}Y_{32}(1)Y_{3-3}(2) - \frac{\sqrt{35}}{14}Y_{31}(1)Y_{3-2}(2) + \frac{\sqrt{42}}{42}Y_{30}(1)Y_{3-1}(2) + \frac{\sqrt{42}}{42}Y_{3-1}(1)Y_{30}(2) - \frac{\sqrt{35}}{14}Y_{3-2}(1)Y_{31}(2) + \frac{5\sqrt{21}}{42}Y_{3-3}(1)Y_{32}(2)$$

$$|20\rangle = \frac{5\sqrt{21}}{42}Y_{33}(1)Y_{3-3}(2) - \frac{\sqrt{21}}{14}Y_{31}(1)Y_{3-1}(2) + \frac{2\sqrt{21}}{21}Y_{30}(1)Y_{30}(2) - \frac{\sqrt{21}}{14}Y_{3-1}(1)Y_{31}(2) + \frac{5\sqrt{21}}{42}Y_{3-3}(1)Y_{33}(2)$$

$$|21\rangle = \frac{5\sqrt{21}}{42}Y_{33}(1)Y_{3-2}(2) - \frac{\sqrt{35}}{14}Y_{32}(1)Y_{3-1}(2) + \frac{\sqrt{42}}{42}Y_{31}(1)Y_{30}(2) + \frac{\sqrt{42}}{42}Y_{30}(1)Y_{31}(2) - \frac{\sqrt{35}}{14}Y_{3-1}(1)Y_{32}(2) + \frac{5\sqrt{21}}{42}Y_{3-2}(1)Y_{33}(2)$$

$$|22\rangle = \frac{\sqrt{210}}{42}Y_{33}(1)Y_{3-1}(2) - \frac{\sqrt{105}}{21}Y_{32}(1)Y_{30}(2) + \frac{\sqrt{14}}{7}Y_{31}(1)Y_{31}(2) - \frac{\sqrt{105}}{21}Y_{30}(1)Y_{32}(2) + \frac{\sqrt{210}}{42}Y_{3-1}(1)Y_{33}(2)$$

The ^{1}G term

$$|4-4\rangle = \frac{\sqrt{33}}{11}Y_{3-1}(1)Y_{3-3}(2) - \frac{\sqrt{55}}{11}Y_{3-2}(1)Y_{3-2}(2) + \frac{\sqrt{33}}{11}Y_{3-3}(1)Y_{3-1}(2)$$

$$|4-3\rangle = \frac{3\sqrt{22}}{22}Y_{30}(1)Y_{3-3}(2) - \frac{\sqrt{11}}{11}Y_{3-1}(1)Y_{3-2}(2) - \frac{\sqrt{11}}{11}Y_{3-2}(1)Y_{3-1}(2) + \frac{3\sqrt{22}}{22}Y_{3-3}(1)Y_{30}(2)$$

$$|4-2\rangle = \frac{3\sqrt{231}}{77}Y_{31}(1)Y_{3-3}(2) + \frac{\sqrt{462}}{154}Y_{30}(1)Y_{3-2}(2) - \frac{2\sqrt{385}}{77}Y_{3-1}(1)Y_{3-1}(2) + \frac{\sqrt{462}}{154}Y_{3-2}(1)Y_{30}(2) + \frac{3\sqrt{231}}{77}Y_{3-3}(1)Y_{31}(2)$$

$$|4-1\rangle = \frac{\sqrt{1155}}{77}Y_{32}(1)Y_{3-3}(2) + \frac{4\sqrt{77}}{77}Y_{31}(1)Y_{3-2}(2) - \frac{\sqrt{2310}}{154}Y_{30}(1)Y_{3-1}(2) - \frac{\sqrt{2310}}{154}Y_{3-1}(1)Y_{30}(2) + \frac{4\sqrt{77}}{77}Y_{3-2}(1)Y_{31}(2) + \frac{\sqrt{1155}}{77}Y_{3-3}(1)Y_{32}(2)$$

$$|40\rangle = \frac{3\sqrt{154}}{154}Y_{33}(1)Y_{3-3}(2) + \frac{\sqrt{154}}{22}Y_{32}(1)Y_{3-2}(2) + \frac{\sqrt{154}}{154}Y_{31}(1)Y_{3-1}(2) - \frac{3\sqrt{154}}{77}Y_{30}(1)Y_{30}(2) + \frac{\sqrt{154}}{154}Y_{3-1}(1)Y_{31}(2) + \frac{\sqrt{154}}{22}Y_{3-2}(1)Y_{32}(2) + \frac{3\sqrt{154}}{154}Y_{3-3}(1)Y_{33}(2)$$

$$|41\rangle = \frac{\sqrt{1155}}{77}Y_{33}(1)Y_{3-2}(2) + \frac{4\sqrt{77}}{77}Y_{32}(1)Y_{3-1}(2) - \frac{\sqrt{2310}}{154}Y_{31}(1)Y_{30}(2) - \frac{\sqrt{2310}}{154}Y_{30}(1)Y_{31}(2) + \frac{4\sqrt{77}}{77}Y_{3-1}(1)Y_{32}(2) + \frac{\sqrt{1155}}{77}Y_{3-2}(1)Y_{33}(2)$$

$$|42\rangle = \frac{3\sqrt{231}}{77}Y_{33}(1)Y_{3-1}(2) + \frac{\sqrt{462}}{154}Y_{32}(1)Y_{30}(2) - \frac{2\sqrt{385}}{77}Y_{31}(1)Y_{31}(2) + \frac{\sqrt{462}}{154}Y_{30}(1)Y_{32}(2) + \frac{3\sqrt{231}}{77}Y_{3-1}(1)Y_{33}(2)$$

$$|43\rangle = \frac{3\sqrt{22}}{22}Y_{33}(1)Y_{30}(2) - \frac{\sqrt{11}}{11}Y_{32}(1)Y_{31}(2) - \frac{\sqrt{11}}{11}Y_{31}(1)Y_{32}(2) + \frac{3\sqrt{22}}{22}Y_{30}(1)Y_{33}(2)$$

$$|44\rangle = \frac{\sqrt{33}}{11}Y_{33}(1)Y_{31}(2) - \frac{\sqrt{55}}{11}Y_{32}(1)Y_{32}(2) + \frac{\sqrt{33}}{11}Y_{31}(1)Y_{33}(2)$$

The ¹I term

$$|6-6\rangle = Y_{3-3}(1)Y_{3-3}(2)$$

$$|6-5\rangle = \frac{1}{\sqrt{2}}\left(Y_{3-2}(1)Y_{3-3}(2) + Y_{3-3}(1)Y_{3-2}(2)\right)$$

$$|6-4\rangle = \frac{\sqrt{110}}{22} Y_{3-1}(1) Y_{3-3}(2) + \frac{\sqrt{66}}{11} Y_{3-2}(1) Y_{3-2}(2) + \frac{\sqrt{110}}{22} Y_{3-3}(1) Y_{3-1}(2)$$

$$|6-3\rangle = \frac{\sqrt{11}}{11} Y_{30}(1) Y_{3-3}(2) + \frac{3\sqrt{22}}{22} Y_{3-1}(1) Y_{3-1}(2) + \frac{3\sqrt{22}}{22} Y_{3-2}(1) Y_{3-1}(2) + \frac{\sqrt{11}}{11} Y_{3-3}(1) Y_{30}(2)$$

$$|6-2\rangle = \frac{\sqrt{33}}{33} Y_{31}(1) Y_{3-3}(2) + \frac{2\sqrt{66}}{33} Y_{30}(1) Y_{3-2}(2) + \frac{\sqrt{55}}{11} Y_{3-1}(1) Y_{3-1}(2) + \frac{2\sqrt{66}}{33} Y_{3-2}(1) Y_{30}(2) + \frac{\sqrt{33}}{33} Y_{3-3}(1) Y_{31}(2)$$

$$|6-1\rangle = \frac{\sqrt{33}}{66} Y_{32}(1) Y_{3-3}(2) + \frac{\sqrt{55}}{22} Y_{31}(1) Y_{3-2}(2) + \frac{5\sqrt{66}}{66} Y_{30}(1) Y_{3-1}(2) + \frac{5\sqrt{66}}{66} Y_{3-1}(1) Y_{30}(2) + \frac{\sqrt{55}}{22} Y_{3-2}(1) Y_{31}(2) + \frac{\sqrt{33}}{66} Y_{3-3}(1) Y_{32}(2)$$

$$|60\rangle = \frac{\sqrt{231}}{462} Y_{33}(1) Y_{3-3}(2) + \frac{\sqrt{231}}{77} Y_{32}(1) Y_{3-2}(2) + \frac{5\sqrt{231}}{154} Y_{31}(1) Y_{3-1}(2) + \frac{10\sqrt{231}}{231} Y_{30}(1) Y_{30}(2) + \frac{5\sqrt{231}}{154} Y_{3-1}(1) Y_{31}(2) + \frac{\sqrt{231}}{77} Y_{3-2}(1) Y_{32}(2) + \frac{\sqrt{231}}{462} Y_{3-3}(1) Y_{33}(2)$$

$$|61\rangle = \frac{\sqrt{33}}{66} Y_{33}(1) Y_{3-2}(2) + \frac{\sqrt{55}}{22} Y_{32}(1) Y_{3-1}(2) + \frac{5\sqrt{66}}{66} Y_{31}(1) Y_{30}(2) + \frac{5\sqrt{66}}{66} Y_{30}(1) Y_{31}(2) + \frac{\sqrt{55}}{22} Y_{3-1}(1) Y_{32}(2) + \frac{\sqrt{33}}{66} Y_{3-2}(1) Y_{33}(2)$$

$$|62\rangle = \frac{\sqrt{33}}{33}Y_{33}(1)Y_{3-1}(2) + \frac{2\sqrt{66}}{33}Y_{32}(1)Y_{30}(2) + \frac{\sqrt{55}}{11}Y_{31}(1)Y_{31}(2) + \frac{2\sqrt{66}}{33}Y_{30}(1)Y_{32}(2) + \frac{\sqrt{33}}{33}Y_{3-1}(1)Y_{33}(2)$$

$$|63\rangle = \frac{\sqrt{11}}{11}Y_{33}(1)Y_{30}(2) + \frac{3\sqrt{22}}{22}Y_{32}(1)Y_{31}(2) + \frac{3\sqrt{22}}{22}Y_{31}(1)Y_{32}(2) + \frac{\sqrt{11}}{11}Y_{30}(1)Y_{33}(2)$$

$$|64\rangle = \frac{\sqrt{110}}{22}Y_{33}(1)Y_{31}(2) + \frac{\sqrt{66}}{11}Y_{32}(1)Y_{32}(2) + \frac{\sqrt{110}}{22}Y_{31}(1)Y_{33}(2)$$

$$|65\rangle = \frac{1}{\sqrt{2}}\left(Y_{32}(1)Y_{33}(2) + Y_{33}(1)Y_{32}(2)\right)$$

$$|66\rangle = Y_{33}(1)Y_{33}(2)$$

Appendix III

Symmetrized Wave Functions of the S, P, D, F, G, H Terms in the Cubic Crystal Fields*

The $|LM\rangle$ notation is used, where L and M stand for the orbital and magnetic quantum numbers of the corresponding term. The first symbol in the brackets stands for one of the irreducible representations of the cubic group, and the second symbol is a letter denoting the orbital term.

S term. *L* = 0, the wave function transforms as A_1.

$$\Psi_1(A_1, S) = |00\rangle .$$

P term. *L* = 1, three wave functions transform as T_1.

$$\Psi_1(T_1, P) = |11\rangle , \ \Psi_2(T_1, P) = |10\rangle , \ \Psi_3(T_1, P) = |1-1\rangle .$$

D term. *L* = 2, five wave functions transform as E + T_2.

$$\Psi_1(E, D) = |20\rangle , \ \Psi_2(E, D) = \frac{1}{\sqrt{2}}\left(|22\rangle + |2-2\rangle\right),$$

$$\Psi_1(T_2, D) = |21\rangle , \Psi_2(T_2, D) = \frac{i}{\sqrt{2}}\left(|22\rangle - |2-2\rangle\right), \Psi_3(T_2, D) = |2-1\rangle .$$

F term. *L* = 3, seven wave functions transform as A_2 + T_1 + T_2.

$$\Psi_1(A_2, F) = \frac{i}{\sqrt{2}}\left(|32\rangle - |3-2\rangle\right),$$

*Leushin AM (1968). Tables of the functions transforming according to the irreducible representations of the crystallographic point groups (in Russian: Таблицы функций, преобразующихся по неприводимым представлениям кристаллографических точечных групп), Moscow, Nauka.

$$\Psi_1(T_1, F) = \sqrt{\frac{3}{8}}|31\rangle + \sqrt{\frac{5}{8}}|3-3\rangle,$$

$$\Psi_2(T_1, F) = -|30\rangle,$$

$$\Psi_3(T_1, F) = \sqrt{\frac{5}{8}}|33\rangle + \sqrt{\frac{3}{8}}|3-1\rangle,$$

$$\Psi_1(T_2, F) = -\sqrt{\frac{5}{8}}|31\rangle + \sqrt{\frac{3}{8}}|3-3\rangle,$$

$$\Psi_2(T_2, F) = \frac{i}{\sqrt{2}}(|32\rangle + |3-2\rangle),$$

$$\Psi_3(T_2, F) = -\sqrt{\frac{3}{8}}|33\rangle + \sqrt{\frac{5}{8}}|3-1\rangle,$$

G term. $L = 4$, nine wave functions transform as $A_1 + E + T_1 + T_2$.

$$\Psi_1(A_1, G) = \sqrt{\frac{5}{24}}(|44\rangle + |4-4\rangle) + \sqrt{\frac{7}{12}}|40\rangle,$$

$$\Psi_1(E, G) = \sqrt{\frac{7}{24}}(|44\rangle + |4-4\rangle) - \sqrt{\frac{5}{12}}|40\rangle,$$

$$\Psi_2(E, G) = \frac{1}{\sqrt{2}}(|42\rangle + |4-2\rangle),$$

$$\Psi_1(T_1, G) = \sqrt{\frac{7}{8}}|41\rangle + \sqrt{\frac{1}{8}}|4-3\rangle,$$

$$\Psi_2(T_1, G) = \frac{1}{\sqrt{2}}(|4-4\rangle - |44\rangle),$$

$$\Psi_3(T_1, G) = -\sqrt{\frac{1}{8}}|43\rangle - \sqrt{\frac{7}{8}}|4-1\rangle,$$

$$\Psi_1(T_2, G) = -\sqrt{\frac{1}{8}}|41\rangle + \sqrt{\frac{7}{8}}|4-3\rangle,$$

$$\Psi_2(T_2, G) = \frac{1}{\sqrt{2}}(|42\rangle - |4-2\rangle),$$

$$\Psi_3(T_2, G) = \sqrt{\frac{7}{8}}|43\rangle - \sqrt{\frac{1}{8}}|4-1\rangle.$$

H term. L = 5, eleven wave functions transform as E + 2T$_1$ + T$_2$.

$$\Psi_1(\mathrm{E, H}) = \frac{1}{\sqrt{2}}\left(|54\rangle - |5-4\rangle\right),$$

$$\Psi_2(\mathrm{E, H}) = \frac{1}{\sqrt{2}}\left(-|52\rangle + |5-2\rangle\right),$$

$$\Psi_1(\mathrm{T_1, H}) = \sqrt{\frac{5}{128}}|55\rangle + \sqrt{\frac{21}{64}}|51\rangle - \sqrt{\frac{81}{128}}|5-3\rangle,$$

$$\Psi_2(\mathrm{T_1, H}) = \frac{1}{\sqrt{2}}\left(|54\rangle + |5-4\rangle\right),$$

$$\Psi_3(\mathrm{T_1, H}) = -\sqrt{\frac{81}{128}}|53\rangle + \sqrt{\frac{21}{64}}|5-1\rangle + \sqrt{\frac{5}{128}}|5-5\rangle,$$

$$\Psi_1(\mathrm{T_1', H}) = \sqrt{\frac{63}{128}}|55\rangle + \sqrt{\frac{15}{64}}|51\rangle + \sqrt{\frac{35}{128}}|5-3\rangle,$$

$$\Psi_2(\mathrm{T_1', H}) = |50\rangle,$$

$$\Psi_3(\mathrm{T_1', H}) = \sqrt{\frac{35}{128}}|53\rangle + \sqrt{\frac{15}{64}}|5-1\rangle + \sqrt{\frac{63}{128}}|5-5\rangle,$$

$$\Psi_1(\mathrm{T_2, H}) = \sqrt{\frac{15}{32}}|55\rangle - \sqrt{\frac{7}{16}}|51\rangle - \sqrt{\frac{3}{32}}|5-3\rangle,$$

$$\Psi_2(\mathrm{T_2, H}) = \frac{i}{\sqrt{2}}\left(|52\rangle + |5-2\rangle\right),$$

$$\Psi_3(\mathrm{T_2, H}) = \sqrt{\frac{3}{32}}|53\rangle + \sqrt{\frac{7}{16}}|5-1\rangle - \sqrt{\frac{15}{32}}|5-5\rangle.$$

I term. L = 6, thirteen wave functions transform as A$_1$ + A$_2$ + E + T$_1$ + 2T$_2$.

$$\Psi_1(\mathrm{A_1, I}) = \sqrt{\frac{7}{16}}\left(|64\rangle + |6-4\rangle\right) - \sqrt{\frac{1}{8}}|60\rangle,$$

$$\Psi_1(\mathrm{A_2, I}) = i\sqrt{\frac{5}{32}}\left(|66\rangle + |6-6\rangle\right) + i\sqrt{\frac{11}{32}}\left(|62\rangle - |6-2\rangle\right),$$

$$\Psi_1(\mathrm{E, I}) = \sqrt{\frac{1}{16}}\left(|64\rangle + |6-4\rangle\right) + \sqrt{\frac{7}{8}}|60\rangle,$$

$$\Psi_1(\mathrm{E}, \mathrm{I}) = \sqrt{\frac{11}{32}}\left(|66\rangle + |6-6\rangle\right) + \sqrt{\frac{5}{32}}\left(|62\rangle + |6-2\rangle\right),$$

$$\Psi_1(\mathrm{T}_1, \mathrm{I}) = \sqrt{\frac{11}{32}}|65\rangle - \sqrt{\frac{3}{16}}|61\rangle + \sqrt{\frac{15}{32}}|6-3\rangle,$$

$$\Psi_2(\mathrm{T}_1, \mathrm{I}) = \frac{1}{\sqrt{2}}\left(-|64\rangle + |6-4\rangle\right),$$

$$\Psi_3(\mathrm{T}_1, \mathrm{I}) = -\sqrt{\frac{15}{32}}|63\rangle + \sqrt{\frac{3}{16}}|6-1\rangle - \sqrt{\frac{11}{32}}|6-5\rangle,$$

$$\Psi_1(\mathrm{T}_2, \mathrm{I}) = \sqrt{\frac{165}{256}}|65\rangle + \sqrt{\frac{5}{128}}|61\rangle - \sqrt{\frac{81}{256}}|6-3\rangle,$$

$$\Psi_2(\mathrm{T}_2, \mathrm{I}) = \frac{i}{\sqrt{2}}\left(|62\rangle - |6-2\rangle\right),$$

$$\Psi_3(\mathrm{T}_2, \mathrm{I}) = -\sqrt{\frac{81}{256}}|63\rangle + \sqrt{\frac{5}{128}}|6-1\rangle + \sqrt{\frac{165}{256}}|6-5\rangle,$$

$$\Psi_1(\mathrm{T}_2', \mathrm{I}) = \sqrt{\frac{3}{256}}|65\rangle + \sqrt{\frac{99}{128}}|61\rangle + \sqrt{\frac{55}{256}}|6-3\rangle,$$

$$\Psi_2(\mathrm{T}_2', \mathrm{I}) = \frac{i}{\sqrt{2}}\left(|66\rangle - |6-6\rangle\right),$$

$$\Psi_3(\mathrm{T}_2', \mathrm{I}) = \sqrt{\frac{55}{256}}|63\rangle + \sqrt{\frac{99}{128}}|6-1\rangle + \sqrt{\frac{3}{256}}|6-5\rangle.$$

Index